D1756466

This book is due for return not later than the
last date stamped below, unless recalled sooner.

Stochastic Processes, Optimization, and Control Theory: Applications in Financial Engineering, Queueing Networks, and Manufacturing Systems

A Volume in Honor of Suresh Sethi

Stochastic Processes, Optimization, and Control Theory: Applications in Financial Engineering, Queueing Networks, and Manufacturing Systems

A Volume in Honor of Suresh Sethi

Houmin Yan, George Yin, and Qing Zhang

Editors

 Springer

Houmin Yan
The Chinese Univ. of HK
Hong Kong

George Yin
Wayne State Univ.
Detroit, MI, USA

Qing Zhang
University of Georgia
Athens, GA, USA

Library of Congress Control Number: 2006924379

ISBN-10: 0-387-33770-9 (HB)   ISBN-10: 0-387-33815-2 (e-book)
ISBN-13: 978-0387-33770-8 (HB)   ISBN-13: 978-0387-33815-6 (e-book)

Printed on acid-free paper.

Printed in the United States of America.

9 8 7 6 5 4 3 2 1

springer.com

Dedicated to

Suresh P. Sethi

On the Occasion of His 60th Birthday

Contents

viii

PREFACE

This edited volume contains 16 research articles and presents recent and pressing issues in stochastic processes, control theory, differential games, optimization, and their applications in finance, manufacturing, queueing networks, and climate control. One of the salient features is that the book is highly multi-disciplinary. It assembles experts from the fields of operations research, control theory and optimization, stochastic analysis, and financial engineering to review and substantially update the recent progress in these fields. Another distinct characteristic of the book is that all papers are motivated by applications in which optimization, control, and stochastics are inseparable. The book will be a timely addition to the literature and will be of interest to people working in the aforementioned fields. All papers in this volume have been reviewed.

This volume is dedicated to Professor Suresh Sethi on the occasion of his 60th birthday. In view of his fundamental contributions, his distinguished career, his substantial achievements, his influence to the control theory and applications, operations research, and management science, and his dedication to the scientific community, we have invited a number of leading experts in the fields of optimization, control, and operation management, to contribute to this volume in honor of him.

Without the help of many individuals, this book could not have come into being. We thank the series editor Professor Frederick S. Hillier for his time and consideration. Our thanks also go to Gary Folven, Carolyn Ford, and the Springer's professionals for their assistance in finalizing the book. Finally, we express our gratitude to all authors for their invaluable contributions.

Hong Kong   Houmin Yan
Detroit, Michigan   George Yin
Athens, Georgia   Qing Zhang

S.P. Sethi's Curriculum Vitae

Suresh P. Sethi, FRSC
Ashbel Smith Professor of Operations Management
Director, Center of Intelligent Supply Networks (C4ISN)
School of Management, University of Texas at Dallas
Richardson, TX 75080

Education

Carnegie Mellon University - Ph.D. Operations Research (1972);
M.S.I.A. (1971)
Washington State University - M.B.A. (1969)
Indian Institute of Technology Bombay - B. Tech. (1967)

Employment

University of Texas at Dallas 1997-present
University of Toronto 1973-97
Rice University 1972-73
Stanford University 1971-72

Visiting Positions

Bilkent University; Helsinki School of Economics and Business Administration; University of Siena; University of Vienna; Curtin University; Chinese University of Hong Kong; University of Lyon; Institut National de Recherche en Informatique et en Automatique; Australian National University; Brown University; Institute of Operations Research Technical University; International Insitute for Applied Systems Analysis; Science Center Berlin; University of Arizona, Georgia Insitute of Technology; Carnegie Mellon University; Brookhaven National Laboratory

Honors and Awards

2005	*POMS* Fellow
2004	*Wickham-Skinner Best Paper Award* at The 2nd World Conference on POM
2003	*INFORMS* Fellow; *AAAS* Fellow
2001	*IEEE* Fellow; Outstanding Contribution in Education Award, Greater Dallas Indo-American Chamber of Commerce

2000   Senior Research Fellow of the IC2 Institute

1999   Fellow of the New York Academy of Sciences

1998   C.Y. O'Connor Fellow, Curtin Univ., Perth,
Australia (June-July)

1997   First listed in Canadian Who's Who,
University of Toronto Press

1996   Award of Merit of Canadian Operational Research Society
(CORS);
Honorary Professor, Zhejiang Univ. of Technology,
Hangzhou, China

1994   Fellow of The Royal Society of Canada (FSRC); a.k.a.
The Canadian Academy of Sciences and Humanities

1992   Best Paper Award at POM-92 (second prize)

1991   Erskine Fellow, University of Canterbury, Christchurch,
New Zealand

1984-85 Connaught Senior Research Fellow, University of Toronto

1971   Phi Kappa Phi (National Honor Society in all fields)

1969   Beta Gamma Sigma
(National Honor Society in Business Administration)

Books and Monographs

1 Sethi, S.P. and Thompson, G.L., *Optimal Control Theory: Applications to Management Science*, Martinus Nijhoff, Boston, 1981.

2 Sethi, S.P. and Thompson, G.L., *Solutions Manual for Optimal Control Theory: Applications to Management Science*, Martinus Nijhoff, Boston, 1981.

3 Sethi, S.P. and Zhang, Q., *Hierarchical Decision Making in Stochastic Manufacturing Systems*, Birkhäuser Boston, Cambridge, MA, 1994.

4 Sethi, S.P., *Optimal Consumption and Investment with Bankruptcy*, Kluwer Academic Publishers, Norwell, MA, 1997.

5 Sethi, S.P. and Thompson, G.L., *Optimal Control Theory: Applications to Management Science and Economics*, Second Edition, Kluwer Academic Publishers, Boston, 2000.

6 Sethi, S.P., Zhang, H., and Zhang, Q., *Average-Cost Control of Stochastic Manufacturing Systems*, Springer, New York, NY, 2005.

7 Sethi, S.P., Yan, H., and Zhang, H., *Inventory and Supply Chain Management with Forecast Updates*, Springer, New York, NY, 2005.

8 Beyer, D., Cheng, F., Sethi, S.P., and Taksar, M.I., *Markovian Demand Inventory Models*, Springer, in press.

9 Dawande, M., Geismar, H.N., Sethi, S.P., and Sriskandarajah, C., *Throughput Optimization in Robotic Cells*, Springer, in press.

Research Publications - Operations Management

1 Bensoussan, A., Liu, R.H., and Sethi, S.P., "Optimality of an (s, S) Policy with Compound Poisson and Diffusion Demands: A QVI Approach," *SIAM Journal on Control and Optimization*, in press.

2 Feng, Q., Gallego, G., Sethi, S.P., Yan, H., and Zhang, H. "Are Base-stock Policies Optimal in Inventory Problems with Multiple Delivery Modes?" *Operations Research*, in press.

3 Presman, E. and Sethi, S.P., "Stochastic Inventory Models with Continuous and Poisson Demands and Discounted and Average Costs," *Production and Operations Management*, in press.

4 Geismar, H.N., Sethi, S.P., Sidney, J.B., and Sriskandarajah, C., "A Note on Productivity Gains in Flexible Robotic Cells," *International Journal of Flexible Manufacturing Systems*, 17, 1, 2005, in press.

5 Bensoussan, A., Cakanyildirim, M., and Sethi, S.P., "Optimal Ordering Policies for Inventory Problems with Dynamic Information Delays," *Production and Operations Management*, conditionally accepted.

6 Hou, Y., Sethi, S.P., Zhang, H., and Zhang, Q., "Asymptotically Optimal Production Policies in Dynamic Stochastic Jobshops with Limited Buffers," *Journal of Mathematical Analysis and Applications*, in press.

7 Bensoussan, A., Cakanyildirim, M., and Sethi, S.P., "Optimality of Base Stock and (s, S) Policies for Inventory Problems with Information Delays," *Journal of Optimization Theory and Applications*, 130, 2, August 2006, in press.

8 Drobouchevitch, I.G., Sethi, S.P., and Sriskandarajah, C., "Scheduling Dual Gripper Robotic Cells: One-unit Cycles," *European Journal of Operational Research*, 171, 2006, 598-631.

9 Bensoussan, A., Cakanyildirim, M., and Sethi, S.P., "Partially Observed Inventory Systems," *Proceedings of 44th IEEE CDC and ECC*, Seville, Spain, Dec. 12-15, 2005.

10 Bensoussan, A., Cakanyildirim, M., and Sethi, S.P., "Optimality of Standard Inventory Policies with Information Delays," *Proceedings of 6th APIEMS 2005*, Manila, Philippines, Dec. 4-7, 2005, CD Article #1221.

11 Gallego, G. and Sethi, S.P., "$\mathcal{K}$-Convexity in $\Re^n$," *Journal of Optimization Theory and Applications*, 127, 1, October 2005, 71-88.

12 Feng, Q., Gallego, G., Sethi, S.P., Yan, H., and Zhang, H. "Optimality and Nonoptimality of the Base-stock Policy in Inventory Problems with Multiple Delivery Modes," *Journal of Industrial and Management Optimization*, 2, 1, 2006, 19-42.

13 Bensoussan, A., Cakanyildirim, M., and Sethi, S.P., "On the Optimal Control of Partially Observed Inventory Systems," *Comptes Rendus de l'Academié des Sciences Paris*, Ser. I 341, 2005, 419-426.

14 Sethi, S.P., Yan, H., Zhang, H., and Zhou, J., "Information Updated Supply Chain with Service-Level Constraints," *Journal of Industrial and Management Optimization*, 1, 4, November 2005, 513-531.

15 Sethi, S.P., Yan, H., Yan, J.H., and Zhang, H., "An Analysis of Staged Purchases in Deregulated Time-Sequential Electricity Markets," *Journal of Industrial Management Optimization*, 1, 4, November 2005, 443-463 .

16 Dawande, M., Geismar, N., and Sethi, S.P., "Dominance of Cyclic Solutions and Challenges in the Scheduling of Robotic Cells," *SIAM Review*, 47, 4, December 2005, 709-721.

17 Beyer, D. and Sethi, S.P., "Average-Cost Optimality in Inventory Models with Markovian Demands and Lost Sales," in *Analysis, Control and Optimization of Complex Dynamic Systems*, E.K. Boukas and R.P. Malhame, (Eds.), Kluwer, 2005, 3-23.

18 Huang, H., Yan, H., and Sethi, S.P., "Purchase Contract Management with Demand Forecast Updates," *IIE Transactions on Scheduling and Logistics*, 37, 8, August 2005, 775-785.

19 Dawande, M., Geismar, H. N., Sethi, S.P., and Sriskandarajah, C., "Sequencing and Scheduling in Robotics Cells: Recent Developments," *Journal of Scheduling*, 8, 2005, 387-426.

20 Sethi, S.P., Yan, H., and Zhang, H., "Analysis of a Duopoly Supply Chain and its Application in Electricity Spot Markets," *Annals of Operations Research*, 135, 1, 2005, 239-259.

21 Gan, X., Sethi, S.P., and Yan, H., "Channel Coordination with a Risk-Neutral Supplier and a Downside-Risk-Averse Retailer," *Production and Operations Management*, 14, 1, 2005, 80-89.

22 Drobouchevitch, I.G., Sethi, S.P., Sidney, J.B., and Sriskandarajah, C., "Scheduling Multiple Parts in Two-Machine Dual Gripper Robot Cells: Heuristic Algorithm and Performance Guarantee," *International Journal of Operations & Quantitative Management*, 10, 4, 2004, 297-314.

23 Feng, Q., Gallego, G., Sethi, S.P., Yan, H., and Zhang, H., "Periodic-Review Inventory Model with Three Consecutive Delivery Modes and Forecast Updates," *Journal of Optimization Theory and Applications*, 124, 1, January 2005, 137-155.

24 Sethi, S.P., Yan, H., and Zhang, H., "Quantity Flexible Contracts: Optimal Decisions with Information Updates," *Decision Sciences*, 35, 4, Fall 2004, 691-712.

25 Gan, X., Sethi, S.P., and Yan, H., "Coordination of a Supply Chain with Risk-Averse Agents," *Production and Operations Management*, 13, 2, 2004, 135-149.

26 Sethi, S.P. and Zhang, Q., "Problem 4.3 Feedback Control in Flowshops," in *Unsolved Problems in Mathematical Systems and Control Theory*, V. D. Blondel and A. Megretski (Eds.), Princeton University Press, Princeton, NJ, 2004, 140-143.

27 Sriskandarajah, C., Drobouchevitch, I.G., Sethi, S.P., and Chandrasekaran, R., "Scheduling Multiple Parts in a Robotic Cell Served by a Dual-Gripper Robot," *Operations Research*, 52, 1, Jan-Feb 2004, 65-82.

28 Sethi, S.P., Sidney, J.B., and Sriskandarajah, C., "Throughput Comparison of Robotic Flowshops Versus Robotic Openshops," in *Operational Research and its Applications: Recent Trends*, M.R. Rao and M.C. Puri (Eds.), Allied Publishers Pvt. Limited, New Delhi, India, *Proceedings of the Sixth International Conference of the Association of Asia-Pacific Operational Research Societies (APORS)*, Dec. 8-10, 2003, New Delhi, India, Vol. I, 10-18.

29 Sethi, S.P., Yan, H., and Zhang, H., "Inventory Models with Fixed Costs, Forecast Updates, and Two Delivery Modes," *Operations Research*, 51, 2, March-April 2003, 321-328.

30 Dawande, M., Sriskandarajah, C., and Sethi, S.P., "On Throughput Maximization in Constant Travel-Time Robotic Cells," *Manufacturing & Service Operations Management*, 4, 4, Fall 2002, 296-312.

31 Yan, H., Sriskandarajah, C., Sethi, S.P., and Yue, X., "Supply-Chain Redesign to Reduce Safety Stock Levels: Sequencing and Merging Operations," *IEEE Transactions on Engineering Management: Special Issue on Supply Chain Management*, 49, 3, Aug. 2002, 243-257.

32 Presman, E., Sethi, S.P., Zhang, H., and Zhang, Q., "On Optimality of Stochastic N-Machine Flowshop with Long-Run Average Cost," in *Stochastic Theory and Control*, B. Pasik-Duncan (Ed.), Lecture Notes in Control and Information Sciences, Vol. 280, Springer-Verlag, Berlin, 2002, 399-417.

33 Sethi, S.P., Yan, H., Zhang, H., and Zhang, Q., "Optimal and Hierarchical Controls in Dynamic Stochastic Manufacturing Systems: A Survey," *Manufacturing & Service Operations Management*, 4, 2, Spring 2002, 133-170.

34 Chand, S., Hsu, V.N., and Sethi, S.P., "Forecast, Solution and Rolling Horizons in Operations Management Problems: A Classified Bibliography," *Manufacturing & Service Operations Management*, 4, 1, Winter 2002, 25-43.

35 Beyer, D., Sethi, S.P., and Sridhar, R., "Stochastic Multiproduct Inventory Models with Limited Storage," *Journal of Optimization Theory and Applications*, 111, 3, Dec. 2001, 553-588.

36 Yeh, D.H.M., Sethi, A., Sethi, S.P., and Sriskandarajah, C., "Scheduling of the Injection Process for a Golf Club Head Fabrication Lines," *International Journal of Operations and Quantitative Management*, 7, 3, Sept. 2001, 149-164.

37 Beyer, D., Sethi, S.P., and Sridhar, R., "Average-Cost Optimality of a Base-Stock Policy for a Multi-Product Inventory Model with Limited Storage," *Proceedings of the International Workshop on Decision and Control in Management Sciences in honor of Professor Alain Haurie*, Montreal, Quebec, Canada, Oct. 19-20, 2000,

Decision and Control in Management Science, Essays in Honor of Alain Haurie, G. Zaccour (Ed.), Kluwer Academic Publishers, Boston, 2002, 241-260.

38 Presman, E., Sethi, S.P., Zhang, H., and Bisi, A., "Average Cost Optimal Policy for a Stochastic Two-Machine Flowshop with Limited Work-In-Process," *Proceedings of the 3rd World Congress of Nonlinear Analysis*, 2001, Nonlinear Analysis, 47, 2001, 5671-5678.

39 Sethi, S.P., Sidney, J.B., and Sriskandarajah, C., "Scheduling in Dual Gripper Robotic Cells for Productivity Gains," *IEEE Transactions on Robotics and Automation*, 17, 3, June 2001, 324-341.

40 Sethi, S.P., Yan, H., Zhang, H., and Zhang, Q., "Turnpike Set Analysis in Stochastic Manufacturing Systems with Long-Run Average Cost," in *Optimal Control and Partial Differential Equations in Honour of Professor Alain Bensoussan's 60th Birthday*, J.L. Menaldi, E. Rofman, and A. Sulem (Eds.), IOS Press, Amsterdam, 2001, 414-423.

41 Sethi, S.P., Yan, H., and Zhang, H., "Peeling Layers of an Onion: Inventory Model with Multiple Delivery Modes and Forecast Updates," *Journal of Optimization Theory and Applications*, 108, 2, Feb. 2001, 253-281.

42 Presman E., Sethi, S.P., Zhang H., and Bisi, Arnab, "Average Cost Optimality for an Unreliable Two-Machine Flowshop with Limited Internal Buffer," in *Optimization Theory and Its Application, Annals of Operations Research*, L. Caccetta and K.L. Teo (Eds.), Kluwer Academic Publishers, The Netherlands, 98, Dec. 2000, 333-351.

43 Presman E., Sethi, S.P., Zhang H., and Zhang Q., "Optimal Production Planning in a Stochastic N-Machine Flowshop with Long-Run Average Cost," in *Mathematics and Its Applications to Industry*, S. K. Malik (Ed.), Indian National Science Academy, New Delhi, India, 2000, 121-140.

44 Sethi, S.P., Sorger, G., and Zhou, X.Y., "Stability of Real-Time Lot-Scheduling Policies and Machine Replacement Policies with Quality Levels," *IEEE Transactions on Automatic Control*, 45, 11, Nov. 2000, 2193-2196.

45 Huang, H., Yan, H., and Sethi, S.P., "Purchase Contract Management with Demand Forecast Updates," *Proceedings of the Second*

West Lake International Conference on Small & Medium Business, Oct. 16-18, 2000, Zhejiang University of Technology, Hangzhou, China, 161-167.

46 Sethi, S.P., Yan, H., Zhang, H., and Zhang, Q., "Optimal and Hierarchical Controls in Dynamic Stochastic Manufacturing Systems: A Review," *ITORMS (Interactive Transactions of OR/MS)*, 3, 2, 2000, available at http://itorms.iris.okstate.edu/.

47 Sethi, S.P., Zhang, H., and Zhang, Q., "Hierarchical Production Control in a Stochastic N-Machine Flowshop with Long-Run Average Cost," *Journal of Mathematical Analysis and Applications*, 251, 1, 2000, 285-309.

48 Sethi, S.P., Zhang, H., and Zhang, Q., "Hierarchical Production Control in Dynamic Stochastic Jobshops with Long-Run Average Cost," *Journal of Optimization Theory and Applications*, 106, 2, Aug. 2000, 231-264.

49 Presman, E., Sethi, S.P., and Zhang, H., "Optimal Production Planning in Stochastic Jobshops with Long-Run Average Cost," in *Optimization, Dynamics, and Economic Analysis, Essays in Honor of Gustav Feichtinger*, E.J. Dockner, R.F. Hartl, M. Luptačik, and G. Sorger (Eds.), Physica-Verlag, Heidelberg, 2000, 259-274.

50 Yan, H., Lou, S., and Sethi, S.P., "Robustness of Various Production Control Policies in Semiconductor Manufacturing," *Production and Operations Management*, 9, 2, Summer, 2000, 171-183.

51 Sethi, S.P., Zhang, H., and Zhang, Q., "Optimal Production Rates in a Deterministic Two-Product Manufacturing System," *Optimal Control Applications and Methods*, 21, 3, May-June, 2000, 125-135.

52 Sethi, S.P., Zhang, H., and Zhang, Q., "Hierarchical Production Control in a Stochastic N-Machine Flowshop with Limited Buffers," *Journal of Mathematical Analysis and Applications*, 246, 1, 2000, 28-57.

53 Sethi, S.P., Sriskandarajah, C., Chu, K.F., and Yan, H., "Efficient Setup/Dispatching Policies in a Semiconductor Manufacturing Facility," *Proceedings of 38th IEEE CDC*, Phoenix, Arizona, Dec. 7-10, 1999, 1368-1372.

54 Cheng, F. and Sethi, S.P., "A Periodic Review Inventory Model with Demand Influenced by Promotion Decisions," *Management Science*, 45, 11, 1999, 1510-1523.

55 Beyer, D. and Sethi, S.P., "The Classical Average-Cost Inventory Models of Iglehart and Veinott-Wagner Revisited," *Journal of Optimization Theory and Applications*, 101, 3, 1999, 523-555.

56 Sethi, S.P. and Zhang, H., "Average-Cost Optimal Policies for a Unreliable Flexible Multiproduct Machine," *The International Journal of Flexible Manufacturing Systems*, 11, 1999, 147-157.

57 Yang, J., Yan, H., and Sethi, S.P., "Optimal Production Planning in Pull Flow Lines with Multiple Products," *European Journal of Operational Research*, 119, 3, 1999, 583-604.

58 Cheng, F.M. and Sethi, S.P., "Optimality of State-Dependent (s, S) Policies in Inventory Models with Markov-Modulated Demand and Lost Sales," *Production and Operations Management*, 8, 2, Summer, 1999, 183-192.

59 Sethi, S.P., Sriskandarajah, C., Van de Velde, S., Wang, M.Y., and Hoogeveen, H., "Minimizing Makespan in a Pallet-Constrained Flowshop," *Journal of Scheduling*, 2, 3, May-June, 1999, 115-133.

60 Sethi, S.P. and Zhang, H., "Hierarchical Production Controls for a Stochastic Manufacturing System with Long-Run Average Cost: Asymptotic Optimality," in *Stochastic Analysis, Control, Optimization and Applications: A Volume in Honor of W.H. Fleming*, W.M. McEneaney, G. Yin, and Q. Zhang (Eds.), Systems & Control: Foundations & Applications, Birkhäuser, Boston, 1999, 621-637.

61 Presman, E., Sethi, S.P., Zhang, H., and Zhang, Q., "Optimality of Zero-Inventory Policies for an Unreliable Manufacturing System Producing Two Part Types," *Dynamics of Continuous, Discrete and Impulsive Systems*, 4, 1998, 485-496.

62 Sethi, S.P., Zhang, H., and Zhang, Q., "Minimum Average-Cost Production Planning in Stochastic Manufacturing Systems," *Mathematical Models and Methods in Applied Sciences*, 8, 7, 1998, 1251-1276.

63 Sethi, S.P. and Zhang, Q., "Asymptotic Optimality of Hierarchical Controls in Stochastic Manufacturing Systems: A Review," *Proceedings of the Conference in Honor of Prof. Gerald L. Thompson on his 70th Birthday*, Pittsburgh, PA. Oct. 17-18, 1993, *Operations Research: Methods, Models, and Applications*, J.E. Aronson and S. Zionts (Eds.), Quoram Books, Westport, CT, 1998, 267-294.

64 Presman, E., Sethi, S.P., Zhang, H., and Zhang, Q., "Analysis of Average Cost Optimality for an Unreliable Two-Machine Flow-shop," *Proceedings of the Fourth International Conference on Optimization Techniques and Applications*, July 1-3, 1998, Curtin University of Technology, Perth, Australia, 1998, 94-112.

65 Beyer, D. and Sethi, S.P., "A Proof of the EOQ Formula Using Quasi-Variational Inequalities," *International Journal of Systems Science*, 29, 11, 1998, 1295-1299.

66 Sethi, S.P., Suo, W., Taksar, M.I., and Yan, H., "Optimal Production Planning in a Multi-Product Stochastic Manufacturing System with Long-Run Average Cost," *Discrete Event Dynamic Systems: Theory and Applications*, 8, 1998, 37-54.

67 Presman E., Sethi S.P., Zhang, H., and Zhang Q., "Optimal Production Planning in a Stochastic N-Machine Flowshop with Long-Run Average Cost" (abridged version), *Proceedings of the Second International Conference on the Applications of Mathematics to Science and Engineering (CIAMASI '98)*, Oct. 27-29, 1998, Université Hassan II Aïn Chock, Casablanca, Morocco, 704-711.

68 Beyer, D., Sethi, S.P., and Taksar, M.I., "Inventory Models with Markovian Demands and Cost Functions of Polynomial Growth," *Journal of Optimization Theory and Applications*, 98, 2, 1998, 281-323.

69 Jennings, L.S., Sethi, S.P., and Teo, K.L., "Computation of Optimal Production Plans for Manufacturing Systems," *Proceedings of the 2nd World Congress of Nonlinear Analysis*, Athens, Greece, July 16-17, 1996, *Nonlinear Analysis, Theory, Methods & Applications*, 30, 7, 1997, 4329-4338.

70 Yeh, D.H.M., Sethi, S.P., and Sriskandarajah, C. "Scheduling of the Injection Process for Golf Club Head Fabrication Lines," *Proceedings of the 22nd International Conference on Computers and Industrial Engineering*, Cairo, Egypt, December 20-22, 1997, 172-175.

71 Wang, Y., Sethi, S.P., Sriskandarajah, C., and Van de Velde, S.L., "Minimizing Makespan in Flowshops with Pallet Requirements: Computational Complexity," *INFOR*, 35, 4, 1997, 277-285.

72 Presman, E., Sethi, S.P., and Suo, W., "Optimal Feedback Controls in Dynamic Stochastic Jobshops," in *Mathematics of Stochastic Manufacturing Systems*, G. Yin and Q. Zhang (Eds.), *Lectures*

in Applied Mathematics, Vol. 33, American Mathematical Society, Providence, RI, 1997, 235-252.

73 Presman, E., Sethi, S.P., and Suo, W., "Existence of Optimal Feedback Production Plans in Stochastic Flowshops with Limited Buffers," *Automatica*, 33, 10, 1997, 1899-1903.

74 Sethi, S.P., Zhang, H., and Zhang, Q., "Hierarchical Production Control in a Stochastic Manufacturing System with Long-Run Average Cost," *Journal of Mathematical Analysis and Applications*, 214, 1, 1997, 151-172.

75 Beyer, D. and Sethi, S.P., "Average Cost Optimality in Inventory Models with Markovian Demands," *Journal of Optimization Theory and Applications*, 92, 3, 1997, 497-526.

76 Samaratunga, C., Sethi, S.P., and Zhou, X., "Computational Evaluation of Hierarchical Production Control Policies for Stochastic Manufacturing Systems," *Operations Research*, 45, 2, 1997, 258-274.

77 Sethi, S.P., Suo, W., Taksar, M.I., and Zhang, Q., "Optimal Production Planning in a Stochastic Manufacturing System with Long-Run Average Cost," *Journal of Optimization Theory and Applications*, 92, 1, 1997, 161-188.

78 Sethi, S.P., Zhang, Q., and Zhou, X., "Hierarchical Production Controls in a Stochastic Two-Machine Flowshop with a Finite Internal Buffer," *IEEE Transactions on Robotics and Automation*, 13, 1, 1997, 1-13.

79 Wang, Y., Sethi, S.P., and Van de Velde, S., "Minimizing Makespan in a Class of Reentrant Shops," *Operations Research*, 45, 5, 1997, 702-712.

80 Sethi, S.P. and Cheng, F., "Optimality of (s, S) Policies in Inventory Models with Markovian Demand," *Operations Research*, 45, 6, 1997, 931-939.

81 Sethi, S.P., "Some Insights into Near-Optimal Plans for Stochastic Manufacturing Systems," *Mathematics of Stochastic Manufacturing Systems*, G. Yin and Q. Zhang (Eds.); *Lectures in Applied Mathematics*, Vol. 33, American Mathematical Society, Providence, RI, 1997, 287-315.

82 Yang, J., Yan, H., and Sethi, S.P., "Optimal Production Control for Pull Flow Lines with Multiple Part Types," *Proceedings of the 35th IEEE CDC*, Kobe, Japan, 1996, 3847-3848.

83 Sethi, S.P., Suo, W., Taksar, M.I., and Yan, H., "Minimum Average-Cost Production Plan in a Multi-Product Stochastic Manufacturing System," *Proceedings of the 1996 IEEE Conference on Emerging Technologies and Factory Automation*, Kauai, HI, November 18-21, 1996, 361-365.

84 Sethi, S.P., Taksar, M.I., and Zhang Q., "A Hierarchical Decomposition of Capacity and Production Decisions in Stochastic Manufacturing Systems: Summary of Results," in *Proceedings: Workshop on Hierarchical Approaches in Forest Management in Public and Private Organizations*, Toronto, Canada, May 25-29, 1992; D. Martell, L. Davis and A. Weintraub (Eds.), *Petawawa National Forestry Institute Information Report PI-X0124*, Canadian Forest Service, 1996, 155-163.

85 Sethi, S.P., Suo, W., Taksar, M.I., and Zhang, Q., "Producing in a Manufacturing System with Minimum Average Cost," *Proceedings of the 2nd World Congress of Nonlinear Analysis*, Athens, Greece, July 16-17, 1996, *Nonlinear Analysis, Theory, Methods & Applications*, 30, 7, 1997, 4357-4363.

86 Yan, H., Lou, S., Sethi, S.P., Gardel, A., and Deosthali, P., "Testing the Robustness of Two-Boundary Control Policies in Semiconductor Manufacturing," *IEEE Transactions on Semiconductor Manufacturing*, 9, 2, 1996, 285-288.

87 Beyer, D. and Sethi, S.P., "Average Cost Optimality in Inventory Models with Markovian Demands: A Summary," *1996 M&SOM Conference Proceedings*, Hanover, NH, June 24-25, 1996, 40-45.

88 Sethi, S.P. and Zhou, X., "Optimal Feedback Controls in Deterministic Dynamic Two-Machine Flowshops," *Operations Research Letters*, 19, 5, 1996, 225-235.

89 Sethi, S.P. and Zhou, X., "Asymptotic Optimal Feedback Controls in Stochastic Dynamic Two-Machine Flowshops," in *Recent Advances in Control and Optimization of Manufacturing Systems*, G. Yin and Q. Zhang (Eds.), Lecture Notes in Control and Information Sciences, 214, Springer-Verlag, New York, 1996, 147-180.

90 Presman, E., Sethi, S.P., and Zhang, Q., "Optimal Feedback Production Planning in a Stochastic N-Machine Flowshop," *Automatica*, 31, 9, 1995, 1325-1332.

91 Kubiak, W., Sethi, S.P., and Sriskandarajah, C., "An Efficient Algorithm for a Job Shop Problem," *Annals of Operations Research*, 57, 1995, 203-216.

92 Krichagina, E., Lou, S., Sethi, S.P., and Taksar, M.I., "Diffusion Approximation for a Controlled Stochastic Manufacturing System with Average Cost Minimization," *Mathematics of Operations Research*, 20, 4, 1995, 895-922.

93 Chu, C., Proth, J.M., and Sethi, S.P., "Heuristic Procedures for Minimizing Makespan and the Number of Required Pallets," *European Journal of Operational Research*, 86, 3, 1995, 491-502.

94 Wang, Y., Sethi, S.P., Sriskandarajah, C., and Van de Velde, S.L., "Minimizing Makespan in Flowshops with Pallet Requirements: Computational Complexity," *Proceedings of the International Workshop on Intelligent Scheduling of Robots and Flexible Manufacturing Systems*, Holon, Israel, July 2, 1995, 105-114.

95 Sethi, S.P., Taksar, M.I., and Zhang, Q., "Hierarchical Capacity Expansion and Production Planning Decisions in Stochastic Manufacturing Systems," *Journal of Operations Management, Special Issue on Economics of Operations Management*, 12, 1995, 331-352.

96 Sethi, S.P. and Zhang, Q., "Hierarchical Production and Setup Scheduling in Stochastic Manufacturing Systems," *IEEE Transactions on Automatic Control*, 40, 5, 1995, 924-930.

97 Cheng, F. and Sethi, S.P., "Periodic Review Inventory Models with Markovian Demands," *Proceedings of the Instrument Society of America Annual Conference*, Toronto, Canada April 25-27, 1995, 95-104.

98 Sethi, S.P. and Zhang, Q., "Multilevel Hierarchical Decision Making in Stochastic Marketing - Production Systems," *SIAM Journal on Control and Optimization*, 33, 2, 1995, 528-553.

99 Sethi, S.P. and Zhang, Q., "Asymptotic Optimal Controls in Stochastic Manufacturing Systems with Machine Failures Dependent on Production Rates," *Stochastics and Stochastics Reports*, 48, 1994, 97-121.

100 Sethi, S.P. and Zhang, Q., "Multilevel Hierarchical Open-Loop and Feedback Controls in Stochastic Marketing-Production Systems," *IEEE Transactions on Robotics and Automation*, 10, 6, 1994, 831-839.

101 Sethi, S.P. and Zhang, Q., "Hierarchical Production and Setup Scheduling in Stochastic Manufacturing Systems," *Proceedings of the 33rd IEEE-CDC*, Lake Buena Vista, FL, Dec. 14-16, Vol. 2 of 4, 1994, 1571-1576.

102 Sethi, S.P. and Zhou, X., "Stochastic Dynamic Job Shops and Hierarchical Production Planning," *IEEE Transactions on Automatic Control*, 39, 10, 1994, 2061-2076.

103 Sethi, S.P., Taksar, M.I., and Zhang, Q., "Hierarchical Decomposition of Capacity Expansion and Production Scheduling Decision over Time under Uncertainty," *Proceedings of the 1994 Meeting of the Slovenian Informatika Society*, Portoroz, Slovena, September 13-15, 1994, 43-53.

104 Sethi, S.P., Taksar, M.I., and Zhang, Q., "Hierarchical Decomposition of Production and Capacity Investment Decisions in Stochastic Manufacturing Systems," Canadian National Contribution at *IFORS '93*, July 12-16, 1993, Lisbon, Portugal, *International Transactions of Operational Research*, 1, 4, 1994, 435-451.

105 Lou, S., Sethi, S.P., and Zhang, Q., "Optimal Feedback Production Planning in a Stochastic Two-Machine Flowshop," *European Journal of Operational Research*, 73, 2, 1994, 331-345.

106 Kubiak, W. and Sethi, S.P., "Optimal Just-in-Time Schedules for Flexible Transfer Lines," *International Journal of Flexible Manufacturing Systems*, 6, 2, 1994, 137-154.

107 Sethi, S.P. and Zhang, Q., "Hierarchical Production Planning in Dynamic Stochastic Manufacturing Systems: Asymptotic Optimality and Error Bounds," *Journal of Mathematical Analysis and Applications*, 181, 2, 1994, 285-319.

108 Sethi, S.P. and Zhang, Q., "Hierarchical Controls in Stochastic Manufacturing Systems," *SIAG/CST Newsletter*, 2, 1, 1994, 1-5.

109 Zhou, X. and Sethi, S.P., "A Sufficient Condition for Near Optimal Stochastic Controls and Its Application to Manufacturing Systems," *Applied Mathematics & Optimization*, 29, 1994, 67-92.

110 Sethi, S.P., Zhang, Q., and Zhou, X., "Hierarchical Controls in Stochastic Manufacturing Systems with Convex Costs," *Journal of Optimization Theory and Applications*, 80, 2, Feb. 1994, 303-321.

111 Krichagina, E., Lou, S., Sethi, S.P., and Taksar, M.I., "Diffusion Approximation for a Manufacturing System with Average Cost Minimization," *Proceedings 12th IFAC World Congress*, Sydney, Australia, 1993.

112 Presman, E., Sethi, S.P., and Zhang, Q., "Optimal Feedback Production Planning in a Stochastic N-machine Flowshop," *Proceedings of 12th World Congress of International Federation of Automatic Control*, Sydney, Australia, July 18-23, 4, 1993, 505-508.

113 Sethi, S.P., Yan, H., Zhang, Q., and Zhou, X., "Feedback Production Planning in a Stochastic Two-Machine Flowshop: Asymptotic Analysis and Computational Results," *International Journal Of Production Economics*, 30-31, 1993, 79-93.

114 Krichagina, E., Lou, S., Sethi, S.P., and Taksar, M.I., "Production Control in a Failure-Prone Manufacturing System: Diffusion Approximation and Asymptotic Optimality," *The Annals of Applied Probability*, 3, 2, 1993, 421-453.

115 Lou, S., Sethi, S.P., and Sorger, G., "Stability of Real-Time Lot Scheduling Policies for an Unreliable Machine," *IEEE Transactions on Automatic Control*, 37, 12, 1992, 1966-1970.

116 Sethi, S.P., Taksar, M.I., and Zhang, Q., "Capacity and Production Decisions in Stochastic Manufacturing Systems: An Asymptotic Optimal Hierarchical Approach," *Production and Operations Management*, 1, 4, 1992, 367-392.

117 Sethi, S.P. and Zhang, Q., "Multilevel Hierarchical Controls in Dynamic Stochastic Marketing-Production Systems," *Proceedings of the 31st IEEE-CDC*, Tucson, Arizona, Dec. 16-18, 1992, 2090-2095.

118 Sethi, S.P., Zhang, Q., and Zhou, X., "Hierarchical Controls in Stochastic Manufacturing Systems with Machines in Tandem," *Stochastics and Stochastics Reports*, 41, 1992, 89-118.

119 Sethi, S.P., Zhang, Q., and Zhou, X., "Hierarchical Production Controls in a Two-Machine Stochastic Flowshop with a Finite Internal Buffer," *Proceedings of the 31st IEEE-CDC*, Tucson, AZ, Dec. 16-18, 1992, 2074-2079.

120 Jaumard, B., Lou, S., Lu, S.H., Sethi, S.P., and Sriskandarajah, C., "Heuristics for the Design of Part-Orienting Systems," *International Journal of Flexible Manufacturing Systems*, 5, 1993, 167-185.

121 Sethi, S.P., Sriskandarajah, C., Sorger, G., Blazewicz, J., and Kubiak, W., "Sequencing of Parts and Robot Moves in a Robotic Cell," *International Journal of Flexible Manufacturing Systems*, 4, 1992, 331-358.

122 Zhou, X. and Sethi, S.P., "A Sufficient Condition for Near Optimal Stochastic Controls and its Application to an HMMS Model under Uncertainty," *Optimization: Techniques and Applications*, Vol. 1, K.H. Phua, C.M. Wang, W.Y. Yeong, T.Y. Leong, H.T. Loh, K.C. Tan, and F.S. Chou (Eds.) World Scientific, Singapore, 1992, 423-432.

123 Sethi, S.P., Taksar, M.I., and Zhang, Q., "Hierarchical Investment and Production Decisions in Stochastic Manufacturing Systems," *Stochastic Theory and Adaptive Control*, T. Duncan and B. Pasik-Duncan (Eds.), Springer-Verlag, New York 1992, 426-435.

124 Bylka, S. and Sethi, S.P., "Existence and Derivation of Forecast Horizons in Dynamic Lot Size Model with Nondecreasing Holding Costs," *Production and Operations Management*, 1, 2, 1992, 212-224.

125 Bylka, S., Sethi, S.P., and Sorger, G., "Minimal Forecast Horizons in Equipment Replacement Models with Multiple Technologies and General Switching Costs," *Naval Research Logistics*, 39, 1992, 487-507.

126 Sethi, S.P., Soner, H.M. Zhang, Q., and Jiang, J., "Turnpike Sets and Their Analysis in Stochastic Production Planning Problems," *Mathematics of Operations Research*, 17, 4, 1992, 932-950.

127 Sethi, S.P. and Zhang, Q., "Asymptotic Optimality in Hierarchical Control of Manufacturing Systems Under Uncertainty: State of the Art," *Operations Research Proceedings 1990*, Aug. 28-31, 1990, Vienna, Austria, W. Buhler, G. Feichtinger, R. Hartl, F. Radermacher, and P. Stahly (Eds.), Springer-Verlag, Berlin, 1992, 249-263.

128 Sethi, S.P., Zhang, Q., and Zhou, X.Y., "Hierarchical Production Planning in a Stochastic Two-Machine Flowshop," *Seventh Inter-*

national Working Seminar on Production Economics, Igls, Austria, Feb. 17-21, 1992, Pre-Print Volume 1, 473-488.

129 Lou, S., Yan, H., Sethi, S.P., Gardel, A., and Deosthali, P., "Using Simulation to Test the Robustness of Various Existing Production Control Policies," *Proceedings of the 1991 Winter Simulation Conference*, B. L. Nelson, W. D. Kelton and G. M. Clark (Eds.), 1991, 261-269.

130 Lehoczky, J.P., Sethi, S.P., Soner H., and Taksar, M.I., "An Asymptotic Analysis of Hierarchical Control of Manufacturing Systems under Uncertainty," *Mathematics of Operations Research*, 16, 3, 1991, 596-608.

131 Hall, N., Sethi, S.P., and Sriskandarajah, C., "On the Complexity of Generalized Due Date Scheduling Problems," *European Journal of Operational Research*, 51, 1, 1991, 101-109.

132 Chand, S., Sethi, S.P., and Sorger, G., "Forecast Horizons in the Discounted Dynamic Lot Size Model," *Management Science*, 38, 7, 1992, 1034-1048.

133 Jiang, J. and Sethi, S.P., "A State Aggregation Approach to Manufacturing Systems Having Machine States with Weak and Strong Interactions," *Operations Research*, 39, 6, 1991, 970-978.

134 Hall, N., Kubiak, W., and Sethi, S.P., "Earliness-Tardiness Scheduling Problems, II: Deviation of Completion Times about a Restrictive Common Due Date," *Operations Research*, 39, 5, 1991, 847-856.

135 Kubiak, W., Lou, S., and Sethi, S.P., "Equivalence of Mean Flow Time Problems and Mean Absolute Deviation Problems," *Operations Research Letters*, 9, 6, 1990, 371-374.

136 Kubiak, W. and Sethi, S.P., "A Note on 'Level Schedules for Mixed-Model Assembly Lines in Just-in-Time Production Systems'," *Management Science*, 37, 1, 1991, 121-122.

137 Lou, S., Sethi, S.P., and Sorger, G., "Analysis of a Class of Real-Time Multiproduct Lot Scheduling Policies," *IEEE Transactions on Automatic Control*, 36, 2, 1991, 243-248.

138 Sethi, A. and Sethi, S.P., "Flexibility in Manufacturing: An Abbreviated Survey," *Proceedings of the 1990 Pacific Conference on*

Manufacturing, Sydney and Melbourne, Australia, Dec. 17-21, 1990.

139 Sethi, S.P., Soner, H.M., Zhang, Q., and Jiang, J., "Turnpike Sets in Shochastic Production Planning Problems," *Proceedings of the 29th IEEE CDC*, Honolulu, HI, Dec. 5-7, 1990, 590-595.

140 Sethi, S.P., Sriskandarajah, C., Tayi, G.K., and Rao, M.R., "Heuristic Methods for Selection and Ordering of Part-Orienting Devices," *Operations Research*, 38, 1, 1990, 84-98.

141 Chand, S. and Sethi, S.P., "A Dynamic Lot Size Model with Learning in Setups," *Operations Research*, 38, 4, July-August 1990, 644-655.

142 Lou, S., Yan, H., Sethi, S.P., Gardel, A., and Deosthali, P., "Hub-Centered Production Control of Wafer Fabrication," *Proceedings of the IEEE Advanced Semiconductor Manufacturing Conference and Workshop*, Danvers, MA, Sept. 1990, 27-32.

143 Chand, S., Sethi, S.P., and Proth, J., "Existence of Forecast Horizons in Undiscounted Discrete Time Lot Size Models," *Operations Research*, 38, 5, Sept.-Oct. 1990, 884-892.

144 Sethi, A. and Sethi, S.P., "Flexibility in Manufacturing: A Survey," *International Journal of Flexible Manufacturing Systems*, 2, 1990, 289-328.

145 Blazewicz, J., Sethi, S.P., and Sriskandarajah, C., "Scheduling of Robot Moves and Parts in a Robotic Cell," *Proceedings of the Third ORSA/TIMS Special Interest Conference on FMS*, Cambridge, MA, August 14-16, 1989, 281-286.

146 Sriskandarajah, C. and Sethi, S.P., "Scheduling Algorithms for Flexible Flowshops: Worst and Average Case Performance," *European Journal of Operational Research*, 43, 2, 1989, 143-160.

147 Erlenkotter, D., Sethi, S.P., and Okada, N., "Planning or Surprise: Water Resources Development under Demand and Supply Uncertainty I: The General Model," *Management Science*, 35, 2, 1989, 149-163.

148 Sriskandarajah, C., Sethi, S.P., and Ladet, P., "Scheduling Methods for a Class of Flexible Manufacturing Systems," *Annals of Operations Research*, 17, 1989, 139-162.

149 Bhaskaran, S. and Sethi, S.P., "The Dynamic Lot Size Model with Stochastic Demands: A Decision Horizon Study," *INFOR*, 26, 3, 1988, 213-224.

150 Soner, H.M., Lehoczky, J.P., Sethi, S.P., and Taksar, M.I., "An Asymptotic Analysis of Hierarchical Control of Manufacturing Systems," *Proceedings of the 27th IEEE Conference on Decision and Control*, Austin, TX, Dec. 7-9, 1988, 1856-1857.

151 Rempala, R. and Sethi, S.P., "Forecast Horizons in Single Product Inventory Models," *Optimal Control Theory and Economic Analysis 3*, G. Feichtinger (Ed.), North-Holland, Amsterdam, 1988, 225-233.

152 Sriskandarajah, C., Ladet, P., and Sethi, S.P., "A Scheduling Method for a Class of Flexible Manufacturing Systems," *Proceedings of the 3rd International Conference on Advances in Production Management Systems (APMS 87)*, Winnipeg, Manitoba, Canada, August 1987, 3-18.

153 Sriskandarajah, C., Ladet, P., and Sethi, S.P., "A Scheduling Method for a Class of Flexible Manufacturing Systems," *Modern Production Management Systems*, A. Kusiak (Ed.), North-Holland, N.Y., 1987, 3-18.

154 Davis, M.H.A., Dempster, M.A.H., Sethi, S.P., and Vermes, D., "Optimal Capacity Expansion Under Uncertainty," *Advances in Applied Probability*, 19, 1987, 156-176.

155 Fleming, W.H., Sethi, S.P., and Soner, H.M., "An Optimal Stochastic Production Planning Problem with Randomly Fluctuating Demand," *SIAM Journal on Control and Optimization*, 25, 6, 1987, 1494-1502.

156 Sethi, S.P. and Sriskandarajah, C., "Optimal Selection and Ordering of Part-Orienting Devices: Series Configurations," *Proceedings of the 2nd International Conference on Production Systems*, Paris, France, April 1987, 1-14.

157 Sethi, S.P., "Forecast Horizons in Operations Management Problems: A Tutorial," *Proceedings of the 2nd International Conference on Production Systems*, Paris, France, April 1987, 37-48.

158 Soner, H.M., Fleming, W.H., and Sethi, S.P., "A Stochastic Production Planning Model with Random Demand," *Proceedings of*

the 24th IEEE Conference on Decision and Control, Fort Lauderdale, FL, Dec. 11-13, 1985, 1345-1346.

159 Sethi, S.P., Thompson, G.L., and Udayabhanu, V., "Profit Maximization Models for Exponential Decay Processes," *European Journal of Operational Research,* 22, 1, 1985, 101-115.

160 Bensoussan, A., Sethi, S.P., Vickson, R., and Derzko, N.A., "Stochastic Production Planning with Production Constraints: A Summary," *Proceedings of the 23rd IEEE Conference on Decision and Control,* Vol. 3, Las Vegas, NV, 1984, 1645-1646.

161 Sethi, S.P., "A Quantity Discount Lot Size Model with Disposals," *International Journal of Production Research,* 22, 1, 1984, 31-39.

162 Thompson, G.L., Sethi, S.P., and Teng, J., "Strong Planning and Forecast Horizons for a Model with Simultaneous Price and Production Decisions," *European Journal of Operational Research,* 16, 3, 1984, 378-388.

163 Sethi, S.P., "Applications of the Maximum Principle to Production and Inventory Problems," *Proceedings Third International Symposium on Inventories,* Budapest, Hungary, August 27-31, 1984, 753-756.

164 Hartl, R.F. and Sethi, S.P., "Optimal Control Problems with Differential Inclusions: Sufficiency Conditions and an Application to a Production-Inventory Model," *Optimal Control Applications & Methods,* 5, 4, Oct.-Dec. 1984, 289-307.

165 Teng, J., Thompson, G.L., and Sethi, S.P., "Strong Decision and Forecast Horizons in a Convex Production Planning Problem," *Optimal Control Applications & Methods,* 5, 4, Oct-Dec 1984, 319-330.

166 Bensoussan, A., Sethi, S.P., Vickson, R., and Derzko, N.A., "Stochastic Production Planning with Production Constraints," *SIAM Journal on Control and Optimization,* 22, 6, Nov. 1984, 920-935.

167 Browne, J., Dubois, D., Rathmill, K., Sethi, S.P., and Stecke, K., "Classification of Flexible Manufacturing Systems," *The FMS Magazine,* April 1984, 114-117.

168 D'Cruz, J., Sethi, S.P., and Sunderji, A.K., "The Role of MNCs in the Transfer of FMS Technology: Implications for Home and Host

Countries," *Proceedings of the First ORSA/TIMS Conference on FMS*, Ann Arbor, MI., 1984.

169 Sambandam, N. and Sethi, S.P., "Heuristic Flow-Shop Scheduling to Minimize Sum of Job Completion Times," *Canadian Journal Of Administrative Sciences*, 1, Dec. 1984, 308-320.

170 Chand, S. and Sethi, S.P., "Finite-Production-Rate Inventory Models with First-and-Second-Shift Setups," *Naval Research Logistics Quarterly*, 30, 1983, 401-414.

171 Chand, S. and Sethi, S.P., "Planning Horizon Procedures for Machine Replacement Models with Several Possible Replacement Alternatives," *Naval Research Logistics Quarterly*, 29, 3, 1982, 483-493.

172 Sethi, S.P. and Chand, S., "Multiple Finite Production Rate Dynamic Lot Size Inventory Models," *Operations Research*, 29, 5, Sept-Oct 1981, 931-944.

173 Sethi, S.P. and Thompson, G.L., "Simple Models in Stochastic Production Planning," in *Applied Stochastic Control in Econometrics and Management Science*, A. Bensoussan, P. Kleindorfer and C. Tapiero (Eds.), North-Holland, New York, NY, 1981, 295-304.

174 Thompson, G.L. and Sethi, S.P., "Turnpike Horizons for Production Planning," *Management Science*, 26, 3, 1980, 229-241.

175 Sethi, S.P. and Chand, S., "Planning Horizon Procedures in Machine Replacement Models," *Management Science*, 25, 2, Feb. 1979, 140-151; "Errata," 26, 3, 1980, 342.

176 Sethi, S.P. and Thompson, G.L., "Christmas Toy Manufacturer's Problem: An Application of the Stochastic Maximum Principle," *Opsearch*, 14, 3, 1977, 161-173.

177 Sethi, S.P. "Simultaneous Optimization of Preventive Maintenance and Replacement Policy for Machines: A Modern Control Theory Approach," *AIIE Transactions*, 5, 2, June 1973, 156-163.

178 Sethi, S.P. and Morton, T.E., "A Mixed Optimization Technique for the Generalized Machine Replacement Problem," *Naval Research Logistics Quarterly*, 19, 3, Sept. 1972, 471-481.

Research Publications - Finance and Economics

1 Sethi, S.P. and Taksar, M.I., "Optimal Financing of a Corporation Subject to Random Returns: A Summary," *Proceedings of 41st IEEE CDC*, Las Vegas, NV, December 10-13, 2002, 395-397.

2 Sethi, S.P., "Optimal Consumption – Investment Decisions Allowing for Bankruptcy: A Brief Survey" in *Markov Processes and Controlled Markov Chains*, H. Zhenting , J. A. Filar, and A. Chen (Eds.), Kluwer Academic Publishers, Dordrecht, 2002, 371-387.

3 Sethi, S.P. and Taksar, M.I., "Optimal Financing of a Corporation Subject to Random Returns," *Mathematical Finance*, 12, 2, April 2002, 155-172.

4 Gordon, M.J. and Sethi, S.P., "Consumption and Investment When Bankruptcy is Not a Fate Worse Than Death," in *Method, Theory and Policy in Keynes, Essays in honor of Paul Davidson: Volume 3*, Philip Arestis (Ed.), Edward Elgar Publishing, Northampton, MA, 1998, 88-108.

5 Sethi, S.P., "Optimal Consumption - Investment Decisions Allowing for Bankruptcy: A Survey" in *Worldwide Asset and Liability Modeling*, W. T. Ziemba and J. M. Mulvey (Eds.), Cambridge University Press, Cambridge, U.K., 1998, 387-426

6 Cadenillas, A. and Sethi, S.P., "Consumption–Investment Problem with Subsistence Consumption, Bankruptcy, and Random Market Coefficients," *Journal of Optimization Theory and Applications*, 93, 1997, 243-272.

7 Cadenillas, A. and Sethi, S.P., "The Consumption-Investment Problem with Subsistence Consumption, Bankruptcy, and Random Market Coefficients," Chapter 12 in Sethi, S.P., *Optimal Consumption and Investment with Bankruptcy*, Kluwer, Norwell, MA, 1997, 247-280.

8 Presman, E. and Sethi, S.P., "Consumption Behavior in Investment/ Consumption Problems with Bankruptcy," Chapter 9 in Sethi, S.P., *Optimal Consumption and Investment with Bankruptcy*, Kluwer, Norwell, MA, 1997, 185-205.

9 Presman, E. and Sethi, S.P., "Equivalence of Objective Functionals in Infinite Horizon and Random Horizon Problems," Chapter 10 in Sethi, S.P., *Optimal Consumption and Investment with Bankruptcy*, Kluwer, Norwell, MA, 1997, 207-216.

10 Gordon, M. and Sethi, S.P., "A Contribution to Micro Foundation for Keynesian Macroeconomic Models," Chapter 11 in Sethi, S.P., *Optimal Consumption and Investment with Bankruptcy*, Kluwer, Norwell, MA, 1997, 217-244.

11 Presman, E. and Sethi, S.P., "Risk-Aversion in Consumption/ Investment Problems with Subsistence Consumption and Bankruptcy," Chapter 8 in Sethi, S.P., *Optimal Consumption and Investment with Bankruptcy*, Kluwer, Norwell, MA, 1997, 155-184.

12 Presman, E. and Sethi, S.P., "Distribution of Bankruptcy Time in a Consumption/Portfolio Problem, Chapter 7 in Sethi, S.P., *Optimal Consumption and Investment with Bankruptcy*, Kluwer, Norwell, MA, 1997, 145-154.

13 Sethi, S.P., Taksar, M.I., and Presman, E., "Explicit Solution of a General Consumption/Investment Problem with Subsistence Consumption and Bankruptcy," Chapter 6 in Sethi, S.P., *Optimal Consumption and Investment with Bankruptcy*, Kluwer, Norwell, MA, 1997, 119-143.

14 Sethi, S.P. and Taksar, M.I., "Infinite-Horizon Investment Consumption Model with a Nonterminal Bankruptcy," Chapter 4 in Sethi, S.P., it Optimal Consumption and Investment with Bankruptcy, Kluwer, Norwell, MA, 1997, 67-84.

15 Presman, E. and Sethi, S.P., "Risk Aversion in Consumption/ Investment Problems," Chapter 5 in Sethi, S.P., *Optimal Consumption and Investment with Bankruptcy*, Kluwer, Norwell, MA, 1997, 85-116.

16 Sethi, S.P. and Taksar, M.I., "A Note on Merton's 'Optimum Consumption and Portfolio Rules in a Continuous-Time Model'," Chapter 3 in Sethi, S.P., *Optimal Consumption and Investment with Bankruptcy*, Kluwer, Norwell, MA, 1997, 59-65.

17 Karatzas, I., Lehoczky, J.P., Sethi, S.P., and Shreve, S.E., "Explicit Solution of a General Consumption/Investment Problem," Chapter 2 in Sethi, S.P., *Optimal Consumption and Investment with Bankruptcy*, Kluwer, Norwell, MA, 1997, 23-56.

18 Lehoczky, J.P., Sethi, S.P. and Shreve, S.E., "A Martingale Formulation for Optimal Consumption/Investment Decision Making," Chapter 15 in Sethi, S.P., *Optimal Consumption and Investment with Bankruptcy*, Kluwer, Norwell, MA, 1997, 379-406.

19 Lehoczky, J.P., Sethi, S.P. and Shreve, S.E., "Optimal Consumption and Investment Policies Allowing Consumption Constraints, Bankruptcy, and Welfare," Chapter 14 in Sethi, S.P., *Optimal Consumption and Investment with Bankruptcy*, Kluwer, Norwell, MA, 1997, 303-378.

20 Sethi, S.P., Gordon, M.J. and Ingham, B., "Optimal Dynamic Consumption and Portfolio Planning in a Welfare State, Chapter 13 in Sethi. S.P., *Optimal Consumption and Investment with Bankruptcy*, Kluwer, Norwell, MA, 1997, 285-302.

21 Sethi, S.P., "Optimal Consumption – Investment Decisions Allowing for Bankruptcy: A Survey" in *Proceedings of The First International Conference on Pacific Basin Business and Economics*, June 15-16, 1996, National Central University, Chungli, Taiwan, 317-355.

22 Sethi, S.P., "When Does the Share Price Equal the Present Value of Future Dividends? - A Modified Dividend Approach," *Economic Theory*, 8, 1996, 307-319.

23 Presman, E. and Sethi, S.P., "Distribution of Bankruptcy Time in a Consumption/Portfolio Problem," *Journal Economic Dynamics and Control*, 20, 1996, 471-477.

24 Sethi, S.P. and Taksar, M.I., "Infinite-Horizon Investment Consumption Model with a Nonterminal Bankruptcy," *Journal of Optimization Theory and Applications*, 74, 2, Aug. 1992, 333-346.

25 Sethi, S.P., Taksar, M.I. and Presman, E., "Explicit Solution of a General Consumption/Portfolio Problem with Subsistence Consumption and Bankruptcy," *Journal of Economic Dynamics and Control*, 16, 1992, 747-768; "Erratum," 19, 1995, 1297-1298.

26 Sethi, S.P., Derzko, N.A. and Lehoczky, J.P., "A Stochastic Extension of the Miller-Modigliani Framework," *Mathematical Finance*, 1, 4, 1991, 57-76; "Erratum," 6, 4, October 1996, 407-408.

27 Presman, E. and Sethi, S.P., "Risk-Aversion Behavior in Consumption/Investment Problems," *Mathematical Finance*, 1, 1, 1991, 101-124; "Erratum," 1, 3, July 1991, p. 86.

28 Sethi, S.P. and Sorger, G., "An Exercise in Modeling of Consumption, Import, and Export of an Exhaustible Resource," *Optimal Control Applications & Methods*, 11, 1990, 191-196.

29 Sethi, S.P. and Taksar, M.I., "A Note on Merton's 'Optimum Consumption and Portfolio Rules in a Continuous-Time Model'," *Journal of Economic Theory*, 46, 1988, 395-401.

30 Sethi, S.P. and Taksar, M.I., "Optimal Consumption and Investment Policies with Bankruptcy Modelled by a Diffusion with Delayed Reflection," *Proceedings of 25th IEEE Conference on Decision and Control*, Athens, Greece, Dec. 1986, 267-269.

31 Sethi, S.P., "Dynamic Optimal Consumption-Investment Problems: A Survey," *Proceedings of the 1986 IFAC Workshop on Modelling, Decision and Game with Applications to Social Phenomena*, Beijing, China, August 11-15, l986, 116-121.

32 Karatzas, I., Lehoczky, J.P., Sethi, S.P. and Shreve, S.E., "Explicit Solution of a General Consumption/Investment Problem," *Mathematics of Operations Research*, 11, 2, May 1986, 261-294.

33 Karatzas, I., Lehoczky, J.P., Sethi, S.P. and Shreve, S.E., "Explicit Solution of a General Consumption/Investment Problem," *Proceedings of the International Conference on Stochastic Optimization*, Kiev, 1984, V. Arkin, A. Shiraev and R. Wets (Eds.), Lecture Notes in Control and Information Sciences, 81, Springer-Verlag, 1986, 59-69.

34 Karatzas, I., Lehoczky, J.P., Sethi, S.P. and Shreve, S.E., "Explicit Solution of a General Consumption/Investment Problem: A Summary," *Proceedings of the 3rd Bad Honnef Conference on Stochastic Differential Systems*, Bonn, West Germany, Lecture Notes on Control and Information Sciences, Springer-Verlag, June 1985.

35 Lehoczky, J.P., Sethi, S.P. and Shreve, S.E., "A Martingale Formulation for Optimal Consumption/Investment Decision Making," *Optimal Control Theory and Economic Analysis 2*, G. Feichtinger (Ed.), North-Holland, Amsterdam, 1985, 135-153.

36 Sethi, S.P., Derzko, N.A. and Lehoczky, J.P., "General Solution of the Stochastic Price-Dividend Integral Equation: A Theory of Financial Valuation," *SIAM Journal on Mathematical Analysis*, 15, 6, Nov. 1984, 1100-1113.

37 Sethi, S.P., "A Note on a Simplified Approach to the Valuation of Risky Streams," *Operations Research Letters*, 3, 1, April 1984, 13-17.

38 Lehoczky, J.P., Sethi, S.P. and Shreve, S.E., "Optimal Consumption and Investment Policies Allowing Consumption Constraints, Bankruptcy, and Welfare," *Mathematics of Operations Research*, 8, 4, Nov. 1983, 613-636.

39 Sethi, S.P. and Thompson, G.L., "Planning and Forecast Horizons in a Simple Wheat Trading Model," *Operations Research in Progress*, G. Feichtinger and P. Kall (Eds.), Reidel Pub. Co., 1982, 203-214.

40 Derzko, N.A. and Sethi, S.P., "General Solution of the Price-Dividend Integral Equation," *SIAM Journal on Mathematical Analysis*, 13, 1, Jan. 1982, 106-111.

41 Lehoczky, J.P., Sethi, S.P. and Shreve, S.E., "Degenerate Diffusion Processes in Portfolio Management," *Proceedings of the 1982 American Control Conference*, Arlington, VA, June 14-16, 1982.

42 Sethi, S.P., Derzko, N.A. and Lehoczky, J.P., "Mathematical Analysis of the Miller-Modigliani Theory," *Operations Research Letters*, 1, 4, Sept. 1982, 148-152.

43 Sethi, S.P. and Lehoczky, J.P., "A Comparison of the Ito and Stratonovich Formulations of Problems in Finance," *Journal of Economic Dynamics and Control*, 3, 1981, 343-356.

44 Derzko, N.A. and Sethi, S.P., "Optimal Exploration and Consumption of a Natural Resource: Stochastic Case," *International Journal of Policy Analysis*, 5, 3, Sept. 1981, 185-200.

45 Derzko, N.A. and Sethi, S.P., "Optimal Exploration and Consumption of a Natural Resource: Deterministic Case," *Optimal Control Applications & Methods*, 2, 1, 1981, 1-21.

46 Bhaskaran, S. and Sethi, S.P., "Planning Horizons for the Wheat Trading Model," *Proceedings of AMS 81 Conference, 5: Life, Men, and Societies*, 1981, 197-201.

47 Sethi, S.P., "Optimal Depletion of an Exhaustible Resource," *Applied Mathematical Modeling*, 3, Oct. 1979, 367-378.

48 Sethi, S.P., Gordon, M.J., and Ingham, B., "Optimal Dynamic Consumption and Portfolio Planning in a Welfare State," *TIMS Studies in the Management Science*, 11, 1979, 179-196.

49 Sethi, S.P., "Optimal Equity and Financing Model of Krouse and Lee Corrections and Extensions," *Journal of Financial and Quantitative Analysis*, 13, 3, Sept. 1978, 487-505.

50 Quirin, G.D., Sethi, S.P. and Todd, J.D., "Market Feedbacks and the Limits to Growth," *INFOR*, 15, 1, Feb. 1977, 1-21. (Also reported by Jean Poulain in the French Canadian Daily *La Presse*, Oct. 8, 1977.)

51 Sethi, S.P. and McGuire, T.W., "Optimal Skill Mix: An Application of the Maximum Principle for Systems with Retarded Controls," *Journal of Optimization Theory and Applications*, 23, 2, Oct. 1977, 245-275.

52 Sethi, S.P., "A Linear Bang-Bang Model of Firm Behavior and Water Quality," *IEEE Transactions on Automatic Control*, AC-22, 5, Oct. 1977, 706-714.

53 Sethi, S.P., "Optimal Investment Policy: An Application of Stoke's Theorem," *Journal of Optimization Theory and Applications*, 18, 2, Feb. 1976, 229-233.

54 Largay, J. and Sethi, S.P., "A Simplified Control Theoretic Approach to Gift and Estate Tax Planning," *Proceedings of the Midwest AIDS Conference*, Detroit, MI, May 6-7, 1976, 217-220.

55 Brooks, R., Luan, P., Pritchett, J. and Sethi, S.P., "Examples of Dynamic Optimal Entry into a Monopoly," *SCIMA: Journal of Management Science and Applied Cybernetics*, 3, 1, 1974, 1-8.

56 Sethi, S.P. and McGuire, T.W., "An Application of the Maximum Principle to a Heterogenous Labor Model with Retarded Controls," *Optimal Control Theory and its Applications, Part II*, in Lecture Notes in Economics and Mathematical Systems, 106, Springer-Verlag, Berlin, 1974, 338-384.

57 Sethi, S.P., "Note on Modeling Simple Dynamic Cash Balance Problems," *Journal of Financial and Quantitative Analysis*, 8, Sept. 1973, 685-687; "Errata," 13, Sept. 1978, 585-586.

58 McGuire, T. and Sethi, S.P., "Optimal and Market Control in a Dynamic Economic System with Endogenous Heterogeneous Labor," *Proceedings of the IFORS/IFAC International Conference*, Coventry, England, July 9-12, 1973, IEE Conference Publication No. 101, 172-185.

59 Sethi, S.P., "The Management of a Polluting Firm," *Proceedings of the Ninth Annual Meeting of TIMS Southeastern Chapter*, Atlanta, GA, Oct. 18-19, 1973, 264-270.

60 Sethi, S.P., "A Useful Transformation of Hamiltonians Occurring in Optimal Control Problems in Economic Analyses," *Journal of Management Science and Applied Cybernetics*, 2, 3, 1973, 115-131.

61 Sethi, S.P., "A Note on a Planning Horizon Model of Cash Management," *Journal of Financial and Quantitative Analysis*, January 1971, 6, 1, 659-664.

62 Sethi, S.P. and Thomson, G.L., "Applications of Mathematical Control Theory to Finance: Modeling Simple Dynamic Cash Balance Problems," *Journal of Financial and Quantitative Analysis*, Dec. 1970, 5, 4&5, 381-394.

Research Publications - Marketing

1 Haruvy, E., Sethi, S.P., and Zhou, J., "Open Source Development with a Commercial Complementary Product or Service," *POM; Special Issue on Managemenr of Technology*, conditionally accepted.

2 Sethi, S.P., Forward in *Handbook of Niche Marketing Principles and Practice*, Tevfik Dalgic (Ed.) The Haworth Reference Press, New York, NY., 2006, xv-xvi.

3 Bass, F.M., Krishnamoorthy, A., Prasad, A., and Sethi, S.P., "Generic and Brand Advertising Strategies in a Dynamic Duopoly," *Marketing Science*, 24, 4, 2005, 556-568.

4 Hartl, R.F., Novak, A.J., Rao, A.G., and Sethi, S.P., "Dynamic Pricing of a Status Symbol," *Proceedings of Fourth World Congress on Nonlinear Analysts*, July 2004, Orlando, FL.

5 Haruvy, E., Prasad, A., Sethi, S.P., and Zhang, R., "Optimal Firm Contributions to Open Source Software," *Optimal Control and Dynamic Games, Applications in Finance, Management Science and Economics*, C. Deissenberg and R.F. Hartl (Eds.), Springer, Netherlands, 2005, 197-212.

6 Bass, F.M., Krishnamoorthy, A., Prasad, A., and Sethi, S.P., "Advertising Competition with Market Expansion for Finite Horizon Firms," *Journal of Industrial and Management Optimization*, 1, 1, February 2005, 1-19.

7 Kumar, S., Li, Y., and Sethi, S.P., "Optimal Pricing and Advertising Policies for Web Services," *Proceedings of 14th Annual Workshop on Information Technologies Systems (WITS'04)*, Washington, DC, December 2004, 104-109.

8 Prasad, A. and Sethi, S.P., "Competitive Advertising under Uncertainty: Stochastic Differential Game Approach," *Journal of Optimization Theory and Applications*, 123, 1, October 2004, 163-185.

9 Haruvy, E., Prasad, A., and Sethi, S.P., "Harvesting Altruism in Open Source Software Development," *Journal of Optimization Theory and Applications*, 118, 2, August 2003, 381-416.

10 Hartl, R.F., Novak, A.J., Rao, A.G., and Sethi, S.P., "Optimal Pricing of a Product Diffusing in Rich and Poor Populations," *Journal of Optimization Theory and Applications*, 117, 2, May 2003, 349-375.

11 Sethi, S.P. and Bass, F.M., "Optimal Pricing in a Hazard Rate Model of Demand," *Optimal Control Applications and Methods*, 24, 2003, 183-196.

12 Feichtinger, G., Hartl, R.F. and Sethi, S.P., "Dynamic Optimal Control Models in Advertising: Recent Developments," *Management Science*, 40, 2, Feb. 1994, 195-226.

13 Seidman, T.I., Sethi, S.P. and Derzko, N.A., "Dynamics and Optimization of a Distributed Sales-Advertising Model," *Journal of Optimization Theory and Applications*, 52, 3, March 1987, 443-462.

14 Sethi, S.P., "Optimal Long-Run Equilibrium Advertising Level for the Blattberg-Jeuland Model," *Management Science*, 29, 12, Dec. 1983, 1436-1443.

15 Sethi, S.P., "Deterministic and Stochastic Optimization of a Dynamic Advertising Model," *Optimal Control Application and Methods*, 4, 2, 1983, 179-184.

16 Caines, P., Keng, C.W. and Sethi, S.P., "Causality Analysis and Multivariate Autoregressive Modelling with an Application to Supermarket Sales Analysis," *Journal of Economic Dynamics and Control*, 3, 1981, 267-298.

17 Sethi, S.P. and Lee, S.C., "Optimal Advertising for the Nerlove-Arrow Model Under a Replenishable Budget," *Optimal Control Applications & Methods*, 2, 1981, 165-173.

18 Bourguignon, F. and Sethi, S.P., "Dynamic Optimal Pricing and (Possibly) Advertising in the Face of Various Kinds of Potential Entrants," *Journal of Economic Dynamics and Control*, 3, 1981, 119-140.

19 Sethi, S.P., "A Note on the Nerlove-Arrow Model Under Uncertainty," *Operations Research*, 27, 4, July-August 1979, 839-842; "Erratum", 28, 4, July-August 1980, 1026-1027.

20 Sethi, S.P., "Optimal Advertising Policy with the Contagion Model," *Journal of Optimization Theory and Applications*, 29, 4, Dec. 1979, 615-627.

21 Deal, K., Sethi, S.P. and Thompson, G.L., "A Bilinear-Quadratic Differential Game in Advertising," *Control Theory in Mathematical Economics*, P.T. Lui & J.G. Sutinen (Eds.), Marcel Dekker, Inc., N.Y., 1979, 91-109.

22 Pekelman, D. and Sethi, S.P., "Advertising Budgeting, Wearout and Copy Replacement," *Journal of Operational Research Society*, 29, 7, 1978, 651-659.

23 Caines, P.E., Sethi, S.P. and Brotherton, T.W., "Impulse Response Identification and Causality Detection for the Lydia-Pinkham Data," *Annals of Economic and Social Measurement*, 6, 2, 1977, 147-163.

24 Sethi, S.P., "Dynamic Optimal Control Models in Advertising: A Survey," SIAM Review, 19, 4, Oct. 1977, 685-725.

25 Sethi, S.P., "Optimal Advertising for the Nerlove-Arrow Model Under a Budget Constraint," *Operational Research Quarterly*, 28, 3 (ii), 1977, 683-693.

26 Sethi, S.P., "Optimal Control of a Logarithmic Advertising Model," *Operational Research Quarterly*, 26, 2(i), 1975, 317-319.

27 Sethi, S.P., "Optimal Institutional Advertising: Minimum-Time Problem," *Journal of Optimization Theory and Applications*, 14, 2, Aug. 1974, 213-231.

28 Sethi, S.P., "Some Explanatory Remarks on Optimal Control for the Vidale-Wolfe Advertising Model," *Operations Research*, 22, 5, Sept.-Oct. 1974, 1119-1120.

29 Sethi, S.P., "Optimal Control Problems in Advertising," *Optimal Control Theory and its Applications, Part II*, in Lecture Notes in

Economics and Mathematical Systems, Vol. 106, Springer-Verlag, Berlin, 1974, 301-337.

30 Sethi, S.P., Turner, R. E. and Neuman, C. P., "Policy Implications of an Inter-temporal Analysis of Advertising Budgeting Models," *Proceedings of Midwest AIDS Conference*, Michigan State University, April 13-14, 1973, A15-A18.

31 Sethi, S.P., "Optimal Control of the Vidale-Wolfe Advertising Model," *Operations Research*, 21, 4, 1973, 998-1013.

Research Publications - Optimization Theory and Miscellaneous Applications

1 Feng, Q., Mookerjee, V. S., and Sethi, S.P., "Optimal Policies for the Sizing and Timing of Software Maintenance Projects," *European Journal of Operational Research*, in press.

2 Dawande, M., Gavirneni, S., Naranpanawe, S., and Sethi. S.P., "Computing Minimal Forecast Horizons: An Integer Programming Approach," *Journal of Mathematical Modeling and Algorithms*, in press.

3 Ji, Y., Mookerjee, V., and Sethi, S.P., "Optimal Software Development: A Control Theoretic Approach," *Information Systems Research*, 16, 3, 2005, 292-306.

4 Ji, Y., Mookerjee, V., and Sethi, S. P., "An Integrated Planning Model of Systems Development and Release," *Proceedings of Workshop on Information Technologies and Systems*, Seattle, WA, December 2003, 55-60.

5 Sethi, S.P. and Zhang, Q., "Near Optimization of Stochastic Dynamic Systems by Decomposition and Aggregation," *Optimal Control Applications and Methods*, 22, 2001, 333-350.

6 Sethi, S.P. and Zhang, Q., "Near Optimization of Dynamic Systems by Decomposition and Aggregation," *Journal of Optimization Theory and Applications*, 99, 1, 1998, 1-22.

7 Hartl, R.F., Sethi, S.P. and Vickson, R.G., "A Survey of the Maximum Principles for Optimal Control Problems with State Constraints," *SIAM Review*, 37, 2, June 1995, 181-218.

8 Krass, D., Sethi, S.P., and Sorger, G., "Some Complexity Issues in a Class of Knapsack Problems: What Makes a Knapsack Problem

'Hard'?" *INFOR*, Special Issue on Knapsack Problems, 32, 3, 1994, 149-162.

9 Sethi, S.P. and Kubiak, W., "Complexity of a Class of Nonlinear Combinatorial Problems Related to Their Linear Counterparts," *European Journal of Operational Research*, 73, 3, 1994, 569-576.

10 Rempala, R. and Sethi, S.P., "Decision and Forecast Horizons for One-Dimensional Optimal Control Problems: Existence Results and Applications," *Optimal Control Applications & Methods*, 13, 1992, 179-192.

11 Sethi, S.P. and Sorger, G., "A Theory of Rolling Horizon Decision Making," *Annals of Operations Research*, 29, 1991, 387-416.

12 Sethi, S.P. and Taksar, M.I., "Deterministic Equivalent for a Continuous-Time Linear-Convex Stochastic Control Problem," *Journal of Optimization Theory and Applications*, 64, 1, Jan. 1990, 169-181.

13 Sethi, S.P., "Dynamic Optimization: Decision and Forecast Horizons," *Systems and Control Encyclopedia Supplementary Volume 1*, Madan G. Singh (Ed.), Pergamon Press, Oxford, U.K., 1990, 192-198.

14 Kulkarni, V.G. and Sethi, S.P., "Deterministic Retrial Times are Optimal in Queues with Forbidden States," *INFOR*, 27, 3, Aug. 1989, 374-386.

15 Bes, C. and Sethi, S.P., "Solution of a Class of Stochastic Linear-Convex Control Problems Using Deterministic Equivalents," *Journal of Optimization Theory and Applications*, 62, 1, July 1989, 17-27.

16 Sethi, S.P. and Sorger, G., "Concepts of Forecast Horizons in Stochastic Dynamic Games," *Proceedings of the 28th IEEE Conference on Decision and Control*, Tampa, FL, Dec. 13-15, 1989, 195-197.

17 Bes, C. and Sethi, S.P., "Concepts of Forecast and Decision Horizons: Applications to Dynamic Stochastic Optimization Problems," *Mathematics of Operations Research*, 13, 2, May 1988, 295-310.

18 Sethi, S.P. and Taksar, M.I., "Deterministic and Stochastic Control Problems with Identical Optimal Cost Functions," *Analysis and Optimization of Systems*, A. Bensoussan and J. L. Lions (Eds.),

Lecture Notes in Control and Information Sciences, Springer-Verlag, New York, 1988, 641-645.

19 Bhaskaran, S. and Sethi, S.P., "Decision and Forecast Horizons in a Stochastic Environment: A Survey," *Optimal Control Applications & Methods*, 8, 1987, 201-217.

20 Sethi, S.P. and Bes, C., "Dynamic Stochastic Optimization Problems in the Framework of Forecast and Decision Horizons," *Proceedings of Optimization Days Conference*, Montreal, Quebec, April 30-May 2, 1986, Advances in Optimization and Control, H.A. Eiselt and G. Pederzoli (Eds.) Lecture Notes in Economics and Mathematical Systems, 302, Springer-Verlag, 1988, 230-246.

21 Hartl, R.F. and Sethi, S.P., "Solution of Generalized Linear Optimal Control Problems Using a Simplex-like Method in Continuous Time I: Theory," *Optimal Control Theory and Economic Analysis 2*, G. Feichtinger (Ed.), Elsevier Science Publishers B.V. (North-Holland), 1985, 45-62.

22 Hartl, R.F. and Sethi, S.P., "Solution of Generalized Linear Optimal Control Problems Using a Simplex-like Method in Continuous Time II: Examples," *Optimal Control Theory and Economic Analysis 2*, G. Feichtinger (Ed.), Elsevier Science Publishers B.V. (North-Holland), 1985, 63-87.

23 Sethi, S.P. and Bhaskaran, S., "Conditions for the Existence of Decision Horizons for Discounted Problems in a Stochastic Environment: A Note," *Operations Research Letters*, 4, 2, July 1985, 61-64.

24 Hartl, R.F. and Sethi, S.P., "Optimal Control of a Class of Systems with Continuous Lags: Dynamic Programming Approach and Economic Interpretations," *Journal of Optimization Theory and Applications*, 43, 1, May 1984, 73-88.

25 Haurie, A., Sethi, S.P. and Hartl, R.F., "Optimal Control of an Age-Structured Population Model with Applications to Social Services Planning," *Journal of Large Scale Systems*, 6, 1984, 133-158.

26 Haurie, A. and Sethi, S.P., "Decision and Forecast Horizons, Agreeable Plans, and the Maximum Principle for Infinite Horizon Control Problems," *Operations Research Letters*, 3, 5, Dec. 1984, 261-265.

27 Derzko, N.A., Sethi, S.P. and Thompson, G.L., "Necessary and Sufficient Conditions for Optimal Control of Quasilinear Partial Differential Systems," *Journal of Optimization Theory and Applications*, 43, 1, May 1984, 89-101.

28 Sethi, S.P., "Applications of Optimal Control to Management Science Problems," *Proceedings of the 1983 World Conference on Systems*, Caracas, Venezuela, July 11-15, 1983.

29 Hartl, R.F. and Sethi, S.P., "A Note on the Free Terminal Time Transversality Condition," *Zeitschrift fur Operations Research, Series: Theory*, 27, 5, 1983, 203-208.

30 Bhaskaran, S. and Sethi, S.P., "Planning Horizon Research - The Dynamic Programming/Control Theory Interface," *Proceedings of International AMSE Winter Symposium*, Bermuda, March 1-3, 1983, 155-160.

31 Sethi, S.P., Drews, W.P. and Segers, R.G., "A Unified Framework for Linear Control Problems with State-Variable Inequality Constraints," *Journal of Optimization Theory and Applications*, 36, 1, Jan. 1982, 93-109.

32 Sethi, S.P. and Thompson, G.L., "A Tutorial on Optimal Control Theory," *INFOR*, 19, 4, 1981, 279-291.

33 Dantzig, G.B. and Sethi, S.P., "Linear Optimal Control Problems and Generalized Linear Programs," *Journal of Operational Research Society*, 32, 1981, 467-476.

34 Derzko, N.A., Sethi, S.P. and Thompson, G.L., "Distributed Parameter Systems Approach to the Optimal Cattle Ranching Problem," *Optimal Control Applications & Methods*, 1, 1980, 3-10.

35 Bookbinder, J.H. and Sethi, S.P., "The Dynamic Transportation Problem: A Survey," *Naval Research Logistics Quarterly*, 27, 1, March 1980, 65-87.

36 Sethi, S.P., "Optimal Pilfering Policies for Dynamic Continuous Thieves," *Management Science*, 25, 6, June 1979, 535-542.

37 Caines, P.E. and Sethi, S.P., "Recursiveness, Causality, and Feedback," *IEEE Transactions on Automatic Control*, AC-24, 1, Feb. 1979, 113-115.

38 Sethi, S.P. and Staats, P. W., "Optimal Control of Some Simple Deterministic Epidemic Models," *Journal of Operational Research Society*, 29, 2, 1978, 129-136.

39 Sethi, S.P., "Optimal Quarantine Programmes for Controlling an Epidemic Spread," *Journal of Operational Research Society*, 29, 3, 1978, 265-268.

40 Sethi, S.P., "A Survey of Management Science Applications of the Deterministic Maximum Principle," *TIMS Studies in the Management Science*, 9, 1978, 33-68.

41 Sethi, S.P. and Ingham, B., "Some Classroom Notes on the Toymaker Example of Howard," *Decision Line*, 8, 4, Sept. 1977, 4-6.

42 Sethi, S.P., "Nearest Feasible Paths in Optimal Control Problems: Theory, Examples, and Counterexamples," *Journal of Optimization Theory and Applications*, 23, 4, Dec. 1977, 563-579.

43 Burdet, C.A. and Sethi, S.P., "On the Maximum Principle for a Class of Discrete Dynamical Systems with Lags," *Journal of Optimization Theory and Applications*, 19, 3, July 1976, 445-454.

44 Sethi, S.P., "Quantitative Guidelines for Communicable Disease Control Program: A Complete Synthesis," *Biometrics*, 30, 4, Dec. 1974, 681-691; a detailed Abstract in Statistical Theory and Method Abstracts, 1975.

45 Sethi, S.P., "Sufficient Conditions for the Optimal Control of a Class of Systems with Continuous Lags," *Journal of Optimization Theory and Applications*, 13, 5, May 1974, 545-552; Errata Corrige, 38, 1, Sept. 1982, 153-154.

46 Sethi, S.P., "An Application of Optimal Control Theory in Forest Management," *Journal of Management Science and Applied Cybernetics*, 2, 1, 1973, 9-16.

47 Sethi, S.P. and Thompson, G.L., "A Maximum Principle for Stochastic Networks," *Proceedings of the TIMS*, XX International Meeting, Tel-Aviv, Israel, June 24-29, 1973, 1, 244-248.

Editorial Boards/Co-editing Activities

Assoc. Editor, Journal of Industrial and Management Optimization	2004-Present
Assoc. Editor, Automatica	2004-Present
Assoc. Editor, Decision Sciences	2004
Departmental Editor, Production and Operations Management	2003-Present
Senior Editor, Manufacturing and Service Operations Management	2001-2005
Assoc. Editor, Journal of Mathematical Analysis & Applications	2000-2004
Assoc. Editor, Journal of Scientific & Industrial Research	1999-2003
Assoc. Editor, Journal of Statistics and Management Systems	1998-2004
Advisor, Production and Operations Management	1990-2003
Assoc. Editor, Optimal Control Applications & Methods	1980-2004
Assoc. Editor, Operations Research	1994-1996
Assoc. Editor, Int. Journal of Flexible Manufacturing Systems	1992-1997
Assoc. Editor, Journal of Canadian Operational Research Society	1981-1987

Guest Editor, Special Issue on Optimal Control Methods in Management Science and Economics of *The Automatica*, in progress.

A Guest Editor, Special Issue on Information Systems and Agile Manufacturing of the *International Journal of Agile Manufacturing*, Vol. 4, No. 2, 2001.

Co-editor, Book Series *Applied Information Technology*, Plenum Pub. Co., New York, 1987-2004.

A Co-editor of the Special Issue on Management Systems of the *Journal of Large Scale Systems Theory and Applications*, Vol. 6, No. 2, Apr. 1984.

Sethi, S.P. and Thompson, G.L. (Eds.), *Special Issue of INFOR on Applied Control Theory*, Vol. 19, No. 4, Nov. 1981.

Chapter 1

TCP-AQM INTERACTION: PERIODIC OPTIMIZATION VIA LINEAR PROGRAMMING

K.E. Avrachenkov
INRIA Sophia Antipolis
k.avrachenkov@sophia.inria.fr

L.D. Finlay
University of South Australia
School of Mathematics and Statistics
luke.finlay@unisa.edu.au

V.G. Gaitsgory
University of South Australia
School of Mathematics and Statistics
vladimir.gaitsgory@unisa.edu.au

Abstract   We investigate the interaction between Transmission Control Protocol (TCP) and an Active Queue Management (AQM) router, that are designed to control congestion in the Internet. TCP controls the sending rate with which the data is injected into the network and AQM generates control signals based on the congestion level. For a given TCP version, we define the optimal strategy for the AQM router as a solution of a nonlinear periodic optimization problem, and we find this solution using a linear programming approach. We show that depending on the choice of the utility function for the sending rate, the optimal control is either periodic or steady state.

Keywords: Transmission Control Protocol (TCP), Active Queue Management (AQM), Deterministic Long-Run Average Optimal Control, Periodic Optimization, Linear Programming Approach.

2

GLOSSARY

TCP/IP   Transmission Control Protocol/Internet Protocol
AIMD   Additive Increase Multiplicative Decrease
MIMD   Multiplicative Increase Multiplicative Decrease
AQM   Active Queue Management
ECN   Explicit Congestion Notification
AQM-ECN   AQM with ECN
AQM-non-ECN   AQM without ECN
RED   Random Early Detection
LPP   Linear Programming Problem

1.   Introduction and statement of the problem

Most traffic in the Internet is governed by the TCP/IP protocol [2], [8]. Data packets of an Internet connection travel from a source node to a destination node via a series of routers. Some routers, particularly edge routers, experience periods of congestion when packets spend a non-negligible time waiting in the router buffers to be transmitted over the next hop. The TCP protocol tries to adjust the sending rate of a source to match the available bandwidth along the path. During the principle Congestion Avoidance phase the current TCP New Reno version uses Additive Increase Multiplicative Decrease (AIMD) binary feedback congestion control scheme. In the absence of congestion signals from the network TCP increases sending rate linearly in time, and upon the reception of a congestion signal TCP reduces the sending rate by a multiplicative factor. Thus, the instantaneous AIMD TCP sending rate exhibits a "saw-tooth" behavior. Congestion signals can be either packet losses or Explicit Congestion Notifications (ECN) [14]. At the present state of the Internet, nearly all congestion signals are generated by packet losses. Packets can be dropped either when the router buffer is full or when an Active Queue Management (AQM) scheme is employed [5]. In particular, AQM RED [5] drops or marks packets with a probability which is a piece-wise linear function of the average queue length. Given an ambiguity in the choice of the AQM parameters (see [3] and [12]), so far AQM is rarely used in practice. In the present work, we study the interaction between TCP and AQM. In particular, we pose and try to answer the question: What should be the optimal dropping or marking strategy in the AQM router? For the performance criterion, we choose the average utility function of the throughput minus either the average cost of queueing or the average cost of losses. This performance criterion with a linear utility function was introduced in [1]. We

have analyzed not only the currently used AIMD congestion control, but
also Multiplicative Increase Multiplicative Decrease (MIMD) congestion
control. In particular, MIMD (or Scalable TCP [9]) is proposed for congestion control in high speed networks. However, since it turns out that
the results for MIMD and AIMD congestion control schemes are similar,
we provide the detailed analysis only for AIMD TCP.

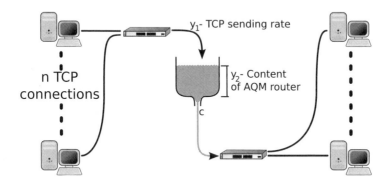

Figure 1.1.   Fluid model for data network.

We restrict the analysis to the single bottleneck network topology
(see Figure 1.1). In particular, we suppose that n TCP connections
cross a single bottleneck router with the AQM mechanism. We take the
fluid approach for modeling the interaction between TCP and AQM [10],
[11], [15]. In such an approach, the variables stand for approximations
of average values and their evolution is described by deterministic differential equations. Since we consider long-run time average criteria, our
TCP-AQM interaction model falls into the framework of the periodic
optimization described as follows (see,e.g., [4]). Consider the control
system

$$\dot{y}(t) = f(u(t), y(t)), \quad t \in [0, T], \quad T > 0 , \qquad (1.1)$$

where the function $f(u, y) : U \times \mathbb{R}^m \to \mathbb{R}^m$ is continuous in (u, y) and
satisfies Lipschitz conditions in y; the controls are Lebesque measurable
functions $u(t) : [0, T] \to U$ and U is a compact subset of $\mathbb{R}^n$.

Let Y be a compact subset of $\mathbb{R}^m$. A pair $(u(t), y(t))$ will be called
admissible on the interval $[0, T]$ if the equation (1.1) is satisfied for almost
all $t \in [0, T]$ and $y(t) \in Y$ $\forall t \in [0, T]$. A pair $(u(t), y(t))$ will be called
periodic admissible on the interval $[0, T]$ if it is admissible on $[0, T]$ and
$y(0) = y(T)$.

Let $g(u, y) : U \times \mathbb{R}^m \to \mathbb{R}^1$ be a continuous function. The following problem is commonly referred to as the *periodic optimization problem*:

$$\sup_{(u(\cdot), y(\cdot))} \frac{1}{T} \int_0^T g(u(t), y(t)) dt \stackrel{def}{=} G_{per} , \tag{1.2}$$

where *sup* is over the length of the time interval $T > 0$ and over the periodic admissible pairs defined on $[0, T]$.

A very special family of periodic admissible pairs is that consisting of constant valued controls and corresponding steady state solutions of (1.1):

$$(u(t), y(t)) = (u, y) \in \mathcal{M} \stackrel{def}{=} \{(u, y) \mid (u, y) \in U \times Y , f(u, y) = 0 \}. \tag{1.3}$$

If *sup* is sought over the admissible pairs from this family, the problem (1.2) is reduced to

$$\sup_{(u,y) \in \mathcal{M}} g(u, y) \stackrel{def}{=} G_{ss} \tag{1.4}$$

which is called a *steady state optimization problem*. Note that

$$G_{per} \geq G_{ss} \tag{1.5}$$

and that, as can be easily verified, $G_{per} = G_{ss}$ if the system (1.1) is linear, the sets U, Y are convex and the function $g(u, y)$ is concave. Note also that in a general case (e.g., the dynamics is non-linear and/or the integrand is not concave), (1.5) can take the form of a strict inequality (examples can be found in [4],[6],[7] and in references therein).

We formulate the problem of optimal control of TCP-AQM interaction as a periodic optimization problem, in which the state space is two dimensional

$$y = (y_1, y_2) , \quad f(u, y) = (f_1(u, y_1), f_2(u, y_1)) ; \tag{1.6}$$

and the control u is a scalar: $u(t) \in U$, with

$$U \stackrel{def}{=} \{u : 0 \leq u \leq 1\} . \tag{1.7}$$

We consider two congestion control schemes: Additive Increase Multiplicative Decrease (AIMD) scheme and Multiplicative Increase Multiplicative Decrease (MIMD) scheme. In both cases the first state component $y_1(t)$ is interpreted as a sending rate at the moment t, while the second state component $y_2(t)$ represents the size of the queue in the router buffer. In the AIMD scheme, the evolution of $y_1(t)$ is defined by the equation

$$\dot{y}_1(t) = f_1(u(t), y_1(t)) \stackrel{def}{=} \alpha(1 - u(t)) - \beta y_1^2(t)u(t), \qquad (1.8)$$

where $\alpha = \alpha_0 n / \tau^2$ and $\beta = 1 - \beta_0/n$. Here n is the number of competing TCP connections, τ is the round trip time. Typical values for α_0 and β_0 are 1 and 0.5, respectively. In the MIMD scheme, the evolution of $y_1(t)$ is defined by the equation

$$\dot{y}_1(t) = f_1(u(t), y_1(t)) \stackrel{def}{=} \gamma y_1(t)(1 - u(t)) - \beta y_1^2(t)u(t), \qquad (1.9)$$

where $\gamma = \gamma_0/\tau$ and β as in the AIMD case. A typical value for γ_0 is 0.01. The control $u(t)$ is interpreted as the dropping/marking probability. A detail derivation of equation (1.8) can be found for instance in [10] and [15]. Note also that if the control is not applied ($u(t) = 0$), the sending rate grows linearly in time if AIMD is used, and the sending rate grows exponentially in time if MIMD is used.

We study active queue management with and without explicit congestion notifications. When AQM with ECN is used, the packets are not dropped from the buffer when control is applied and the buffer is not full. In the case of AQM with ECN (AQM-ECN scheme), the evolution of the router buffer content $y_2(t)$ is described by

$$\dot{y}_2(t) = f_2(u(t), y_1(t)) = f_2(y_1(t)) \stackrel{def}{=} \begin{cases} y_1(t) - c, & 0 < y_2(t) < B, \\ [y_1(t) - c]_+, & y_2(t) = 0, \\ [y_1(t) - c]_-, & y_2(t) = B, \end{cases}$$

$$(1.10)$$

where c is the router capacity, B is the buffer size, $[a]_+ = \max(a, 0)$ and $[a]_- = \min(a, 0)$. In the case of AQM without ECN (AQM-non-ECN scheme), AQM signals congestion by dropping packets with rate $u(t)y_1(t)$. Consequently, the dynamics of the router buffer content $y_2(t)$ is described by

$$\dot{y}_2(t) = f_2(u(t), y_1(t)) \stackrel{def}{=} \begin{cases} (1 - u(t))y_1(t) - c, & 0 < y_2(t) < B, \\ [(1 - u(t))y_1(t) - c]_+, & y_2(t) = 0, \\ [(1 - u(t))y_1(t) - c]_-, & y_2(t) = B. \end{cases}$$

$$(1.11)$$

The function $g(u, y)$ in the objective (1.2) will be defined as follows

$$g(u, y) = \psi(y_1) - \kappa y_2 - Me^{K(y_2 - B)}y_1, \qquad (1.12)$$

where $\psi(\cdot)$ is the utility function for the sending rate value, κy_2 is the cost of delaying the data in the buffer, and $Me^{K(y_2 - B)}y_1$ is the penalty function for losing data when the buffer is full. Examples of the utility functions we will be dealing with are:

$$\psi(y_1) = y_1 \ , \quad \psi(y_1) = \log(1 + y_1) \ , \quad \psi(y_1) = (y_1)^2 \ . \qquad (1.13)$$

The rest of the paper is organized as follows: In Section 2 we give an overview of the linear programming approach to periodic optimization. Then, in Sections 3 and 4 we apply the general technique of Section 2 to the problem of interaction between TCP and AQM. We show that depending on the utility function for the sending rate, we obtain either periodic or steady state optimal solution. Some technical proofs are postponed to the Appendix. We conclude the paper with Section 5.

2.   Linear programming approach

In [7] it has been shown that the periodic optimization problem (1.2) can be approximated by a family of finite dimensional Linear Programming Problems (LPPs) (called in the sequel as approximating LPP). This approximating LPP is constructed as follows.

Let y_j $(j = 1, ..., m)$ stand for the jth component of y and let $\phi_i(y)$ be the monomomial:

$$\phi_i(y) \overset{def}{=} y_1^{i_1}...y_m^{i_m} \ ,$$

where i is the multi-index: $i \overset{def}{=} (i_1, ..., i_m)$. Let us denote by I_N the set of multi-indices

$$I_N \overset{def}{=} \{i \ : \ i = (i_1, ..., i_m) \ , \quad i_1, ..., i_m = 0, 1, ..., N \ , \quad i_1 + ... + i_m \geq 1 \ \}.$$

Note that the number of elements in I_N is $(N+1)^m - 1$. Assume that, for any $\Delta > 0$, the points $(u_l^\Delta, y_k^\Delta) \in U \times Y$, $l = 1, ..., L^\Delta$, $k = 1, ..., K^\Delta$, are being chosen in such a way that, for any $(u, y) \in U \times Y$, there exists (u_l^Δ, y_k^Δ) such that $||(u, y) - (u_l^\Delta, y_k^\Delta)|| \leq c\Delta$, where c is a constant.

Define the polyhedral set $W_N^\Delta \subset \mathbb{R}^{L^\Delta + K^\Delta}$

$$W_N^\Delta \overset{def}{=} \left\{ \gamma = \{\gamma_{l,k}\} \geq 0 \ : \ \sum_{l,k} \gamma_{l,k} = 1 \ , \right.$$

$$\left. \sum_{l,k} (\phi_i'(y_k^\Delta))^T f(u_l^\Delta, y_k^\Delta) \gamma_{l,k} = 0 \ , i \in I_N \right\}, \qquad (1.14)$$

where $\phi_i'(\cdot)$ is the gradient of $\phi_i(\cdot)$. Define the approximating LPP as follows

$$\max_{\gamma \in W_N^\Delta} \sum_{l,k} \gamma_{l,k} g(u_l^\Delta, y_k^\Delta) \overset{def}{=} G_N^\Delta \ , \qquad (1.15)$$

where $\sum_{l,k} \overset{def}{=} \sum_{l=1}^{L^\Delta} \sum_{k=1}^{K^\Delta}$.

As shown in [7] (under certain natural and easily verifiable conditions), *there exists the limit of the optimal value* $G^{N,\Delta}$ *of the LPP (1.15) and this limit is equal to the optimal value* G_{per} *of the periodic optimization problem (1.2):*

$$\lim_{N\to\infty} \lim_{\Delta\to 0} G_N^\Delta = G_{per} . \qquad (1.16)$$

Also, for any fixed N,

$$\lim_{\Delta\to 0} G_N^\Delta \overset{def}{=} G_N \geq G_{per} . \qquad (1.17)$$

Thus, G_N^Δ can be used as an approximation of G_{per} if N is large and Δ is small enough.

Let $(u^*(\cdot), y^*(\cdot))$ be the solution of the periodic optimization problem (1.2) defined on the optimal period $T = T^*$ (assuming that this solution exists and is unique) and let $\gamma^{N,\Delta} \overset{def}{=} \{\gamma_{l,k}^{N,\Delta}\}$ be an optimal basic solution of the approximating LPP (1.15). From the consideration in [7] it follows that an element $\gamma_{l,k}^{N,\Delta}$ of $\gamma^{N,\Delta}$ can be interpreted as an estimate of the "proportion" of time spent by the optimal pair $(u^*(\cdot), y^*(\cdot))$ in a Δ-neighborhood of the point (u_l, y_k), and in particular, the fact that $\gamma_{l,k}^\Delta$ is positive or zero can be interpreted as an indication of that whether or not the optimal pair attends the Δ-neighborhood of (u_l, y_k).

Define the set Θ by the equation

$$\Theta \overset{def}{=} \{(u, y) : (u, y) = (u^*(\tau), y^*(\tau)) \ for \ some \ \tau \in [0, T^*] \} . \qquad (1.18)$$

This Θ is the graph of the of the optimal feedback control function, which is defined on the optimal state trajectory $\mathcal{Y} \overset{def}{=} \{y : (u, y) \in \Theta\}$ by the equation $\psi(y) \overset{def}{=} u \ \forall \ (u, y) \in \Theta$. For the definition of $\psi(\cdot)$ to make sense, it is assumed that the set Θ is such that from the fact that $(u', y) \in \Theta$ and $(u'', y) \in \Theta$ it follows that $u' = u''$ (this assumption being satisfied if the closed curve defined by $y^*(\tau)$, $\tau \in [0, T^*]$ does not intersect itself).

Define also the sets:

$$\Theta_N^\Delta \overset{def}{=} \{(u_l^\Delta, y_k^\Delta) : \gamma_{l,k}^{N,\Delta} > 0\}, \qquad (1.19)$$

$$\mathcal{Y}_N^\Delta \overset{def}{=} \{y : (u, y) \in \Theta_N^\Delta\} , \qquad (1.20)$$

$$\psi_N^\Delta(y) \overset{def}{=} u \ \forall \ (u, y) \in \Theta_N^\Delta, \qquad (1.21)$$

where again it is assumed that from the fact that $(u', y) \in \Theta_N^\Delta$ and $(u'', y) \in \Theta_N^\Delta$ it follows that $u' = u''$. Note that the set Θ_N^Δ (and the set

$\mathcal{Y}_N^\Delta$) can contain no more than $(N+1)^m$ elements since $\gamma^{N,\Delta}$, being a basic solution of the LPP (1.15), has no more than $(N+1)^m$ positive elements (the number of the equality type constraints in (1.15)).

As mentioned above, the fact that $\gamma_{l,k}^{N,\Delta}$ is positive or zero can be interpreted as an indication of that whether or not the optimal pair attends the Δ-neighborhood of (u_l^Δ, y_k^Δ), and thus, one may expect that Θ_N^Δ can provide some approximation for Θ if N is large and Δ is small enough. Such an approximation has been formalized in [7], where it has been established that:

(i) *Corresponding to an arbitrary small $r > 0$, there exists N_0 such that, for $N \geq N_0$ and $\Delta \leq \Delta_N$ (Δ_N is positive and small enough),*

$$\Theta \subset \Theta_N^\Delta + r\mathcal{B} . \tag{1.22}$$

(ii) *Corresponding to an arbitrary small $r > 0$ and arbitrary small $\delta > 0$, there exists N_0 such that, for $N \geq N_0$ and $\Delta \leq \Delta_N$ (Δ_N being positive and small enough),*

$$\Theta_N^{\Delta,\delta} \subset \Theta + r\mathcal{B} , \tag{1.23}$$

where $\Theta_N^{\Delta,\delta} \overset{def}{=} \{(u_l^\Delta, y_k^\Delta) : \gamma_{l,k}^\Delta \geq \delta \} .$
Note that in both (1.22) and (1.23), $\mathcal{B}$ is the closed unit ball in $\mathbb{R}^{n+m}$.

The fact that Θ_N^Δ "approximates" Θ for N large and Δ small enough leads to the fact that $\mathcal{Y}_N^\Delta$ approximates $\mathcal{Y}$ and to the fact that $\psi_N^\Delta(y)$ approximates (in a certain sense) $\psi(y)$. This gives rise to the following algorithm for construction of near-optimal periodic admissible pair [7]:

1) Find an optimal basic solution γ_N^Δ and the optimal value G_N^Δ of the approximating LPP (1.15) for N large and Δ small enough; the expression "N large and Δ small enough" is understood in the sense that a further increment of N and/or a decrement of Δ lead only to insignificant changes of the optimal value G_N^Δ and, thus, the latter can be considered to be approximately equal to G_{per} (see (1.16)).

2) Define Θ_N^Δ, $\mathcal{Y}_N^\Delta$, $\psi_N^\Delta(y)$ as in (1.19). By (1.22) and (1.23), the points of $\mathcal{Y}_N^\Delta$ will be concentrated around a closed curve being the optimal periodic state trajectory while $\psi_N^\Delta(y)$ will give a point wise approximation to the optimal feedback control.

3) Extrapolate the definition of the function $\psi_N^\Delta(y)$ to some neighborhood of $\mathcal{Y}_N^\Delta$ and integrate the system (1.1) starting from an initial point $y(0) \in \mathcal{Y}_N^\Delta$ and using $\psi_N^\Delta(y)$ as a feedback control. The end point of the integration period, T^Δ, is identified by the fact that the solution "returns" to a small vicinity of the starting point $y(0)$.

4) Adjust the initial condition and/or control to obtain a periodic admissible pair $(u^\Delta(\tau), y^\Delta(\tau))$ defined on the interval $[0, T^\Delta]$. Calculate the integral $\frac{1}{T^\Delta} \int_0^{T^\Delta} g(u^\Delta(\tau), y^\Delta(\tau)) d\tau$ and compare it with G_N^Δ. If the value of the integral proves to be close to G_N^Δ, then, by (1.16), the constructed admissible pair is a "good" approximation to the solution of the periodic optimization problem (1.2).

In conclusion of this section, let us consider the following important special case. Assume that, for all N large and Δ small enough, the optimal basic solution $\gamma^{N,\Delta}$ of the approximating LPP (1.15) has the property that

$$\gamma_{l^*,k^*}^{N,\Delta} = 1 , \qquad \gamma_{l,k}^{N,\Delta} = 0 \quad \forall \ (l, k) \neq (l^*, k^*) , \qquad (1.24)$$

which is equivalent to that the set Θ_N^Δ consists of only one point

$$\Theta_N^\Delta = \{(u_{l^*}^\Delta, y_{k^*}^\Delta)\} . \qquad (1.25)$$

Note that that the indexes l^*, k^* in (1.24) and (1.24) may depend on N and Δ.

Assume that there exists a limit

$$\lim_{\Delta \to 0} (u_{l^*}^\Delta, y_{k^*}^\Delta) = (\bar{u}, \bar{y}) , \qquad (1.26)$$

(the same for all sufficiently large N). Then, as follows from results of [7], the pair $(\bar{u}, \bar{y})$ is the steady state solution of the periodic optimization problem (1.2) and, in particular,

$$G_{per} \ = \ G_{ss} \ = \ g(\bar{u}, \bar{y}) . \qquad (1.27)$$

3.   Optimal periodic solution

In this and the next sections it is always assumed that

$$Y = \{(y_1, y_2) \mid y_i \in [0, 4], \ i = 1, 2\} \qquad (1.28)$$

and that U is defined by (1.7); it is also assumed everywhere that $c = 1$ and $B = 4$ (see the equations describing the dynamics of the buffer's content (1.10) and (1.11)).

Let us consider the interaction between AIMD TCP (1.8) and the AQM-ECN router (1.10), the former being taken with $\alpha = 1/98$ and $\beta = 1/2$ (such a choice of these parameters corresponds to the case of a single TCP connection and a typical value of the round trip time).

Let us use the objective function (1.12), with the following values of the parameters: $\kappa = 0$, $M = 20$ and $K = 5$; and with the utility function being defined by the equation

$$\psi(y_1) = y_1^2. \tag{1.29}$$

Note that a conventional choice of the sending rate utility function is a concave function. However, the analysis of a convex utility function makes the present work complete and leads to some interesting observations. The concave utility functions are analyzed in the next section.

Define the grid of $U \times Y$ by the equations (with U and Y mentioned as above)

$$u_i^\Delta \stackrel{def}{=} i\Delta, \quad y_{1,j}^\Delta \stackrel{def}{=} j\Delta, \quad y_{2,k}^\Delta \stackrel{def}{=} k\Delta. \tag{1.30}$$

Here $i = 0, 1, \ldots, \frac{1}{\Delta}$ and $j, k = 0, 1, \ldots, \frac{4}{\Delta}$ (Δ is chosen in such a way that $\frac{1}{\Delta}$ is an integer). The approximating LPP (1.15) can be written in this specific case as

$$G_N^\Delta \stackrel{def}{=} \max_{\gamma \in W_N^\Delta} \sum_{i,j,k} \left((y_{1,j}^\Delta)^2 - 20e^{5(y_{2,k}-4)} y_{1,j}^\Delta \right) \gamma_{i,j,k}, \tag{1.31}$$

where W_N^Δ is a polyhedral set defined by the equation

$$W_N^\Delta \stackrel{def}{=} \left\{ \gamma = \{\gamma_{i,j,k}\} \geq 0 : \sum_{i,j,k} \gamma_{i,j,k} = 1 , \right.$$

$$\left. \sum_{i,j,k} (\phi'_{i_1,i_2}(y_{1,j}^\Delta, y_{2,k}^\Delta))^T f(u_i^\Delta, y_{1,j}^\Delta, y_{2,k}^\Delta) \gamma_{i,j,k} = 0, \ (i_1, i_2) \in I_N \right\}, \tag{1.32}$$

in which $\phi_{i_1,i_2}(y_1, y_2) \stackrel{def}{=} y_1^{i_1} y_2^{i_2}$.

The problem (1.31) was solved using the CPLEX LP solver [16] for $N = 5$ and $N = 7$ with Δ varying from 0.0125 to 0.2. We have obtained the following optimal values of the LPP (1.31):

$$G_5^{0.1} \approx 1.0152, \quad G_5^{0.05} \approx 1.0174, \quad G_5^{0.025} \approx 1.0179, \quad G_5^{0.0125} \approx 1.0180,$$
$$G_7^{0.2} \approx 1.0156, \quad G_7^{0.1} \approx 1.0174, \quad G_7^{0.025} \approx 1.0175, \quad G_7^{0.0125} \approx 1.0175.$$

From this data one may conclude that $G_7 = \lim_{\Delta \to 0} G_7^\Delta \approx 1.0175$. Since $G_7 \geq G_{per}$, it follows that, if for some admissible periodic pair $(u(\tau), y(\tau))$,

$$\frac{1}{T} \int_0^T \left(y_1^2(\tau) - y_1(\tau)100e^{20(y_2(\tau)-4)} \right) d\tau \approx 1.0175 , \tag{1.33}$$

then this pair is an approximate solution of the periodic optimization problem (1.2).

Let $\left\{\gamma_{i,j,k}^{N,\Delta}\right\}$ stand for the solution of (1.31) and define the sets

$$\Theta_N^\Delta \overset{def}{=} \left\{(u_i, y_{1,j}, y_{2,k}) \; : \; \gamma_{i,j,k}^{N,\Delta} \neq 0\right\},$$

$$\mathcal{Y}_N^\Delta \overset{def}{=} \left\{(y_{1,j}, y_{2,k}) \; : \; \sum_i \gamma_{i,j,k}^{N,\Delta} \neq 0\right\}. \tag{1.34}$$

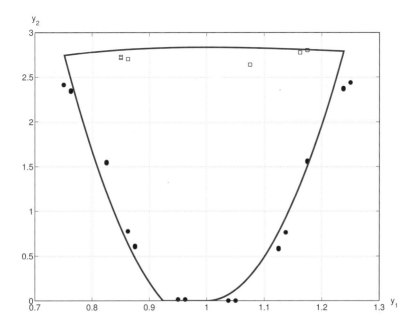

Figure 1.2.   Optimal state trajectory approximation

Let us mark with dots the points on the plane (y_1, y_2) which belong to $\mathcal{Y}_N^\Delta$ for $N = 7$ and $\Delta = 0.0125$. The result is depicted in Figure 1.2. The points are represented with $\square$ or $\bullet$ and have an associated $u = 1$ or $u = 0$, respectively. It is possible to construct a feedback control by using two thresholds for the buffer content. As the queue length y_2 is decreasing (in the region where $y_1 < 1$), we have a certain threshold for when the control should be dropped and allow the data rate y_1 to grow. The same can be said for the opposite case when the queue is increasing and $y_1 \geq 1$. The threshold values in our numerical example can be chosen as 2.7466 and 2.7939, respectively. Thus, the feedback control is defined as

$$u = \begin{cases} 1 & , \quad y_1 < 1 \text{ and } y_2 > 2.7466 \\ 1 & , \quad y_1 \geq 1 \text{ and } y_2 > 2.7939 \\ 0 & , \quad \text{otherwise} \end{cases} \tag{1.35}$$

Using this feedback control, we can integrate the system with the initial point $y_1 = 1, y_2 = 0$. The optimal state trajectory is plotted as a solid line in Figure 1.2. In Figure 1.3 we show the evolution of the state variables and the optimal control.

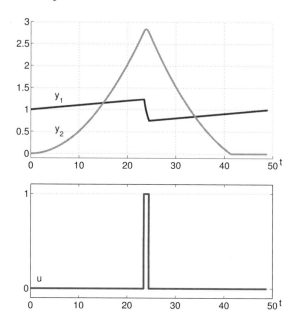

Figure 1.3. Approximated periodic solution and optimal control

The value of the objective function calculated on this pair is approximately 1.0083. Comparing it with (1.33), one can conclude that the admissible pair which has been constructed is an approximation to the solution of (1.2).

Curiously enough, the evolution of the optimal sending rate y_1 resembles a "saw-tooth" behavior of the "instantaneous" TCP sending rate. This is despite the fact that the variables in our fluid model stand for average values and a convex utility function does not seem to correspond to the commonly used utility concept for elastic traffic.

We have also tested the current objective function for the interaction between AIMD and AQM-non-ECN and for the MIMD congestion control. For those cases, we have also detected similar periodic optimal solutions.

4.   Optimal steady state solution

As in the previous section, let us consider the interaction between AIMD TCP (1.8) and the AQM-ECN router (1.10). However, in contrast to the above consideration, let us choose the following objective function

$$g(u, y) = y_1 - y_2. \tag{1.36}$$

That is, take $\psi(y_1) = y_1$, $\kappa = 1$, and $M = 0$ in (1.12). Note that, as can be easily verified, in this case, the solution of the steady state optimization problem (1.4) is

$$\bar{u} = 0.02 , \quad \bar{y}_1 = 1 , \quad \bar{y}_2 = 0 \tag{1.37}$$

and, in particular, $G_{ss} = \bar{y}_1 - \bar{y}_2 = 1$.

As in section 3, define the grid of $U \times Y$ by the equations

$$u_i^{\Delta} \stackrel{def}{=} i\Delta, \quad y_{1,j}^{\Delta} \stackrel{def}{=} j\Delta, \quad y_{2,k}^{\Delta} \stackrel{def}{=} k\Delta, \tag{1.38}$$

where $i = 0, 1, \ldots, \frac{1}{\Delta}$ and $j, k = 0, 1, \ldots, \frac{4}{\Delta}$ (Δ being chosen in such a way that $\frac{1}{\Delta}$ is an integer). The approximating LPP (1.15) in this specific case is of the form

$$\max_{\gamma \in W_N^{\Delta}} \sum_{i,j,k} \left(y_{1,j}^{\Delta} - y_{2,k}^{\Delta} \right) \gamma_{i,j,k} = G_N^{\Delta}, \tag{1.39}$$

where W_N^{Δ} has exactly the same form as in (1.31). That is,

$$W_N^{\Delta} \stackrel{def}{=} \left\{ \gamma = \{\gamma_{i,j,k}\} \geq 0 \, : \, \sum_{i,j,k} \gamma_{i,j,k} = 1 , \right.$$

$$\left. \sum_{i,j,k} (\phi'_{i_1,i_2}(y_{1,j}^{\Delta}, y_{2,k}^{\Delta}))^T f(u_i^{\Delta}, y_{1,j}^{\Delta}, y_{2,k}^{\Delta})\gamma_{i,j,k} = 0, \, (i_1, i_2) \in I_N \right\}, \tag{1.40}$$

with $\phi_{i_1,i_2}(y_1, y_2) \stackrel{def}{=} y_1^{i_1} y_2^{i_2}$.

Proposition 1. *For any $N = 1, 2, \ldots$, and any $\Delta > 0$ such that $\frac{0.02}{\Delta} \stackrel{def}{=} i^*$ is integer, there exists a basic optimal solution $\gamma^{N,\Delta} \stackrel{def}{=} \{\gamma_{i,j,k}^{N,\Delta}\}$ of the LPP (1.39) defined by the equations*

$$\gamma_{i^*,j^*,k^*}^{N,\Delta} = 1 , \quad \gamma_{i,j,k}^{N,\Delta} = 0 \, \forall \, (i, j, k) \neq (i^*, j^*, k^*) , \tag{1.41}$$

where i^* is as above and $j^* = \frac{1}{\Delta}$, $k^* = 0$.

Proof of the proposition is given in the Appendix.

Since, by definition, $u_{i^*}^\Delta = 0.02$, $y_{1,j^*}^\Delta = 1$, $y_{2,k^*}^\Delta = 0$, that is, $(u_{i^*}^\Delta, y_{1,j^*}^\Delta, y_{2,k^*}^\Delta)$ coincides with the optimal steady state regime $(\bar{u}, \bar{y}_1, \bar{y}_2)$ defined in (1.37), one obtains the following corollary of Proposition 1 (see (1.26) and (1.27)).

Corollary 2. *The periodic optimization problem (1.2) has a steady state solution and this steady state solution is defined by (1.37). In particular,*

$$G_{per} = G_{ss} = \bar{y}_1 - \bar{y}_2 = 1 . \tag{1.42}$$

We have also checked numerically the other criteria with concave utility functions for the sending rate. It appears that if the utility function for the sending rate is concave, the optimal solution is steady state. The same conclusion holds for the case of interaction between AIMD TCP and AQM-non-ECN and when MIMD is used instead of AIMD.

5. Conclusions

We have analyzed the interaction between TCP and AQM using the fluid model approach. The fluid model approach leads to a periodic optimization problem. We have shown that depending on the choice of the utility function for the sending rate, the optimal solution is either periodic or steady state. In particular, we have obtained steady state solution for all concave utility functions and periodic solutions for some convex utility functions. Even though a convex utility function does not seem to correspond to the commonly used utility concept for elastic traffic, the optimal periodic solution resembles strikingly the "saw-tooth" behavior of the instantaneous TCP sending rate evolution. With the help of linear programming approach for periodic optimization we have succeeded to prove that the steady state solution is indeed an optimal solution for the given non-linear periodic optimization problem.

Acknowledgments

The work was supported by the Australian Research Council Discovery-Project Grants DP0346099, DP0664330 and Linkage International Grant LX0560049.

Appendix. Proof of Proposition 1.

Proof of Proposition 1. First of all let us note that from the fact that $(u_{i^*}^\Delta, y_{1,j^*}^\Delta, y_{2,k^*}^\Delta)$ coincides with $(\bar{u}, \bar{y}_1, \bar{y}_2)$ and the latter is a steady state solution, it follows that $f(u_{i^*}^\Delta, y_{1,j^*}^\Delta, y_{2,k^*}^\Delta) = 0$. This implies that the vector $\gamma^{N,\Delta}$ with the components defined in (1.41) is feasible, the value of the objective function obtained on this solution is equal to

$$y_{1,j^*}^{\Delta} - y_{2,k^*}^{\Delta} = \bar{y}_1 - \bar{y}_2 = 1 . \tag{1.43}$$

To prove that $\gamma^{N,\Delta}$ is is optimal, it is enough to show that the optimal value of the problem dual to (1.39) is equal to 1 (that is, the same as in (1.43)). The set of feasible solutions $\omega \in \mathbb{R}^{(N+1)^2}$ of the problem dual to (1.39) is described by the inequalities

$$\omega_1 + \sum_{(i_1,i_2)\in I_N} \omega_{i_1,i_2}(\phi'_{i_1,i_2}(y_{1,j}^{\Delta}, y_{2,k}^{\Delta}))^T f(u_i^{\Delta}, y_{1,j}^{\Delta}, y_{2,k}^{\Delta}) \geq y_{1,j}^{\Delta} - y_{2,k}^{\Delta}$$

$$\forall (i, j, k) , \tag{1.44}$$

where, for convenience, the first component of ω is denoted as ω_1 and the other $(N+1)^2 - 1$ components are denoted as ω_{i_1,i_2}, $(i_1, i_2) \in I_N$. Note that , the objective function of the dual problem is

$$F_{dual}(\omega) = \omega_1 \tag{1.45}$$

Define the vector $\bar{\omega}$ by the equations

$$\bar{\omega}_1 = 1 , \quad \bar{\omega}_{0,1} = 1 + \Delta , \quad \bar{\omega}_{i_1,i_2} = 0 \ \forall \ (i_1, i_2) \neq (0, 1). \tag{1.46}$$

It is obvious that

$$F_{dual}(\bar{\omega}) = \bar{\omega}_1 = 1 . \tag{1.47}$$

Thus, to prove the desired result, one needs to show that the vector $\bar{\omega}$ satisfies the inequalities (1.44) or equivalently (having in mind (1.46) and the fact that $(\phi'_{0,1}(y_{1,j}^{\Delta}, y_{2,k}^{\Delta}))^T = (0, 1)$) to prove that

$$1 + (1 + \Delta)f_2(y_{1,j}^{\Delta}) \geq y_{1,j}^{\Delta} - y_{2,k}^{\Delta} \ \forall (i, j, k) , \tag{1.48}$$

where $f_2(y_1)$ is defined in (1.10). Consider the following three cases:
(1) $\Delta \leq y_{2,k}^{\Delta} < 4 \ (1 \leq k < \frac{4}{\Delta})$
(2) $y_{2,k}^{\Delta} = 0 \ (k = 0)$
(3) $y_{2,k}^{\Delta} = 4 \ (k = \frac{4}{\Delta})$

By (1.10), in case (1) for any $j = 0, 1, ...\frac{4}{\Delta}$,

$$1 + (1 + \Delta)f_2(y_{1,j}^{\Delta}) = 1 + (1 + \Delta)(y_{1,j}^{\Delta} - 1) \geq y_{1,j}^{\Delta} - \Delta \geq y_{1,j}^{\Delta} - y_{2,k}^{\Delta} .$$

This proves the validity of (1.48) in the given case.

In case (3) (see (1.10)), $f_2(y_{1,j}^{\Delta}) \leq 0$ and, hence, for any $j = 0, 1, \dots \frac{4}{\Delta}$,

$$1 + (1 + \Delta) f_2(y_{1,j}^{\Delta}) \geq 1 \geq y_{1,j}^{\Delta} - 4 = y_{1,j}^{\Delta} - y_{2,k}^{\Delta} \ .$$

To deal with case (2), consider two situations:

(2.a) $0 \leq y_{1,j}^{\Delta} \leq 1$ $(0 \leq j \leq \frac{1}{\Delta})$

(2.b) $1 < y_{1,j}^{\Delta} \leq 4$ $(\frac{1}{\Delta} < j \leq \frac{4}{\Delta})$.

In case (2.a),

$$1 + (1 + \Delta) f_2(y_{1,j}^{\Delta}) = 1 + (1 + \Delta) \max(y_{1,j}^{\Delta} - 1, 0) = 1 \geq y_{1,j}^{\Delta} = y_{1,j}^{\Delta} - y_{2,k}^{\Delta} \ ;$$

and in case (2.b),

$$1 + (1 + \Delta) f_2(y_{1,j}^{\Delta}) = 1 + (1 + \Delta) \max(y_{1,j}^{\Delta} - 1, 0) = 1 + (1 + \Delta)(y_{1,j}^{\Delta} - 1)$$

$$> y_{1,j}^{\Delta} = y_{1,j}^{\Delta} - y_{2,k}^{\Delta} \ .$$

This completes the proof of Proposition 1.

References

[1] K. Avrachenkov, U. Ayesta, and A. Piunovsky, "Optimal Choice of the Buffer Size in the Internet Routers", in Proceedings of IEEE CDC-ECC 2005.

[2] M. Allman, V. Paxson and W. Stevens, TCP congestion control, *RFC 2581*, April 1999, available at http://www.ietf.org/rfc/rfc2581.txt.

[3] M. Christiansen, K. Jeffay, D. Ott and F. Donelson Smith, "Tuning RED for Web Traffic", *IEEE/ACM Trans. on Networking*, v.9, no.3, pp.249-264, June 2001. An earlier version appeared in Proc. of ACM SIGCOMM 2000.

[4] F. Colonius, "Optimal Periodic Control", Lecture Notes in Mathematics, Springer-Verlag, Berlin, 1988.

[5] S. Floyd and V. Jacobson, "Random Early Detection Gateways for Congestion Avoidance", *IEEE/ACM Trans. on Networking*, v.1, no.4, pp.397-413, 1993.

[6] V. Gaitsgory, "Suboptimization of Singularly Perturbed Control Problems", *SIAM J. Control and Optimization*, 30 (1992), No. 5, pp. 1228 - 1240.

[7] V. Gaitsgory and S. Rossomakhine "Linear Programming Approach to Deterministic Long Run Average Problems of Optimal Control", *SIAM J. Control and Optimization*, To appear.

[8] V. Jacobson, Congestion avoidance and control, *ACM SIGCOMM'88*, August 1988.

[9] T. Kelly, "Scalable TCP: Improving performance in highspeed wide area networks", *Computer Comm. Review*, v.33, no.2, pp.83-91, 2003.

[10] S. Low, F. Paganini and J. Doyle, "Internet Congestion Control", *IEEE Control Systems Magazine*, v.22, no.1, pp.28-43, February 2002.

[11] V. Misra, W. Gong and D. Towsley, "A Fluid-based Analysis of a Network of AQM Routers Supporting TCP Flows with an Application to RED", in Proceedings of ACM SIGCOMM 2000.

[12] M. May, J. Bolot, C. Diot and B. Lyles, "Reasons Not to Deploy RED", in Proceedings of 7th International Workshop on Quality of Service (IWQoS'99), June 1999, London, UK.

[13] J. Postel, User Datagram Protocol, *RFC 768*, August 1980, available at `http://www.ietf.org/rfc/rfc0768.txt`.

[14] K. Ramakrishnan, S. Floyd and D. Black, The Addition of Explicit Congestion Notification (ECN) to IP, *RFC 3168*, September 2001, available at `http://www.ietf.org/rfc/rfc3168.txt`.

[15] R. Srikant, *The Mathematics of Internet Congestion Control*, Birkhaüser, Boston, 2004.

[16] ILOG CPLEX `http://ilog.com/products/cplex/`

Chapter 2

EXPLICIT SOLUTIONS OF LINEAR QUADRATIC DIFFERENTIAL GAMES

A. Bensoussan

International Center for Decision and Risk Analysis
University of Texas at Dallas
School of Management, P.O. Box 830688
Richardson, Texas 75083-0688

Alain-Bensoussan@utdallas.edu

Abstract   The theory of linear quadratic differential games is in principle known. An excellent reference for management and economics applications is Dockner et al. (2000). We review here the results, showing that in useful simple cases, explicit solutions are available. This treatment is not included in the previous reference and seems to be original. In non-stationary cases, explicit solutions are not available, we prove the existence of solutions of coupled Riccati equations, which provide a complete solution of the Nash equilibrium problem.

1.   Introduction

Differential games is attracting a lot of interest in the management and economics literature. This is because many players appear in most situations, and traditional optimization techniques for a single decision maker are not sufficient. However the treatment of differential games is much more complex than that of control theory, especially as far as obtaining explicit solutions is concerned. In this article, we complete a presentation of Dockner et al. (2000), a main reference for management and economics applications of differential games, to derive explicit solutions of linear quadratic differential games in a fairly general context. However, when data depend on time, non-stationary situation, we do not have explicit solutions anymore. The problem reduces to solving coupled Riccati equations. We prove existence of the solution of this pair of equations, which provide control strategies for the players.

2. Open Loop Differential Games

Finite Horizon

We consider the following model, [Dockner et al. (2000), Section 7.1]. We have two players, whose controls are denoted by v^1, v^2. The state equation is described by

$$\dot{x} = ax + b^1 v^1 + b^2 v^2, \ x(0) = x_0.$$

The payoffs of player $i = 1, 2$ are given by

$$J^i(x, \mathbf{v}) = \frac{1}{2} \int_0^T [\alpha^i x^2 + \beta^i (v^i)^2] dt.$$

We apply the theory of necessary conditions. We first define the Hamiltonians by the formulas

$$H^i(x, \mathbf{v}, q^i) = \frac{1}{2}(\alpha^i x^2 + \beta^i (v^i)^2) + q^i(ax + b^1 v^1 + b^2 v^2),$$

where $\mathbf{v} = (v^1, v^2)$. Writing the adjoint equations, we obtain

$$-\dot{p}^i = \alpha^i y + ap^i, \ p^i(T) = 0.$$

Writing next the optimality conditions

$$H^i_{v^i}(y, \mathbf{u}, p^i) = 0,$$

we obtain

$$\beta^i u^i + p^i b^i = 0.$$

We then notice an important simplification, namely

$$\frac{p^i}{\alpha^i} = p,$$

with

$$-\dot{p} = y + ap, \ p(T) = 0.$$

We next define

$$M = \sum_{i=1}^{2} \frac{\alpha^i (b^i)^2}{\beta^i}.$$

Collecting results, we obtain the maximum principle conditions

$$\dot{y} = ay - Mp$$

$$-\dot{p} = y + pa$$

$$y(0) = x_0, \ p(T) = 0,$$

and the optimal controls of both players are given by

$$u^i = -\frac{\alpha^i b^i p}{\beta^i}.$$

We can compute $y(t), p(t)$ following a decoupling argument. We postulate

$$p(t) = P(t)y(t).$$

It is easy to show that $P(t)$ is the solution of the Riccati equation

$$-\dot{P} - 2aP + MP^2 - 1 = 0, \quad P(T) = 0,$$

and that furthermore

$$\frac{1}{P(t)} = -a + s\frac{\exp 2s(T-t) + 1}{\exp 2s(T-t) - 1},$$

where

$$s = \sqrt{a^2 + M}.$$

We then obtain the explicit solution

$$y(t) = x_0 \frac{s(\exp s(T-t) + \exp -s(T-t)) - a(\exp s(T-t) - \exp -s(T-t))}{s(\exp sT + \exp -sT) - a(\exp sT - \exp -sT)}.$$

Infinite Horizon

We consider the infinite horizon version of the basic model above. We introduce a discount factor r. The maximum principle leads to the following relations (usual changes with respect to the finite horizon case)

$$\dot{y} = ay - Mp,$$

$$-\dot{p} + rp = y + pa,$$

$$y(0) = x_0.$$

There is no final condition on $p(T)$, but we require the integration conditions

$$y \in L_r^2(0, \infty; R), p \in L_r^2(0, \infty; R).$$

We can check that the solution of the infinite horizon problem is obtained as follows

$$p = Py,$$

with

$$P = \frac{2a - r + \sqrt{(r - 2a)^2 + 4M}}{2M},$$

and y, p satisfy the integrability conditions.

Non-Zero Final Cost

We consider again the finite horizon case, with non-zero final cost. So the payoffs are given by

$$J^i(x, \mathbf{v}) = \frac{1}{2} \int_0^T [\alpha^i x^2 + \beta^i (v^i)^2] dt + \frac{1}{2} \gamma^i (x(T))^2.$$

The adjoint variables p^1, p^2 are then the solutions of

$$-\dot{p}^i = \alpha^i y + a p^i, \; p^i(T) = \gamma^i y(T).$$

We do not have anymore the property

$$\frac{p^i}{\alpha^i} = p.$$

However we shall be able again to derive an explicit solution.

We can see indeed that

$$p^i = \alpha^i p + \gamma^i \pi,$$

where p, π satisfy

$$-\dot{p} = y + ap, \; p(T) = 0,$$

$$-\dot{\pi} = a\pi, \; \pi(T) = y(T).$$

We introduce a number analogous to M

$$N = \sum_{i=1}^{2} \frac{\gamma^i (b^i)^2}{\beta^i},$$

and define

$$\varpi = Mp + N\pi.$$

We deduce the equation for y

$$\dot{y} = ay - \varpi, \; y(0) = x_0,$$

and we check that

$$-\dot{\varpi} = My + a\varpi, \; \varpi(T) = Ny(T).$$

We can decouple the two-point boundary value problem in y, ϖ, by writing

$$\varpi(t) = Q(t)y(t),$$

and Q is the solution of the Riccati equation

$$-\dot{Q} - 2aQ + Q^2 = M, \ Q(T) = N.$$

We deduce easily

$$p = Ry, \ \pi = \rho y,$$

with

$$-\dot{R} = 1 + 2aR - QR, \ R(T) = 0$$

$$-\dot{\rho} = 2a\rho - Q\rho, \ \rho(T) = 1.$$

We next check that

$$Q(t) = a + s\frac{(N - a - s) + (N - a + s)\exp 2s(T - t)}{(N - a + s)\exp 2s(T - t) - (N - a - s)},$$

and that

$$p^1(t) = P^1(t)y(t), \ p^2(t) = P^2(t)y(t),$$

where P^1, P^2 are solutions of the system

$$-\dot{P}^1 - 2aP^1 + \frac{(b^1)^2}{\beta^1}(P^1)^2 + \frac{(b^2)^2}{\beta^2}P^1P^2 = \alpha^1$$

$$-\dot{P}^2 - 2aP^2 + \frac{(b^2)^2}{\beta^2}(P^2)^2 + \frac{(b^1)^2}{\beta^1}P^1P^2 = \alpha^2$$

$$P^1(T) = \gamma^1, \ P^2(T) = \gamma^2,$$

which is a system of Riccati equations. we then assert that

$$P^1(t) = \alpha^1 R(t) + \gamma^1\rho(t),$$

$$P^2(t) = \alpha^2 R(t) + \gamma^2\rho(t).$$

To complete the explicit solution, we check the following formulas

$$\rho(t) = \frac{2s\exp a(T - t)}{(N - a + s)\exp s(T - t) - (N - a - s)\exp -s(T - t)},$$

$$R(t) = \frac{1}{M}[-2sN\exp a(T-t)+(a+s)(N-a+s)\exp s(T-t)$$
$$+(N-a-s)(s-a)\exp -s(T-t)]$$
$$/[(N-a+s)\exp s(T-t)-(N-a-s)\exp -s(T-t)]$$

Furthermore,

$$y(t) = \frac{2sx_0}{(N - a + s)\exp s(T - t) - (N - a - s)\exp -s(T - t)}.$$

3.   Non-Stationary Mode

Maximum Principle

We consider the same problem with non-constant parameters, namely

$$\dot{x}(t) = a(t)x(t) + b^1(t)v^1(t) + b^2(t)v^2(t), \ x(0) = x_0,$$

and

$$J^i(v^1(.), v^2(.)) = \frac{1}{2} \int_0^T (\alpha^i(t)x^2(t) + \beta^i(t)(v^i)^2(t))dt + \frac{1}{2}\gamma^i x^2(T).$$

We can write the maximum principle in a way similar to the stationary case. To save notation, we shall not explicitly write the argument t. This leads to the system

$$\dot{y}(t) = ay - \frac{(b^1)^2}{\beta^1}p^1 - \frac{(b^2)^2}{\beta^2}p^2, \ y(0) = x_0$$

$$-\dot{p}^i = \alpha^i y + ap^i, \ p^i(T) = \gamma^i y(T).$$

Unfortunately the simplifications of the stationary case do not carry over. However, one can use the fact that the system obtained from the maximum principle arguments is linear. So if we set

$$A(t) = \begin{pmatrix} a(t) & -\frac{(b^1)^2}{\beta^1}(t) & -\frac{(b^2)^2}{\beta^2}(t) \\ -\alpha^1(t) & -a(t) & 0 \\ -\alpha^2(t) & 0 & -a(t) \end{pmatrix}$$

and

$$\mathbf{z}(t) = \begin{pmatrix} y(t) \\ p^1(t) \\ p^2(t). \end{pmatrix},$$

then the system of conditions from the maximum principle reads

$$\dot{\mathbf{z}}(t) = A(t)\mathbf{z}(t).$$

Fundamental Matrix

The solution of this non-stationary linear differential system is obtained as follows

$$\mathbf{z}(t) = \Phi(t, \tau)\mathbf{z}(\tau), \forall t > \tau,$$

where $\Phi(t, \tau)$ is called the *fundamental matrix*. In the stationary case, where A does not depend on t we can find the eigenvalues and the eigenvectors of A. We can check that these eigenvalues are $s, -s, -a$. We find next the corresponding eigenvectors, w^1, w^2, w^3. We can show that

$$W = (w^1, w^2, w^3) = \begin{pmatrix} 1 & 1 & 0 \\ \dfrac{\alpha^1}{a+s} & \dfrac{\alpha^1}{a-s} & \dfrac{(b^2)^2}{\beta^2} \\ -\dfrac{\alpha^2}{a+s} & -\dfrac{\alpha^2}{a-s} & -\dfrac{(b^1)^2}{\beta^1} \end{pmatrix}.$$

Let Λ be the diagonal matrix with eigenvalues on the diagonal, we can show that the fundamental matrix is given by

$$\Phi(t, \tau) = W \exp \Lambda(t - \tau) W^{-1}.$$

The fundamental matrix satisfies the matrix differential equation

$$\frac{\partial}{\partial t} \Phi(t, \tau) = A(t)\Phi(t, \tau), \quad \Phi(\tau, \tau) = I.$$

Moreover, this matrix is invertible, with inverse $\Psi(t, \tau) = (\Phi(t, \tau))^{-1}$ given by

$$\frac{\partial}{\partial t} \Psi(t, \tau) = -\Psi(t, \tau) A(t), \quad \Psi(\tau, \tau) = I.$$

Since the Maximum principle leads to a two-point boundary value problem, one must find the values

$$\varpi^1 = p^1(0), \quad \varpi^2 = p^2(0).$$

They are obtained from the conditions

$$p^1(T) = \gamma^1 y(T), \quad p^2(T) = \gamma^2 y(T),$$

which amounts to solving the linear system of algebraic equations

$$(\Phi_{22}(T, 0) - \gamma^1 \Phi_{12}(T, 0))\varpi^1 + (\Phi_{23}(T, 0) - \gamma^1 \Phi_{13}(T, 0))\varpi^2$$
$$= (\gamma^1 \Phi_{11}(T, 0) - \Phi_{21}(T, 0))x_0,$$

$$(\Phi_{32}(T, 0) - \gamma^2 \Phi_{12}(T, 0))\varpi^1 + (\Phi_{33}(T, 0) - \gamma^2 \Phi_{13}(T, 0))\varpi^2$$
$$= (\gamma^2 \Phi_{11}(T, 0) - \Phi_{31}(T, 0))x_0,$$

where $\Phi_{ij}(T, 0)$ represents the element of line i and column j of the matrix $\Phi(T, 0)$.

We can show the formulas

$$\varpi^1 = \frac{\Psi_{21}(T, 0) + \gamma^1 \Psi_{22}(T, 0) + \gamma^2 \Psi_{23}(T, 0)}{\Psi_{11}(T, 0) + \gamma^1 \Psi_{12}(T, 0) + \gamma^2 \Psi_{13}(T, 0)} x_0,$$

$$\varpi^2 = \frac{\Psi_{31}(T, 0) + \gamma^1 \Psi_{32}(T, 0) + \gamma^2 \Psi_{33}(T, 0)}{\Psi_{11}(T, 0) + \gamma^1 \Psi_{12}(T, 0) + \gamma^2 \Psi_{13}(T, 0)} x_0.$$

We can show more generally that

$$p^1(t) = P^1(t)y(t), \quad p^2(t) = P^2(t)y(t),$$

with the formulas

$$P^1(t) = \frac{\Psi_{21}(T,t) + \gamma^1 \Psi_{22}(T,t) + \gamma^2 \Psi_{23}(T,t)}{\Psi_{11}(T,t) + \gamma^1 \Psi_{12}(T,t) + \gamma^2 \Psi_{13}(T,t)},$$

$$P^2(t) = \frac{\Psi_{31}(T,t) + \gamma^1 \Psi_{32}(T,t) + \gamma^2 \Psi_{33}(T,t)}{\Psi_{11}(T,t) + \gamma^1 \Psi_{12}(T,t) + \gamma^2 \Psi_{13}(T,t)}.$$

We can check directly that $P^1(t), P^2(t)$ are solutions of the system of Riccati differential equations

$$-\dot{P}^1 - 2aP^1 + \frac{(b^1)^2}{\beta^1}(P^1)^2 + \frac{(b^2)^2}{\beta^2}P^1P^2 = \alpha^1,$$

$$-\dot{P}^2 - 2aP^2 + \frac{(b^2)^2}{\beta^2}(P^2)^2 + \frac{(b^1)^2}{\beta^1}P^1P^2 = \alpha^2,$$

$$P^1(T) = \gamma^1, \quad P^2(T) = \gamma^2,$$

already mentioned in the stationary case. This time the coefficients depend on time.

4. Closed-Loop Nash Equilibrium

System of PDE

We proceed with the Dynamic Programming formulation. The Hamiltonians are defined by

$$H^i(x, \mathbf{v}, q^i) = \frac{1}{2}(\alpha^i x^2 + \beta^i (v^i)^2) + q^i(ax + b^1 v^1 + b^2 v^2).$$

We look for Nash point equilibriums of the Hamiltonians $H^i(x, \mathbf{v}, q^i)$. We obtain easily

$$u^i(\mathbf{q}) = -q^i \frac{b^i}{\beta^i}.$$

We next write

$$H^1(x, \mathbf{q}) = \frac{1}{2}\alpha^1 x^2 + q^1 ax - \frac{1}{2}(q^1)^2 \frac{(b^1)^2}{\beta^1} - q^1 q^2 \frac{(b^2)^2}{\beta^2},$$

$$H^2(x, \mathbf{q}) = \frac{1}{2}\alpha^1 x^2 + q^2 ax - \frac{1}{2}q^2 \frac{(b^2)^2}{\beta^2} - q^1 q^2 \frac{(b^1)^2}{\beta^1}.$$

Dynamic Programming leads to the following system of partial differential equations

$$\frac{\partial\Psi^1}{\partial t}+\frac{1}{2}\alpha^1 x^2+\frac{\partial\Psi^1}{\partial x}ax-\frac{1}{2}\left(\frac{\partial\Psi^1}{\partial x}\right)^2\frac{(b^1)^2}{\beta^1}-\frac{\partial\Psi^1}{\partial x}\frac{\partial\Psi^2}{\partial x}\frac{(b^2)^2}{\beta^2}=0,$$

(2.1)
$$\frac{\partial\Psi^2}{\partial t}+\frac{1}{2}\alpha^1 x^2+\frac{\partial\Psi^2}{\partial x}ax-\frac{1}{2}\left(\frac{\partial\Psi^2}{\partial x}\right)^2\frac{(b^2)^2}{\beta^2}-\frac{\partial\Psi^1}{\partial x}\frac{\partial\Psi^2}{\partial x}\frac{(b^1)^2}{\beta^1}=0,$$

$$\Psi^1(x,T)=\frac{1}{2}\gamma^1 x^2,\ \Psi^2(x,T)=\frac{1}{2}\gamma^2 x^2,\ \text{a.e.}$$

System of Riccati Equations

The solutions are given by

$$\Psi^i(x,t)=\frac{1}{2}Q^i(t)x^2,$$

where $Q^i(t)$ are solutions of the system of Riccati equations

$$-\dot{Q}^1-2aQ^1+\frac{(b^1)^2}{\beta^1}(Q^1)^2+2\frac{(b^2)^2}{\beta^2}Q^1Q^2=\alpha^1,$$

(2.2)
$$-\dot{Q}^2-2aQ^2+\frac{(b^2)^2}{\beta^2}(Q^2)^2+2\frac{(b^1)^2}{\beta^1}Q^1Q^2=\alpha^2Q^1(T)=\gamma^1,$$
$$Q^2(T)=\gamma^2.$$

These equations are different from those of open-loop control. The coupling term is different, reflecting the coupling through the state.

Stationary Case

The above Riccati equations cannot be solved as easily as in the open loop case. To simplify we consider the stationary case (it corresponds to an infinite horizon problem with no discount). The Riccati equations reduce to the algebraic equations

$$-2aQ^1+\frac{(b^1)^2}{\beta^1}(Q^1)^2+2\frac{(b^2)^2}{\beta^2}Q^1Q^2=\alpha^1,$$

(2.3)
$$-2aQ^2+\frac{(b^2)^2}{\beta^2}(Q^2)^2+2\frac{(b^1)^2}{\beta^1}Q^1Q^2=\alpha^2.$$

To simplify notation, we set

$$\nu^i=\frac{(b^i)^2}{\beta^i}.$$

We can write these equations as

(2.4)
$$(\nu^1 Q^1 + \nu^2 Q^2 - a)^2 = \nu^1 \alpha^1 + (\nu^2 Q^2 - a)^2,$$
$$(\nu^1 Q^1 + \nu^2 Q^2 - a)^2 = \nu^2 \alpha^2 + (\nu^1 Q^1 - a)^2.$$

Set $\rho = \nu^1 Q^1 + \nu^2 Q^2 - a$. We check that

$$\nu^1 Q^1 - \nu^2 Q^2 = \frac{\nu^1 \alpha^1 - \nu^2 \alpha^2}{\rho - a},$$

and that ρ must be solution of the equation

$$\phi(\rho) = (\rho - a)^2(-3\rho^2 - 2a\rho + a^2 + 2M) + (\nu^1 \alpha^1 - \nu^2 \alpha^2)^2 = 0.$$

Assume that

$$a < \frac{s}{\sqrt{3}},$$

we can show that there exists only one solution such that $\rho > -a$. This solution is larger than $s/\sqrt{3}$ and $\sqrt{M/2}$. If

$$a > \frac{s}{\sqrt{3}},$$

then

$$\sqrt{\frac{M}{2}} < \frac{s}{\sqrt{3}},$$

we can show that for a sufficiently large, the equation may have 3 solutions larger than $\sqrt{M/2}$. It has only one larger than a.

As an example, consider the following model of "knowledge as a public good", see [Dockner et al. (2000), Section 9.5]. Two players contribute by investing in accumulating knowledge as a capital, whose evolution is governed by

$$\dot{x} = -\delta x + v^1 + v^2, \ x(0) = x_0.$$

Each player faces an investment cost given by

$$\rho v^i + \frac{1}{2}(v^i)^2,$$

and benefits from the common knowledge according to an individual revenue given by

$$x(t)(a^i - x(t)).$$

Note that in this model the individual profit declines when the collective knowledge is sufficiently large, and can become negative. There is a saturation effect. The pay-off for each player (to be minimized) is given by

$$J^i(v^1, v^2) = \int_0^\infty e^{-rt}[\rho v^i + \frac{1}{2}(v^i)^2 - x(t)(a^i - x(t))]dt.$$

We begin with the closed-loop Nash equilibrium. We write the Dynamic Programming equations . We first consider the Hamiltonians

$$H^i(x, \mathbf{v}, q^i) = -x(a^i - x) + \rho v^i + \frac{1}{2}(v^i)^2 + q^i(-\delta x + v^1 + v^2)$$

and look for a Nash equilibrium in (v^1, v^2).We obtain easily

$$u^i(\mathbf{q}) = -\rho - q^i$$

and it follows that

$$H^i(x, \mathbf{q}) = -x(a^i - x) - \frac{1}{2}\rho^2 - \frac{1}{2}(q^i)^2 - q^i(2\rho + \delta x) - q^1 q^2.$$

So the DP equations read

$$r\Psi^i = -x(a^i - x) - \frac{1}{2}\rho^2 - \frac{1}{2}(\Psi^i_x)^2 - \Psi^i_x(2\rho + \delta x) - \Psi^1_x \Psi^2_x.$$

We look for quadratic solutions

$$\Psi^i = \frac{1}{2}Px^2 - \beta^i x - \gamma^i$$

and we obtain, by identification

$$\frac{3}{2}P^2 + P(\delta + \frac{r}{2}) - 1 = 0,$$

$$\beta^i(P + \delta + r) + P(\beta^1 + \beta^2) = a^i + 2P\rho,$$

$$r\gamma^i = \frac{1}{2}\rho^2 + \frac{1}{2}(\beta^i)^2 - 2\rho\beta^i + \beta^1\beta^2.$$

We take the positive solution

$$P = -\frac{r + 2\delta}{6} + \sqrt{\left(\frac{r + 2\delta}{6}\right)^2 + \frac{2}{3}}$$

and the closed-loop controls are given by

$$u^i(x, t) = -Px + \beta^i - \rho.$$

Applying these controls we get the trajectory

$$\dot{y} = -(\delta + 2P)y + \beta^1 + \beta^2 - 2\rho,$$

which has a stable solution since $P > 0$. The stable solution is

$$\bar{y} = \frac{\beta^1 + \beta^2 - 2\rho}{2P + \delta}.$$

We next consider open-loop controls. We assume to simplify that $a^1 = a^2 = a$. We write the Maximum Principle necessary conditions. Thanks to our simplification, the two adjoint variables coincide. We get the system

$$\dot{y} = -\delta y - 2(\rho + p), \ y(0) = x_0,$$

$$-\dot{p} + (r + \delta)p = -a + 2y,$$

$$u^i = -(\rho + p).$$

The solution of this system is easily obtained as follows

$$p = Qy - q,$$

where Q is a solution of

$$2Q^2 + (r + 2\delta)Q - 2 = 0.$$

Note that $Q > P$. Next q is the solution of

$$-\dot{q} + q(r + \delta + 2Q) = a + 2Q\rho.$$

The corresponding trajectory is defined by

$$\dot{y} + y(\delta + 2Q) = -2\rho + 2q, \ y(0) = x_0.$$

It has also a stable solution, given by

$$\hat{y} = \frac{2q - 2\rho}{\delta + 2Q},$$

where q is given by

$$q = \frac{a + 2Q\rho}{r + \delta + 2Q}.$$

We can show that

$$\hat{y} = \frac{2(a - \rho(r + \delta))}{\delta(r + \delta) + 4},$$

$$\bar{y} = \frac{2(a - \rho(r + \delta) - \rho P)}{\delta(r + \delta) + 4 + P\delta},$$

and conclude that

$$\bar{y} < \hat{y}.$$

The economic interpretation of this inequality is the following. Each player has interest to benefit from the other player's investment and contribute the less possible. In the closed-loop case, one can make more use of the common education than in the open-loop case, resulting in a lower steady state.

From the economic considerations one can conjecture that the steady state should improve in the case of a cooperative game. We shall verify this property. In the cooperative game formulation we take as common objective function the sum of each player objective function

$$J(v^1, v^2) = \int_0^\infty e^{-rt}[\rho(v^1 + v^2) + \frac{1}{2}((v^1)^2 + (v^2)^2) - 2x(t)(a - x(t))]dt,$$

with the trajectory

$$\dot{x} = -\delta x + v^1 + v^2, \ x(0) = x_0.$$

We can write the Maximum Principle for the cooperative game, and obtain

$$\dot{y} = -\delta y - 2(\rho + p), \ y(0) = x_0,$$

$$-\dot{p} + (r + \delta)p = -2a + 4y,$$

$$u^i = -(\rho + p).$$

Let us check simply the steady state, given by the relations

$$\delta y = 2(\rho + p),$$

$$(r + \delta)p = -2a + 4y.$$

This leads to a steady state defined by

$$y^* = 2\frac{2a - (r + \delta)}{8 + \delta(r + \delta)}$$

and it is easy to check that

$$y^* > \hat{y}.$$

Existence Result

We go back to equations (2.2). To simplify, we assume that the coefficients do not depend on time. Recalling the notation ν^i, the equations are

$$-\dot{Q}^1 - 2aQ^1 + \nu^1(Q^1)^2 + 2\nu^2Q^1Q^2 = \alpha^1,$$

(2.5)

$$-\dot{Q}^2 - 2aQ^2 + \nu^2(Q^2)^2 + 2\nu^1Q^1Q^2 = \alpha^2$$

$$Q^1(T) = \gamma^1, \quad Q^2(T) = \gamma^2.$$

It is of interest to mimic the stationary case and introduce

$$\rho = \nu^1Q^1 + \nu^2Q^2 - a,$$

$$\sigma = \nu^1Q^1 - \nu^2Q^2.$$

We obtain the equations

$$-2\dot{\rho} + 3\rho^2 + 2a\rho = 2M + a^2 + \sigma^2, \ \rho(T) = N - a,$$

$$-\dot{\sigma} = \nu^1\alpha^1 - \nu^2\alpha^2 - \sigma(\rho - a), \ \sigma(T) = \nu^1\gamma^1 - \nu^2\gamma^2,$$

which reduce to the algebraic equation for ρ in the stationary case. We want to prove the following result

THEOREM 4.1 *There exist a positive solution of equations (2.2)*

We have seen in the stationary case that there may be several solutions. So we do not claim uniqueness.

Proof. It is better to work with $z = \rho + a$. Thus we got the system

$$-\dot{z} + \frac{3}{2}z^2 - 2az = M + \sigma^2, \ z(T) = N$$

$$-\dot{\sigma} = \nu^1\alpha^1 - \nu^2\alpha^2 - \sigma(z - 2a), \ \sigma(T) = \nu^1\gamma^1 - \nu^2\gamma^2.$$

Note that we recover Q^1, Q^2 from z, σ, by the formulas

$$Q^1 = \frac{z + \sigma}{2}, \ Q^2 = \frac{z - \sigma}{2}.$$

We prove a priori estimates. We first interpret z as the Riccati equation of a control problem. Indeed, consider the control problem

$$\dot{x} = ax + \frac{3}{2}v, \ x(t) = x$$

$$K_{x,t}(v(.)) = \frac{1}{2}\left[\int_t^T \left((M + \frac{\sigma^2(s)}{2})x^2(s) + \frac{3}{2}v^2(s) \right) ds + Nx^2(T) \right]$$

then it is easy to check that

$$\frac{1}{2}z^2(t)x^2 = \min_{v(.)} K_{x,t}(v(.)).$$

It follows immediately that $z(t) > 0$. In addition we can write

$$\frac{1}{2}z^2(t)x^2 \le K_{x,t}(0).$$

Calling

$$\sigma_\infty = \sup_{\{0 \le t \le T\}} |\sigma(t)|,$$

we get the following inequality

$$0 \le z(t) \le \exp 2aT(\frac{M}{2a} + \frac{\sigma^2}{4a} + N).$$

Now $\sigma(t)$ is the solution of a linear equation. So we have the explicit formula

$$\sigma(t) = (\gamma^1\nu^1 - \gamma^2\nu^2)\exp -\int_t^T (z - 2a)(s)ds + (\alpha^1\nu^1 - \alpha^2\nu^2)$$
$$\times \int_t^T \left(\exp -\int_t^s ((z - 2a)(\tau)d\tau\right) ds.$$

Since $z > 0$, it follows easily that

$$\sigma(t) = (\gamma^1\nu^1 - \gamma^2\nu^2)\exp -\int_t^T (z - 2a)(s)ds + (\alpha^1\nu^1 - \alpha^2\nu^2)$$
$$\times \int_t^T \left(\exp -\int_t^s ((z - 2a)(\tau)d\tau\right) ds.$$

We obtain

$$\sigma_\infty \le \exp 2aT \left(|\gamma^1\nu^1 - \gamma^2\nu^2| + \frac{|\alpha^1\nu^1 - \alpha^2\nu^2|}{2a}\right).$$

So we have obtained a priori bounds on $\sigma(t), z(t)$. They have been obtained provided $z(t)$ can be interpreted as the infimum of a control problem. Now if we consider a local solution $\sigma(t), z(t)$, near T, i.e. defined in $t \in (T - \epsilon, T]$ for ϵ sufficiently small, we obtain a positive solution for z, since $z(T) > 0$. Moreover the Control interpretation can be easily obtained, and thus the bounds are valid on this small interval. Since the bounds do not depend on ϵ, we can expand the solution beyond $T - \epsilon$, and in fact in $[0, T]$. We deduce Q^1, Q^2. They are positive near T, and by extension, they are positive. This concludes the proof of Theorem 4.1.

References

E. Dockner, S. Jørgensen, N. van Long,G. Sorger, *Differential Games in Economics and Management Science*, Cambridge University Press, Cambridge, 2000.

Chapter 3

EXTENDED GENERATORS OF MARKOV PROCESSES AND APPLICATIONS

Tomasz R. Bielecki
Applied Mathematics Department
Illinois Institute of Technology
10 W. 32nd Str.
Chicago, IL 60616
bielecki@iit.edu

Ewa Frankiewicz
Faculty of Mathematics and Information Science
Warsaw University of Technology
Pl. Politechniki 1
00-661 Warsaw
e_frankiewicz@o2.pl

Abstract   An extended generator for a semigroup of linear contractions corresponding to a Markov process on a Banach space is introduced and its fundamental properties are examined. It is argued that the extended generator is a better tool to analyze the behavior of the semigroup in various time scales (discrete and continuous) than the "classical" generator of a semigroup. Probabilistic interpretation of extended generators in the context of Markov processes and corresponding martingale process is also provided. A controlled martingale problem is examined. Suitable version of Bellman's optimality equations are introduced, and the corresponding verification theorem, which extends the classical optimality results for continuous time and discrete time controls, is established.

Keywords: Markov processes, infinitesimal operator, martingale problem, stochastic control

1.   Introduction

The paper is devoted to a study of Markov processes of a specific two-component structure. The first component corresponds to a continuous-time type, and in a typical case it can be interpreted as a diffusion-type component. The second component, corresponding to random jumps at discrete times is, loosely speaking, of discrete-time type, in the sense that it can be analyzed using discrete-time techniques. Although the features of a semigroup of linear contractions on Banach space corresponding to processes of that type can be revealed by the strong, weak (in the sense of [6]) and the full generators (in the sense of [7]), the domains of these three "classical" generators turn out to be quite awkward to work with. The extended generator of a process with jumps occurring spontaneously in a Poisson-like fashion, is considered in [5].

In this paper, we propose to analyze the behavior of a semigroup in various time scales by means of a finite family of generating operators; this family is referred to as the *extended generator* for the semigroup (the definition was first given in [2]). We show that this generator is an extension of both strong and weak generators, and it provides an explicit description of the transitions of the process in various time scales. Mimicking the standard probabilistic approach to semigroups of linear contractions, we also present a result on the martingale problem related to the extended generator. Let us stress that in the present paper we are mainly concerned with a special case of an extended generator, called the *CD-extended generator*, that is, the case when we have a continuous-time component and a single discrete-time component.

The second part of the paper is concerned with application of the extended generator to modelling and analysis of control systems with complete information, incorporating both continuous and discrete time scales. We introduce (after [3]) the definition of controlled martingale problem. As expected, our definitions of martingale problems are natural extensions of standard definitions, that can be found, for instance, in [1] and [10]. The case of the relaxed form of a controlled martingale problem is investigated in [3] and [9].

The paper is organized as follows: In Section 2, we present an example of the semigroup corresponding to a uniform motion along the real line with jumps of the size β or $-\beta$ at the deterministic moments and we analyze strong and weak generators for this semigroup. We show by direct calculations that the extended generator is a more convenient tool to analyze a semigroup of this kind than "classical" generators. In this section, we also give the formal definition of the extended generator (in particular, the definition of the CD-extended generator). We also prove that the continuous part of the CD-extended generator is an extension of both strong and weak generators. In Section 3, we discuss a martingale process related to the Markov process in various time scales via the corresponding CD-extended generator Finally, in Section 4, we define the controlled martingale problem of a mixed type. Bellman's equations for the value function are derived, and it is shown that they furnish an efficient way to derive the optimal solution to a martingale control problem of a mixed type. Verification theorem given in this paper is straightforward extension of classical result, which can be found, for instance, in [8].

It should be noted that examples presented in the paper are of an introductory nature. However, we envision several possible applications of the concept of the CD generator to deal with problems where at least two different time scales appear that are singular with respect to each other. Several practically relevant applications of extended generators in the area of finance are studied in [9]. In particular, a finite time horizon problem of optimal consumption-investment for a single agent subject to continuous and discrete time scales is solved there using the concept of the CD generator. In addition, in [9] there is given an approach to the valuation of a defaultable coupon bond based on the concept of the extended generator.

Notation. The following notation is used throughout the text:

$$\lim_{s \downarrow t} = \lim_{s \to t, \, s < t}$$

$$T_{t-}f = \lim_{\epsilon \downarrow 0} T_{(t-\epsilon)}f \quad \text{(if the limit exists)}$$

$$[x] - \text{ the integer part of a real number } x$$

$$[t]_\rho = \begin{cases} 0, & 0 \le t < \rho, \\ k\rho, & k\rho \le t < (k+1)\rho, \text{ for } k = 1, 2, \dots \end{cases}$$

$$\int_{s+}^{t} = \int_{(s,t]} .$$

Moreover $\frac{\partial_+ f}{\partial x}$ denotes the right-hand side partial derivative of the function f with respect to x. Finally, A and $\tilde{A}$ denote strong and weak generators for a contraction semigroup T_t, with respective domains $D(A)$ and $D(\tilde{A})$ (in a sense of [6]).

2. Extended Generator of a Markov Process

Before stating a formal definition of an extended generator, let us consider a simple, but motivating, example of a process, which can be effectively analyzed using various time scales.

2.1 Motivation

Let $(\Omega, \mathbb{F}, \mathbb{P})$ be a probability space. To provide some justification of the concept of an extended generator, we shall consider a simple example of a piecewise deterministic Markov process $X. = (X_t, t \geq 0)$ taking values in $(\mathbb{R}, \mathcal{B})$, where $\mathcal{B}$ is the σ-algebra of Borel subsets in the real line. Specifically, the process X is governed by the following equation

$$dX_t = \alpha \, dt + \sum_{n \leq t} \xi_n, \quad t \in [0, \infty), \qquad (2.1)$$

with $X_0 = 0$, where $\{\xi_n\}_{n=1}^{\infty}$ is a sequence of i.i.d. random variables on $(\Omega, \mathcal{F}, \mathbb{P})$, such that $\mathbb{P}(\xi_n = \beta) = p = 1 - \mathbb{P}(\xi_n = -\beta)$ and $\alpha, \beta > 0$ are constants. We take $\mathbb{F}$ to be the natural filtration generated by the process $X.$ so that $\mathbb{F} = (\mathcal{F}_t^X)_{t \geq 0}$ where $\mathcal{F}_t^X = \sigma(X_s, 0 \leq s \leq t)$. It is clear that the process $X.$ describes a uniform motion with the speed α along the real line, combined with random jumps of the size β or $-\beta$ at deterministic moments $1, 2, \ldots$.

Remarks. The restrictive assumption that the jump component of the considered process is driven by a sequence of random variables taking only two values can be relaxed. Indeed, the jump size can be given as an arbitrary random variable (see [9]). Notice also that for the sake of simplicity we consider here only a continuous deterministic motionmodulated by random jumps; a diffusion type component will be added in Section 4.

Transition semigroup. To analyze the process $X.$ given by (2.1), we enlarge the original state space $\mathbb{R}$ to the product space $\bar{E} = \mathbb{R} \times [0, \infty)$, and we introduce a suitably modified process $Y. = (Y_t, t \geq 0)$ taking values in $\bar{E}$, such that $Y_t = (X_t, t)$. Observe that the process $Y.$ defined above is time-homogeneous. The semigroup T_t of linear contractions corresponding to the process $Y.$ has the form (we denote $q = 1 - p$):

$$T_t f(x, s) = \sum_{k=0}^{[t+s]-[s]} \binom{[t+s] - [s]}{k} p^k q^{[t+s]-[s]-k} \cdot$$

$$\cdot \, f\Big(x + \alpha t + \beta k - \beta \left([t+s] - [s] - k\right), s + t\Big) \qquad (2.2)$$

for any function $f \in B(\bar{E})$, where $B(\bar{E})$ is the space of real-valued, bounded, measurable functions on $\bar{E}$.

LEMMA 2.1 *Suppose that $T_t f$ is given by (2.2). Then, for $t \in (0, 1)$ we have that:*
If $n - 1 \leq s < n - t$, then $T_t f(x, s) = f(x + \alpha t, s + t)$,

If $n-t \le s < n$, then $T_t f(x, s) = p f(x + \alpha t + \beta, s + t) + q f(x + \alpha t - \beta, s + t)$ for $n = 1, 2, \ldots$.

Proof. The result follows from the explicit form of the semigroup (2.2), combined with the fact that for $t \in (0, 1)$ we have

$$[s + t] - [s] = \begin{cases} 0, & \text{if} \quad n - 1 \le s < n - t, \\ 1, & \text{if} \quad n - t \le s < n, \end{cases}$$

for every $n = 1, 2, \ldots$ ∎

Strong and weak generators. It is easily seen that for any $f \in D(A)$ the strong generator of Y is given by

$$A f(x, t) = \frac{\partial f}{\partial t}(x, t) + \alpha \frac{\partial f}{\partial x}(x, t).$$

By virtue of Lemma 2.1, the space $B_0(\bar{E})$ of strong continuity of the semigroup consists of all bounded, measurable functions f characterized by the following condition:

$$\lim_{t \downarrow 0} \sup_{x \in \mathbb{R}} \left[\sup_{\substack{n - t \le s < n \\ n = 1, 2, \ldots}} |p f(x + \alpha t + \beta, s + t) + q f(x + \alpha t - \beta, s + t) - f(x, s)| \right.$$

$$\left. + \sup_{\substack{n - 1 \le s < n - t \\ n = 1, 2, \ldots}} |f(x + \alpha t, s + t) - f(x, s)| \right] = 0. \tag{2.3}$$

The domain $D(A)$ contains all functions $f \in B_0(\bar{E})$ such that the following limit exists:

$$\lim_{t \downarrow 0} \sup_{x \in \mathbb{R}} \left[\sup_{\substack{n - t \le s < n \\ n = 1, 2, \ldots}} \frac{1}{t} |p f(x + \alpha t + \beta, s + t) + q f(x + \alpha t - \beta, s + t) - f(x, s)| \right. \tag{2.4}$$

$$+ \sup_{\substack{n-1 \le s < n-t \\ n=1,2,\dots}} \frac{1}{t} |f(x+\alpha t, s+t) - f(x,s)| \Bigg] .$$

Remarks. Notice that several smooth functions, like $f(x,s) = \sin x$ do not belong to $B_0\left(\bar{E}\right)$, and thus they do not belong to the domain $D(A)$. Since jumps of the semigroup can not be captured by the differential operator $\mathcal{L} = f'_d$, they have to be somehow encoded in the domain $D(A)$, in order to be able to recover the semigroup from $(A, D(A))$. Unfortunately, this encoding, formally done via conditions (2.4), eliminates many nice functions from $D(A)$ and makes the domain quite awkward to work with.

Let us now examine the weak generator $\tilde{A}$ of Y. For any $f \in D(\tilde{A})$ we have

$$\tilde{A}f(x,t) = \frac{\partial_+ f}{\partial t}(x,t) + \alpha \frac{\partial_+ f}{\partial x}(x,t).$$

The space $\tilde{B}_0(\bar{E})$ of weak continuity of the semigroup consists of all bounded, measurable functions which are right-continuous in t and continuous in x. The domain $D(\tilde{A})$ contains all functions $f \in \tilde{B}_0(\bar{E})$ for which following quantities stay bounded as $t \downarrow 0$:

$$\sup_{x \in \mathbb{R}} \Bigg[\sup_{\substack{n-t \le s < n \\ n=1,2,\dots}} \tag{2.5}$$

$$\frac{1}{t} |pf(x+\alpha t + \beta, s+t) + qf(x+\alpha t - \beta, s+t) - f(x,s)| +$$

$$+ \sup_{\substack{n-1 \le s < n-t \\ n=1,2,\dots}} \frac{1}{t} |f(x+\alpha t, s+t) - f(x,s)| \Bigg] .$$

Remarks. Note that in this case the function $f(x,s) = \sin x$ belongs to the class $B_0\left(\bar{E}\right)$, but it does not belong to the domain $D(\tilde{A})$. Again, similarly as for the strong generator, the domain $D(\tilde{A})$ is rather inconvenient to work with. For this reason, the idea of using instead the extended generator for the semigroups of this type seems to be natural and advantageous.

Extended generator. We shall now introduce an extended generator of a semigroup, and we shall show that it can be conveniently used to analyze behavior of the semigroup considered in the present example. To this end, we denote by $C^1_\delta(\bar{E})$ the class of all functions $f \in B(\bar{E})$ which satisfy the following three conditions: (i) f has a continuous right-hand-side derivative with respect to t, (ii) f is of class C^1 in x, (iii) the jumps of the function $h(t) = f(x + \alpha t, s + t)$ may occur only in points $\delta(s) + k$ for $k = 0, 1, 2, \ldots$, where

$$\delta(s) = 1 - s + [s] \text{ for all } s \geq 0. \tag{2.6}$$

Observe first that both domains, $D(A)$ and $D(\tilde{A})$, are clearly subsets of the space $C^1_\delta(\bar{E})$. Moreover, we have the following result:

PROPOSITION 2.1 *For any function $f \in C^1_\delta(\bar{E})$ the semigroup (2.2) admits the following integral representation*

$$T_t f(x, s) = f(x, s) + \int_0^t T_r A_0 f(x, s)\, dr + \int_{0+}^t T_{r-} A_1 f(x, s)\, d\gamma(r, s),$$
$$\tag{2.7}$$

where the operators A_0 and A_1 are given by

$$A_0(x, t) = \frac{\partial_+ f}{\partial t}(x, t) + \frac{\partial f}{\partial x}(x, t), \tag{2.8}$$

$$A_1 f(x, t) = p f(x + \beta, t) + q f(x - \beta, t) - f(x, t), \tag{2.9}$$

$$\gamma(r, s) = \begin{cases} 0, & 0 \leq r < \delta(s) \\ n, & \delta(s) + (n - 1) \leq r < \delta(s) + n \text{ for } n = 1, 2, \ldots, \end{cases}$$

and $\delta(s)$ is given by (2.6).

Proof. For any function $f \in C^1_\delta(\bar{E})$ the right-hand side of (2.7) can be rewritten as follows:

$$R = f(x, s) + \int_0^{\delta(s)} T_r A_0 f(x, s)\, dr + \int_{\delta(s)}^{\delta(s)+1} T_r A_0 f(x, s)\, dr + \ldots$$

$$+ \int_{\delta(s)+(n-1)}^{\delta(s)+n} T_r A_0 f(x, s)\, dr + \int_{\delta(s)+n}^t T_r A_0 f(x, s)\, dr$$
$$+ \quad T_{\delta(s)-} A_1 f(x, s) + T_{(\delta(s)+1)-} A_1 f(x, s) + \ldots$$
$$+ \quad T_{(\delta(s)+n-1)-} A_1 f(x, s) + T_{(\delta(s)+n)-} A_1 f(x, s).$$

It easy to show that for semigroup considered in our example we have

$$A_1 T_{(\delta(s)+n)-} f(x,s) = T_{\delta(s)+n} f(x,s) - T_{(\delta(s)+n)-} f(x,s).$$

Therefore,

$$
\begin{aligned}
R & = T_t f(x,s) - (T_{\delta(s)} - T_{\delta(s)-}) f(x,s) - \ldots \\
& - (T_{(\delta(s)+n-1)} - T_{(\delta(s)+n-1)-}) f(x,s) - (T_{(\delta(s)+n)} - T_{(\delta(s)+n)-}) f(x,s) \\
& + A_1 T_{\delta(s)-} f(x,s) + \ldots + A_1 T_{(\delta(s)+n-1)-} f(x,s) \\
& + A_1 T_{(\delta(s)+n)-} f(x,s) = T_t f(x,s).
\end{aligned}
$$

This proves the proposition. ∎

Remarks. The process given by (2.1) is an example of a Markov process of a two-component structure. The first component (continuous-time type component) corresponds to a uniform motion along the real line. The second component (discrete-time type component) consists of sum of the jumps.

The triple $(A_0, A_1, C_\delta^1(\bar{E}))$ introduced in Proposition 2.1 may serve as an elementary example of the so-called *extended generator* for a semigroup of linear operators (in the present case, for the semigroup T given in (2.2)). In particular, the space $C_\delta^1(\bar{E})$ may be treated as a domain of the extended generator. This class of functions is much more convenient to work with than the domains $D(A)$ and $D(\tilde{A})$. Observe that the considered semigroup is uniquely determined on $B(\bar{E})$ by the function δ and the corresponding triple $(A_0, A_1, C_\delta^1(\bar{E}))$. Specifically, the operator A_0 uniquely determines the continuous part of the semigroup. The operator A_1, corresponding to the discrete part of the semigroup, is uniquely determined by the function δ. In the next paragraph, we provide a formal definition of an abstract extended generator for a semigroup of linear operators.

2.2 Extended Generator

Let E be a locally compact topological space with countable base, endowed with the Borel σ-algebra $\mathcal{B}$. Let $(\Omega, \mathcal{F}, \mathbb{P})$ be an underlying probability space, and let $X = (X_t, t \geq 0)$ be a Markov process of a two-component structure on this space, with values in $(E, \mathcal{B})$ and the transition probability $P(s, x, s+t, \Gamma)$. Let us define the process $Y = (Y_t, t \geq 0)$ with values in an extended state space $\bar{E} = E \times [0, \infty)$, where $Y_t = (X_t, t)$. As a σ-algebra $\bar{\mathcal{B}}$ we take a σ-algebra of all sets $\bar{\Gamma} \in \bar{E}$ such that $\bar{\Gamma}_t = \{x : (x,t) \in \bar{\Gamma}\}$ is $\mathcal{B}$-measurable. Then the process Y is a time-homogeneous Markov process with the transition probability P' given by $P'(t, y, \bar{\Gamma}) = P(s, x, s+t, \bar{\Gamma}_{s+t})$, where $y = (x,s) \in \bar{E}$ and $\bar{\Gamma} \in \bar{\mathcal{B}}$. Let

$B(\bar{E})$ be the space of all real-valued, bounded and measurable functions on $\bar{E}$. Then for every $t, s \geq 0$ and $f \in B(\bar{E})$ the semigroup of linear contractions corresponding to Y is given by

$$T_t f(Y_s) = \mathbf{E}[f(Y_{t+s}) \mid \mathcal{F}_s^Y] = \int_{\bar{E}} f(y) P'(t, Y_s, dy), \quad \mathbb{P}\text{-a.s.}$$

Remarks. Notice that $\mathcal{F}_t^Y = \mathcal{F}_t^X$. To simplify notation we shall write from now on $\mathcal{F}_t$ instead of $\mathcal{F}_t^X$.

The space $B(\bar{E})$ endowed with the sup norm

$$\|f\| = \sup_{y \in \bar{E}} |f(y)| = \sup_{x \in E} \sup_{t \in [0,\infty)} |f(x,t)|$$

is a Banach space. Let $B_0(\bar{E})$ and $\tilde{B}_0(\bar{E})$ denote the spaces of strong and weak continuity of the semigroup T_t on $B(\bar{E})$, respectively. We restrict our attention to processes Y whose trajectories are right-continuous and have left-hand limits. We will also assume that the process Y is normal, that is, $\lim_{t \downarrow 0} P'(t, y, \bar{E}) = 1$ for all $y \in \bar{E}$. This implies that the transition function $P'(t, y, \bar{\Gamma})$ is stochastically continuous (see [6], II 2.8 and Lemma III 3.2), that is, $\lim_{t \downarrow 0} P'(t, y, \bar{\Gamma}) = 1$ for all $y \in \bar{\Gamma}$.

To define the extended generator for this semigroup, we need to introduce some notation: We fix $k \in \mathbb{N}$, and we let $\gamma_i : [0, \infty) \times [0, \infty) \longrightarrow [0, \infty)$ for $i = 1, 2, \ldots, k$ be functions given by

$$\gamma_i(t, s) = \begin{cases} 0, & 0 \leq t < \delta_i(s) \\ n, & \delta_i(s) + (n-1)\rho_i \leq t < \delta_i(s) + n\rho_i \\ & \text{for } n = 1, 2, \ldots, \end{cases} \quad (2.10)$$

where $\delta_i(s) = \rho_i - s + [s]_{\rho_i}$ for $s \geq 0$, $\rho_i > 0, i = 1, \ldots, k$ are fixed constants.

Next, let $Y = ((X_t, t), t \geq 0)$ be a Markov process of a two-component structure, with state space $\bar{E}$. Let T_t be a semigroup of linear contractions corresponding to process Y, and let $\tilde{A}$ be a weak infinitesimal operator corresponding to this process. Then we have the following definition.

DEFINITION 2.2 The sequence of operators $A_0, A_1, \ldots, A_k$ is called the *extended generator* for T_t with domain D if the following conditions are satisfied:
(i) D is any set of functions $f \in B(\bar{E})$ such that $D(\tilde{A}) \subseteq D$, and for

every $f \in D$

$$T_t f(x,s) = f(x,s) + \int_0^t T_r A_0 f(x,s) dr + \sum_{i=1}^k \int_{0+}^t T_{r-} A_i f(x,s) d\gamma_i(r,s);$$

(2.11)

(ii) A_0 is a weak infinitesimal operator corresponding to the continuous-time type component of Y;

(iii) $A_0 f \in \tilde{B}_0(\bar{E})$ for every function $f \in D$;

(iv) A_i, $i = 1, \ldots, k$ are bounded operators on $B(\bar{E})$ such that $A_i T_t f = T_t A_i f$.

Remark. It is essential to observe that the domain D of the extended generator is not unique.

CD-extended generator. In this paper, we are only concerned with the specific type of extended generator, corresponding to two time scales: a continuous scale and a discrete scale (the general case is examined in Frankiewicz [9]). Namely, we postulate that $k = 1$ and we denote $\rho \equiv \rho_1$. Let

$$\gamma_1(t,s) = \gamma(t,s) = \begin{cases} 0, & 0 \le t < \delta(s) \\ n, & \delta(s) + (n-1)\rho \le t < \delta(s) + n\rho \\ & \text{for } n = 1, 2, \ldots \end{cases} \qquad (2.12)$$

and for every $s \ge 0$

$$\delta(s) = \rho - s + [s]_\rho. \qquad (2.13)$$

Observe, that the discrete-time type component of the process Y is a sum of the jumps at the moments $\rho, 2\rho, \ldots$.

Then the fundamental equation (2.11) becomes

$$T_t f(x,s) = f(x,s) + \int_0^t T_r A_0 f(x,s) \, dr + \int_{0+}^t T_{r-} A_1 f(x,s) \, d\gamma(r,s).$$

(2.14)

We shall call the triple (A_0, A_1, D) the CD-extended generator of T (for continuous/discrete time). The operators A_0 and A_1 will be called, respectively, the continuous and discrete parts of the CD-extended generator.

In the remaining part of this section, we shall show that the continuous component of the CD-extended generator is an extension of both strong and weak generators. We first prove the following auxiliary result related to the operator A_1.

LEMMA 2.3 *For any $f \in B(\bar{E})$ the operator A_1 satisfies the equality*

$$A_1 T_{(\delta(s)+\rho n)-} f(x,s) = T_{\delta(s)+\rho n} f(x,s) - T_{(\delta(s)+\rho n)-} f(x,s), \qquad (2.15)$$

for every $n = 0, 1, 2, \ldots$, *with the function* $\delta(\cdot)$ *given by (2.13).*

Proof. By virtue of equation (2.14), we have for every $f \in B(\bar{E})$

$$
T_{\delta(s)+\rho n} f(x, s) - T_{(\delta(s)+\rho n)-} f(x, s) = \int_0^{\delta(s)+\rho n} T_r A_0 f(x, s) dr
$$

$$
+ \int_{0+}^{\delta(s)+\rho n} T_{r-} A_1 f(x, s) d\gamma(r, s) - \int_0^{\delta(s)+\rho n} T_r A_0 f(x, s) dr
$$

$$
- \int_{0+}^{(\delta(s)+\rho n)-} T_{r-} A_1 f(x, s) d\gamma(r, s) = A_1 T_{(\delta(s)+\rho n)-} f(x, s). \quad \blacksquare
$$

We are in the position to prove the following result.

LEMMA 2.4 (A_0, D) *is an extension of both* $(A, D(A))$ *and* $(\tilde{A}, D(\tilde{A}))$.

Proof. We first prove that (A_0, D) is an extension of $(\tilde{A}, D(\tilde{A}))$. To this end, observe that, by assumption, $D(\tilde{A}) \subseteq D$. Therefore, it suffices to show that for $f \in D(\tilde{A})$ we have $A_0 f = \tilde{A} f$.

By the definition of the weak infinitesimal operator and Definition 2.2, we have (with the variables suppressed) that

$$
\tilde{A} f = w \lim_{h \downarrow 0} \frac{1}{h} \left(\int_0^h T_t A_0 f dt + \int_{0+}^h T_{t-} A_1 f d\gamma \right).
$$

Let us examine the second integral. Let $0 < \delta < \delta + \rho < \ldots < \delta + n\rho \leq h$ for $n = 0, \ldots$. Then

$$
\int_{0+}^h T_{t-} A_1 f d\gamma = T_{\delta-} A_1 f + T_{(\delta+\rho)-} A_1 + \ldots + T_{(\delta+n\rho)-} A_1 f.
$$

From Lemma 2.3 and the fact that the operators A_1 and T_t commute, it follows that $T_{(\delta+n\rho)-} A_1 f = A_1 T_{(\delta+n\rho)-} f = T_{\delta+n\rho} f - T_{(\delta+n\rho)-} f$. Recall that if $f \in D(\tilde{A})$ then $T_t f$ is weakly continuous (see [6]), which implies that $T_{(\delta+n\rho)-} f = T_{\delta+n\rho} f$ for every $n \geq 0$. Consequently, $\tilde{A} f = w \lim_{h \downarrow 0} \frac{1}{h} \int_0^h T_t A_0 f dt = A_0 f$. This completes the proof of the first part of the lemma. In an analogous way, we can show that (A_0, D) is also an extension of $(A, D(A))$. $\quad \blacksquare$

3. Martingale Problem

In this section, we discuss a martingale problem associated to the process Y via the corresponding CD-extended generator. The discussion below is motivated by classical results on martingale problems (see [7]

46

for example). Let $Y.$ be the process introduced in Section 3.1. Then we have the following proposition:

PROPOSITION 3.2 *For any $f \in D$ and $r \geq 0$, we define the process $M. = (M_t^{r,f}, t \geq r)$ by setting*

$$M_t^{r,f} = f(Y_t) - f(Y_r) - \int_r^t A_0 f(Y_s)\, ds - \int_{r+}^t A_1 f(Y_{s-})\, d\gamma(s,0). \quad (3.16)$$

Then the process $M.$ is an $\mathbb{F}$-martingale.

Proof. In the proof, we shall write M_t instead of $M_t^{r,f}$. Since $M.$ is an adapted and integrable process, we need only to show that

$$\mathbf{E}[M_{t+s} \mid \mathcal{F}_t] = M_t \quad (3.17)$$

for all $t \geq r$ and $s \geq 0$. Towards this end, we first observe that

$$\mathbf{E}\left[\int_r^{s+t} A_0 f(Y_u)\, du \mid \mathcal{F}_t\right] = \int_r^{s+t} \mathbf{E}[A_0 f(Y_u) \mid \mathcal{F}_t]\, du \quad (3.18)$$

$$= \int_r^t A_0 f(Y_u)\, du + \int_t^{s+t} T_{u-t} A_0 f(Y_t)\, du =$$

$$= \int_r^t A_0 f(Y_u)\, du + \int_0^s T_u A_0 f(Y_t)\, du.$$

We also have that

$$\mathbf{E}\left[\int_{r+}^{s+t} A_1 f(Y_{u-})\, d\gamma(u,0) \mid \mathcal{F}_t\right] =$$

$$= \int_{r+}^t A_1 f(Y_{u-})\, d\gamma(u,0) + \int_{t+}^{s+t} T_{(u-t)-} A_1 f(Y_t)\, d\gamma(u,0).$$

Observe that $\int_{t+}^{s+t} T_{(u-t)-} A_1 f(Y_t)\, d\gamma(u,0) = \int_{0+}^s T_{u-} A_1 f(Y_t)\, d\gamma(u,t)$, and

$$\mathbf{E}\left[\int_{r+}^{s+t} A_1 f(Y_{u-})\, d\gamma(u,0) \mid \mathcal{F}_t\right] = \quad (3.19)$$

$$= \int_{r+}^t A_1 f(Y_{u-})\, d\gamma(u,0) + \int_{0+}^s T_{u-} A_1 f(Y_t)\, d\gamma(u,t).$$

Combining (3.18) with (3.19), we obtain

$$\mathbf{E}[M_{t+s} \mid \mathcal{F}_t] = \mathbf{E}[f(Y_{t+s}) - f(Y_r) - \int_r^t A_0 f(Y_s)\, ds$$

$$-\int_{r+}^{t} A_1 f(Y_{s-})\, d\gamma(s,0)\mid \mathcal{F}_t] = T_s f(Y_t) - \int_0^s T_u A_0 f(Y_t)\, du$$

$$-\int_{0+}^{s} T_{u-} A_1 f(Y_t)\, d\gamma(u,t) - f(Y_r) - \int_r^t A_0 f(Y_u)\, du$$

$$-\int_{r+}^{t} A_1 f(Y_{u-})\, d\gamma(u,0) = f(Y_t) - f(Y_r) - \int_r^t A_0 f(Y_u)\, du$$

$$-\int_{r+}^{t} A_1 f(Y_{u-})\, d\gamma(u,0) = M_t$$

and this in turn implies (3.17). ∎

Remark. Relationship between the extended generator for a Markov semigroup T corresponding to the Markov process Y and the extended generator and random generator for Y in the sense of Kunita (see [4], p.208) is currently under investigation.

4. Controlled Martingale Problem of a Mixed Type

In this section we will be concerned with modelling and analysis of controlled systems with continuous and discrete time scales. For the sake of analytical tractability, the corresponding continuous-time optimization problem will be reformulated as a controlled martingale problem of a mixed type, i.e., a controlled martingale problem associated with a suitable CD-extended generator.

4.1 Cost Functions

Let T be a fixed finite terminal time. Let $\rho > 0$ be fixed constant such that $\rho < T$.

DEFINITION 4.5 Let $X = (X_t, t \in [0,T])$ be a two-component Markow process with values in E, given on a probability space $(\Omega, \mathcal{F}, \mathbb{F}, \mathbb{P})$, with the jumps of the discrete-time type component at the moments $\rho, 2\rho, \ldots$. Process X is called the *controlled process* if the dynamic of X depends on two other stochastic processes:
(i) Process $(u_t, t \in [0,T])$ with values in $U \subseteq \mathbb{R}$, called the *continuous control*;
(ii) Process $(v_t, t \in [0,T])$, with values in $V \subseteq \mathbb{R}$, such that $v_t = v_{j\rho}$ for $t \in [j\rho, (j+1)\rho \wedge T)$, $j = 0, 1, 2, \ldots$, called the *discrete control*.

At any instant $t \in [0,T]$, in order to control the process X_t, we can choose value of u_t. In addition, at any time $t = 0, \rho, 2\rho, \ldots, N(T)$, where

$\rho > 0$, $N(t) = t - \rho$ if $[t]_\rho = t$ and $N(t) = [t]_\rho$ otherwise, we are allowed to choose the values of v_t.

Assumption. We assume throughout that the filtration $\mathbb{F}$ is generated by the observations of the controlled process $X_.$, that is, we deal with the case of complete information.

Our goal is to analyze a family of the following optimization problems parameterized by $(x, t) \in E \times [0, T]$:

Optimization problem $\mathcal{O}(x, t)$**:** Find a controlled process X^{u^*, v^*} and a pair of admissible (in the sense of Definitions 4.8 and 4.9) controls $u^*_., v^*_.$ on the time interval $[t, T]$, which maximize the following functional:

$$J(x, t, u_., v_.) = \mathbb{E}^{u_., v_.}_{x, t} \left[\int_t^T L^0(X_s, s, u_s, v_s) ds + \tag{4.20} \right.$$

$$\left. + \sum_{i=n(t)}^{N(T)} L^1(X_{i-}, i-, v_i) + L^1(X_0, 0, v_0) \mathbb{1}_{\{t=0\}} + \Phi(X_T, T) \right],$$

where $u_., v_.$ are $\mathbb{F}$-adapted stochastic process and $n(t) = [t]_\rho + \rho$ if $t = 0$ or $t \neq [t]_\rho$, and $n(t) = t$ otherwise.

The notation $\mathbb{E}^{u_., v_.}_{x, t}$ is meant to emphasize that we deal with the controlled process X which depends on controls $u_., v_.$ and satisfies the initial condition $X_t = x$. The exact meaning of this statement will be clarified in the next subsection.

The functions $L^0 : E \times [0, T] \times U \times V$ and $L^1 : E \times [0, T] \times V$ are the running cost functions corresponding to continuous and discrete time, respectively, and $\Phi(x, T)$ is the terminal cost function. They are supposed to satisfy suitable technical conditions which guarantee that the above functional is well defined.

Remarks.

1. To simplify notation, we shall write from now on X^*, rather than X^{u^*, v^*}.

2. The controls $u_.$ and $v_.$ are not treated in the same way. The values of u_s are chosen at every time $s \in [t, T]$. The values of v_s are chosen only at the times $s = k\rho, 2k\rho, \ldots, N(T)$ where $k\rho \geq t$, and kept constant between these times.

To solve the Optimization problem $\mathcal{O}(x, t)$ we shall apply backward induction method. Suppose that n is such that $(n + 1)\rho < T$. Let us consider two cases.

If $t \in (n\rho, (n + 1)\rho)$, then we first find optimal continuous control u_s for $s \in [N(T), T]$, treating $v_{N(T)}$ as a parameter. Then, we find the optimal value of $v_{N(T)}$. Then we find optimal continuous control u_s for s in the interval $[N(T) - \rho, N(T))$, treating $v_{N(T)-\rho}$ as a parameter. And so on. We end by finding optimal continuous control u_s for $s \in [t, (n + 1)\rho)$, treating $v_{n\rho}$ as parameter. Therefore the optimization problem $\mathcal{O}(x, t)$ for $t \in (n\rho, (n+1)\rho)$ is parameterized by the value of the discrete control chosen at time $n\rho$. For this reason, given $t \in (n\rho, (n+1)\rho)$, the optimization problem $\mathcal{O}(x, t)$ is in fact solved not only with given initial condition (x, t) but also with given value $v = v_t = v_{n\rho}$.

If $t = n\rho$ for any $n \in \mathbb{N}$, we first solve the Optimization problem $\widetilde{\mathcal{O}}(x, n\rho; v_{n\rho})$.

Optimization problem $\widetilde{\mathcal{O}}(x, n\rho; v_{n\rho})$: Treating $v_{n\rho}$ as a parameter find a controlled process X^{u^*, v^*} and a pair of controls u^*, v^* on the time interval $(n\rho, T]$, which maximize the functional (4.20) for $t = n\rho$.

Solution to the Optimization problem $\widetilde{\mathcal{O}}(x, n\rho; v_{n\rho})$ is found as follows: We start by finding optimal continuous control u_s for $s \in [N(T), T]$, treating $v_{N(T)}$ as a parameter, and we end by finding optimal continuous control u_s for $s \in [n\rho, (n + 1)\rho)$, treating $v_{n\rho}$ as parameter.

Then we return to the Optimization problem $\mathcal{O}(x, n\rho)$ and we find the optimal value of $v_{n\rho}$.

4.2 Bellman's Equations and Verification Theorem

In this subsection, we shall deal with the case of controlled Markov process in a formal way. We postulate that to each fixed controls parameters $u \in U$ and $v \in V$ we may associate a Markov process $X^{u,v}$ with the CD-extended generator denoted as $(A_0^{u,v}, A_1^v, D^{u,v})$. We know already that the domain $D^{u,v}$ is not uniquely specified. Therefore, it will be convenient to assume that we may find a sufficiently large class, denoted by $\tilde{D}$ such that $\tilde{D} \subseteq D^{u,v}$ for each $u \in U$ and $v \in V$. Moreover, in a typical application, it will be clear what is meant by a process with no controls. Thus, we will be in the position to impose the following condition: $D(\tilde{A}) \subseteq D^{u,v}$, where $\tilde{A}$ is the weak generator of the process with no controls.

Let L^0 and L^1 be given functions which are right-continuous in s and right-hand-side differentiable with respect to v and u.

As usual, we are allowed to know the history of X_s for $s \leq t$ when the controls u_t and v_t are chosen. We assume from now on that we are given a family of CD-extended generators $\{(A_0^{u,v}, A_1^v, D^{u,v}), u \in U, v \in V\}$, as well as the class of functions $\tilde{D}$ such that $\tilde{D} \subseteq D^{u,v}$ for each $u \in U$ and $v \in V$. We shall search for a solution of the Optimization Problem

$\mathcal{O}(x,t)$ among processes, which are solutions of the controlled martingale problem of a mixed type, in the sense of the following definitions.

Let $n \in \mathbb{N}$ be a fixed number such that $n\rho < T$.

DEFINITION 4.6 We say that an $E \times U \times V$-valued process (X_s, u_s, v_s), $s \in [t,T]$, where $t \in (n\rho, (n+1)\rho)$, defined on a probability space $(\Omega, \mathbb{F}, \mathbb{P})$ is a solution to the *controlled martingale problem* with the initial conditions $(x,t) \in E \times [0,T]$ and $v_t \in V$ if the following conditions are satisfied:

(i) X_s, $s \in [t,T]$, is corlol (i.e., it is right continuous, with finite left limits) and $\mathbb{F}$-adapted,

(ii) u_s, $s \in [t,T]$, and v_s, $s \in [t,T]$, are $\mathbb{F}$-adapted processes, where $v_s = v_{j\rho}$ for $s \in [j\rho, (j+1)\rho \wedge T)$, $j = n+1, n+2, \ldots$,

(iii) for any function $f(x, s, \bar{v})$ such that $f(\cdot, \cdot, \bar{v}) \in \tilde{D}$ for every $\bar{v} \in V$ the process $M^{t,f} = (M^{t,f}_s, s \in [t,T])$, defined as

$$
\begin{aligned}
M^{t,f}_s = {}& f(X_s, s, v_s) - f(x, t, v_t) \qquad\qquad (4.21) \\
& - \int_t^s A_0^{u_r, v_r} f(X_r, r, v_r)\, dr - \int_{t+}^s A_1^{v_r} f(X_{r-}, r-, v_r)\, d\gamma(r, 0)
\end{aligned}
$$

is an $\mathbb{F}$-martingale.

We shall say in short that $(X., u., v.)$ solves the controlled martingale problem with the initial conditions $(x,t) \in E \times [0,T]$ and $v \in V$.

DEFINITION 4.7 We say that an $E \times U \times V$-valued process (X_s, u_s, v_s), $s \in [t,T]$, where $t = n\rho$, defined on a probability space $(\Omega, \mathbb{F}, \mathbb{P})$ is a solution to the *controlled martingale problem* with the initial condition $(x,t) \in E \times [0,T]$ if the condition (i) from Definition 4.6 is satisfied and:

(ii') u_s, $s \in [t,T]$, and v_s, $s \in [t,T]$, are $\mathbb{F}$-adapted processes, where $v_s = v_{j\rho}$ for $s \in [j\rho, (j+1)\rho \wedge T)$, $j = n, n+1, \ldots$,

(iii') for any function $f(x, s, \bar{v})$ such that $f(\cdot, \cdot, \bar{v}) \in \tilde{D}$ for every $\bar{v} \in V$ the process $M^{t,f} = (M^{t,f}_s, s \in [t,T])$, defined as

$$
\begin{aligned}
M^{t,f}_s = {}& f(X_s, s, v_s) - f(x, t, v_t) \qquad\qquad (4.22) \\
& - \int_t^s A_0^{u_r, v_r} f(X_r, r, v_r)\, dr - \sum_{k=t}^{N(T)} A_1^{v_k} f(X_{k-}, k-, v_k) \\
& - A_1^{v_0} f(X_0, 0, v_0) \mathbb{1}_{\{t=0\}}
\end{aligned}
$$

is an $\mathbb{F}$-martingale.

DEFINITION 4.8 Given a pair $(x,t) \in E \times (0,T]$, where $t \neq n\rho$, and $v \in V$, we say that a pair of controls $u., v.$ is *admissible* for the Optimization problem $\mathcal{O}(x,t)$, if there exists a solution to the controlled martingale problem with the initial conditions (x,t) and $v \in V$.

DEFINITION 4.9 Given a pair $(x, n\rho)$ where $x \in E$, we say that a pair of controls $u_\cdot, v_\cdot$ is *admissible* for the Optimization problem $\mathcal{O}(x, n\rho)$, if there exists a solution to the controlled martingale problem with the initial condition $(x, n\rho)$.

Notations. We write $\mathcal{A}_{(x,t;v)}$ to denote the set of all admissible controls $u_\cdot, v_\cdot$ for the Optimization problem $\mathcal{O}(x,t)$ with the initial conditions (x,t) and v. Also, we write $\mathcal{A}_{(x,t)}$ to denote the set of all admissible controls $u_\cdot, v_\cdot$ for the Optimization problem $\mathcal{O}(x,t)$ with the initial condition (x,t).

DEFINITION 4.10 We say that a function $\hat{V} : E \times [0,T] \to \mathbb{R}$ is the *value function* of the Optimization problem $\mathcal{O}(x,t)$ if

$$\hat{V}(x,t) = \sup_{(u_\cdot,v_\cdot) \in \mathcal{A}_{(x,t;v)}} J(x,t,u_\cdot,v_\cdot), \quad \text{for } t \neq n\rho,$$

$$\hat{V}(x,t) = \sup_{(u_\cdot,v_\cdot) \in \mathcal{A}_{(x,t)}} J(x,t,u_\cdot,v_\cdot), \quad \text{for } t = n\rho.$$

Since the controlled process is Markov, it is natural to expect that the optimal control processes will be of a feedback form. We say that the control processes $u_\cdot$ and $v_\cdot$ are of the *feedback form* if

$$u_s = \bar{u}(X_s, s), \quad v_s = \bar{v}(X_s, s) \tag{4.23}$$

where $\bar{u} : E \times [0,T] \to U$ and $\bar{v} : E \times [0,T] \to V$ are measurable functions. In particular, if the feedback controls $u_\cdot, v_\cdot$ are in $\mathcal{A}_{(x,t;v)}$ or in $\mathcal{A}_{(x,t)}$ then equality (4.23) is assumed to hold for $s \in [t, T]$.

Bellman's equations and verification theorem.
We have $J(x, t, u_\cdot^*, v_\cdot^*) = \hat{V}(x,t)$, provided that the optimal controls $u_\cdot^*$ and $v_\cdot^*$ exist. It is obvious that $J(x, T, u_\cdot^*, v_\cdot^*) = \hat{V}(x,T) = \Phi(x,T)$ for every $x \in E$. To determine the value function $\hat{V}$, we introduce the following pair of coupled Bellman's equations, that have to be solved for the function $W(\cdot, \cdot)$ for every $t \in [(k-1)\rho, (k\rho) \wedge T)$, $k = 1, 2, \ldots$

$$\sup_{u \in U} \left(L^0(x,t,u,v) + A_0^{u,v} W(x,t,v) \right) = 0, \tag{4.24}$$

$$\sup_{v \in V} \left(L^1(x,(k\rho)-,v) + A_1^v W(x,(k\rho)-,v) \right) = 0, \tag{4.25}$$

$$\sup_{v \in V} \left(L^1(x, ((k-1)\rho)-, v) + A_1^v W(x, ((k-1)\rho)-, v) \right) = 0, \quad (4.26)$$

$$W^v(x, T) = \Phi(x, T). \quad (4.27)$$

If $t \in [0, \rho \wedge T)$, instead of the equation (4.26) we have

$$\sup_{v_0 \in V} \left(L^1(x, 0, v_0) + A_1^v W(x, 0, v_0) \right) = 0. \quad (4.28)$$

Then we have the following theorems.

THEOREM 4.11 (**Verification theorem 1**) *Let* $t \in (n\rho, (n+1)\rho)$ *and let a function* $W \in \tilde{D}$ *be a solution to the Bellman's equations (4.24), (4.25), (4.27). Then:*
(a) $W(x, t, v_t) \geq J(x, t, u_., v_.)$ *for every admissible controls* $u_., v_. \in \mathcal{A}_{(x,t;v)}$ *and for every* $v \in V$,
(b) If there exists an admissible solution $(X^*, u^*_., v^*_.)$ *to the controlled martingale problem with the initial conditions* (x, t) *and* v_t *such that*

$$u^*_s \in \arg\max_{u \in U} \left(L^0(X^*_s, s, u, v) + A_0^{u,v} W(X^*_s, s, v) \right) \quad (4.29)$$

and

$$v^*_{k\rho} \in \arg\max_{v \in V} \left(L^1(X^*_{(k\rho)-}, (k\rho)-, v) + A_1^v W(X^*_{(k\rho)-}, (k\rho)-, v) \right) \quad (4.30)$$

for all $s \in [(k-1)\rho \vee t, (k\rho) \wedge T)$, $k = 1, 2, \ldots$, *then* $W(x, t, v) = \hat{V}(x, t)$ *and the pair* $u^*_., v^*_.$ *is an optimal control.*

Proof. (a) Assume that $W \in \tilde{D}$ and $u_., v_. \in \mathcal{A}_{(x,t;v)}$. Then by virtue of Definitions 4.6 and 4.8 we have

$$W(x, t, v_t) = \mathbb{E}_{x,t}^{u.,v.} \left\{ W(X_T, T) - \right. \quad (4.31)$$

$$\left. - \int_t^T A_0^{u_s,v_s} W(X_s, s, v_s) \, ds - \int_{t+}^T A_1^{v_s} W(X_{s-}, s-, v_s) \, d\gamma(s, 0) \right\}.$$

Now, by the Bellman's equations

$$W(x, t, v_t) \geq \mathbb{E}_{x,t}^{u.,v.} \left\{ \int_t^T L^0(X_s, s, u_s, v_s) ds + \right.$$

$$+ \int_{t+}^{T} L^1(X_s, s, v_s)\, d\gamma(s,0) + \Phi(X_T, T) \Big\} =$$

$$= \mathbb{E}_{x,t}^{u_.,v_.} \Big\{ \int_t^T L^0(X_s, s, u_s, v_s)\, ds + \sum_{i=n(t)}^{N(T)} L^1(X_{i-}, i-, v_i) + \Phi(X_T, T) \Big\}$$

$$= J(x, t, u_., v_.),$$

which proves the first part of the theorem.

(b) Assume that $(X^*_., u^*_., v^*_.)$ is a solution to the controlled martingale problem such that (4.29) and (4.30) are satisfied. Then the inequality in part (a) of the theorem becomes equality, and thus $W(x, t, v_t) = J(x, t, u^*_., v^*_.) = \hat{V}(x, t)$. Since we assume that $u^*_., v^*_. \in \mathcal{A}_{(x,t;v)}$ then the pair $u^*_., v^*_.$ is an optimal control. This completes the proof. ∎

THEOREM 4.12 (**Verification theorem 2**) *Let $t = n\rho$ and let a function $W \in \hat{D}$ be a solution to the Bellman's equations (4.24)-(4.28). Then:*

(a) $W(x, t, v_t) \geq J(x, t, u_., v_.)$ for every admissible controls $u_., v_. \in \mathcal{A}_{(x,t)}$,
(b) If there exists an admissible solution $(X^*, u^*_., v^*_.)$ to the controlled martingale problem with the initial condition (x, t) such that

$$u^*_s \in \arg\max_{u \in U} \Big(L^0(X^*_s, s, u, v) + A_0^{u,v} W(X^*_s, s, v) \Big), \qquad (4.32)$$

$$v^*_{k\rho} \in \arg\max_{v \in V} \Big(L^1(X^*_{(k\rho)-}, (k\rho)-, v) + A_1^v W(X^*_{(k\rho)-}, (k\rho)-, v) \Big) \quad (4.33)$$

$$v^*_{n\rho} \in \arg\max_{v \in V} \Big(L^1(X^*_{(n\rho)-}, (n\rho)-, v) + A_1^v W(X^*_{(n\rho)-}, (n\rho)-, v) \Big) \quad (4.34)$$

$$v^*_0 \in \arg\max_{v \in V} \Big(L^1(X^*_0, 0, v) + A_1^v W(X^*_0, 0, v) \Big) \qquad (4.35)$$

for all $s \in [(k-1)\rho \vee n\rho, (k\rho) \wedge T)$, $k = 1, 2, \ldots$, then $W(x, t, v_t) = \hat{V}(x, t)$ and the pair $u^*_., v^*_.$ is an optimal control.

Theorem 4.12 can be proved in the same way as Theorem 4.11, and therefore its proof will be skipped.

Remark. The results of the present section may be used, for example, to solve a consumption-investment problem for a single agent on a finite horizon. The investor starting at time t with a initial endowment $x > 0$, can consume and invest in such a way, that the expected discounted utility from consumption is maximized. This problem is investigated in [9]).

References

[1] Bhatt, A.G., Borkar, V.S. (1996) *Occupation Measures for Controlled Markov Processes: Characterization and Optimality.* Annals of Probability, vol. 24, no. 3, pp. 1531-1562.

[2] Bielecki, T. (1995a) *On Extended Generator for Semigroup of Linear Contractions, Markov Processes and Related Martingales.* Unpublished manuscript.

[3] Bielecki, T. (1995b) *On Time Hybrid Controlled Martingale Problem.* Unpublished manuscript.

[4] Cinlar, E., Jacod, J., Protter, P., Sharpe, M.J. (1980) *Semimartingales and Markov Processes.* Zeitschrift fur Wahrscheinlichkeitstheorie und verwandte Gebiete, 54, pp.161-219.

[5] Davis, M.H.A. (1984) *Piecewise-Deterministic Markov Processes: A General Class of Non-Diffusion Stochastic Models.* J. R. Statist. Soc. B, 46, no. 3.

[6] Dynkin, E.B. (1965) *Markov Processes.* Springer-Verlag.

[7] Ethier, S.E., Kurtz, T.G. (1986) *Markov Processes: Characterization and Convergence.* Wiley, New York.

[8] Fleming, W.H., Soner, H.M. (1992) *Controlled Markov Processes and Viscosity Solutions.* Springer-Verlag.

[9] Frankiewicz, E. (2003) *Extended Generators of Markov Processes and Their Applications in the Mathematics of Finance.* Doctoral dissertation, Warsaw University of Technology. In preparation.

[10] Kurtz, T.G., Stockbridge, R.H. (1998) *Existence of Markov Controls and Characterization of Optimal Markov Controls.* SIAM J. Control Optim. 36, no. 2, pp. 609-653.

Chapter 4

CONTROL OF MANUFACTURING SYSTEMS WITH DELAYED INSPECTION AND LIMITED CAPACITY

E. K. Boukas
Mechanical Engineering Department
École Polytechnique de Montréal
P.O. Box 6079, station "Centre-ville"
Montréal, Québec, Canada H3C 3A7
el-kebir.boukas@polymtl.ca

Abstract   This work deals with the control of manufacturing systems with limited capacity. The production rate of the system for each part is assumed to be bounded by both lower (equal to zero) and upper bounds. The model we propose in this work considers the inspection of the produced parts and a small rate of these parts is rejected. The model considers also the fact that a rate of delivered parts can be returned to the stock level. It is assumed that these rates are not known exactly and some uncertainties are considered. The control problem treated in this work consists of computing the state feedback control law that guarantees the asymptotic stability of our model despite the uncertainties and constraints on the production rates.

Keywords: Manufacturing system, system delay, product inspection.

1.   Introduction

Manufacturing systems are the key success of the industrial countries. Nowadays, most of the big companies in the world use the most advanced technology in control, transportation, machining, etc. to produce their goods. With the trends of globalization of business, we are living in a world where the increasing competition between companies is dictating the business rules. Therefore, to survive, most of the companies are forced to focus seriously on how to produce high quality products at low cost and on how to respond quickly to rapid changes in the demand. The key competitive factors are the new technological advances and the ability to use them to quickly respond to rapid changes in the market. Production planning is one of the key ingredients that has a direct effect on the ability to quickly respond to rapid changes in the market. It is concerned with the optimal allocation of the production capacity of the system to meet the demand efficiently. In general this problem is not easy and requires significant attention.

Production systems belong to the class of large scale systems that are in general more complicated to model and to control. In the last decades, this kind of systems have attracted many researchers from operations research and control communities. Different problems like routing, scheduling, and production and maintenance planning have been tackled and interesting results have been reported in the literature. For more details on what it has been done on the subject, we refer the reader to Sethi and Zhang (1994), Gershwin (1994), Boukas and Liu (2001), Sethi et al. (2005), Simon (1952), Grubbstrom and Wikner (1996), Hennet (2003), Towill (1982), Towill et al. (1997), Wiendahl and Breithaupt (2000), Axsater (1985), Ridalls, C. E. and Bennett (2002), Gavish and Graves (1980), Disney et a. (2000) and the references therein.

Different mathematical model for manufacturing systems have been proposed in the literature and models that use the state space representation (deterministic and stochastic) are among the list. In this direction of research, either in the deterministic or the stochastic frameworls, approaches based on optimal control theory have been used to establish interesting results for solving the previously mentioned problems. For more details on the subject we refer the reader to Boukas and Liu (2001), Sethi et al. (2005) and the references therein.

Recently, an approach based on robust control theory has been developed for the control of the manufacturing systems. Boukas (2005) has formulated the production planning problem as an $\mathcal{H}_\infty$ tracking problem where the solution is obtained by solving a set of linear matrix inequalities that can be successfully solved by the interior point methods. In Rodrigues and Boukas (2005) the production systems are modeled as constrained switched linear systems and the inventory control problem is formulated as a constrained switched $\mathcal{H}_\infty$ problem with a piecewise-affine (PWA) control law. The switching variable for the production systems modeled in this work is the stock level. When the stock level is positive, some of the perishable produced parts are being stored and will deteriorate with time at a given rate. When the stock level is negative it leads to backorders, which means that orders for production of parts are coming in and there are no stocked parts to immediately meet the demand. A state feedback controller that forces the stock level to be kept close to zero (sometimes called a just-in-time policy), even when there are fluctuations in the demand, will be designed in this work using $\mathscr{H}_\infty$ control theory. The synthesis of the state feedback controller that quadratically stabilizes the production dynamics and at the same time rejects the external demand fluctuation (treated as a disturbance) is cast as a set of linear matrix inequalities (LMIs).

In this chapter, our objective is to consider the production planning in the continuous-time framework and extend the model used in Boukas (2005) and Rodrigues and Boukas (2005) to include the inspection process of the produced parts and also the inclusion of a rate of the return of a fraction of the delivered parts. The inspection time is assumed to be constant for all produced part type and it is included in the model as a delay in the dynamics. It is also assumed that in the rest of the chapter that the rejected rate of the inspected parts and the returned rate of the delivered parts are not known exactly and some uncertainties are introduced to count for these variations.

The rest of the chapter is organized as follows. In Section 2, the production planning problem for manufacturing systems is formulated and the required assumptions are given. Section 3 presents the main results of the chapter and which can be summarized in the synthesis of the state feedback controller that will guarantee that the closed-loop will be asymptotically stable and respecting all the constraints of the control for all admissible uncertainties. In Section 4 a numerical example is provided to show the effectiveness of the proposed results.

Notation. The notation used in this chapter is quite standard. $\mathbb{R}^n$ and $\mathbb{R}^{n \times m}$ denote, respectively, the set of the n dimensional space of real vectors and the set of all $n \times m$ real matrices. The superscript "⊤" denotes the transpose of a given matrix. $P > 0$ means that P is symmetric and positive-definite. $\mathbb{I}$ is the identity matrix with compatible dimension. $\ell_2[0, \infty]$ is the space of square summable vector sequence over $[0, \infty]$. $\lambda_{min}(A)$ and $\lambda_{max}(A)$ represent respectively the minimum and maximum eigenvalues of the matrix A. $\|A\|$ represents the norm of of the matrix A $(\|A\| = \sqrt{\lambda_{max}(A^\top A)})$. $\| \cdot \|$ will refer to the Euclidean vector norm whereas $\| \cdot \|_{[0,\infty]}$ denotes the $\ell_2[0, \infty]$-norm over $[0, \infty]$ defined as $\|f\|^2_{[0,\infty]} = \sum_0^\infty \|f_k\|^2$. We define $\mathbf{x}_s(t) = x_{s+t}, t - \tau \leq s \leq t$, noted in the sequel by $\mathbf{x}(t)$.

2. Problem statement

To have an idea on the model under consideration, let us consider a simple production system that consists of one machine producing one part type. Let us assume that the produced parts are inspected and the inspection time is the same for all the produced parts and it is not known exactly and denoted by τ. After the inspection the part is either accepted (added to the stock level) or rejected and that we have to subtract it from the stock. We assume that a rate ρ_{d1} of the inspected parts is rejected and also a rate ρ_{r1} of delivered parts may be returned to the stock. If, we denote by $x_1(t)$, $u_1(t)$ and d_1 respectively the stock level, the production rate and the demand rate (that we assume constant) at time t, the dynamics of the stock is then described by the following differential equation:

$$\begin{cases} \dot{x}_1(t) = [\rho_{r1} + \Delta\rho_{r1}]\, x_1(t) + [-\rho_{d1} + \Delta\rho_{d1}]\, x_1(t - \tau) + u_1(t) - d_1, \\ x(0) = 0 \end{cases} \quad (4.1)$$

where τ represents the inspection time, x_0 is the initial stock level and $\Delta\rho_{r1}$ and $\Delta\rho_{d1}$ represent respectively the uncertainties of the returned rate and rejected rate of parts. These rates are assumed to satisfy the following constraints:

$$\begin{cases} |\Delta\rho_{r1}| \leq \alpha_1, \\ |\Delta\rho_{d1}| \leq \beta_1, \end{cases}$$

where α_1 and β_1 are known positive scalars.

REMARK 2.1 *The bounds on the uncertainties on the rates $\Delta\rho_{r1}$ and $\Delta\rho_{d1}$ are realistic since the rates are always bounded even when the uncertainties are present. This comes from the fact one can either reject none or all the inspected parts or take back none or all the delivered parts.*

The production rate is limited by the capacity of the machine and therefore we have the following constraints:

$$0 \leq u_1(t) \leq \bar{u}_1. \quad (4.2)$$

The problem we will address in the rest of this chapter is to force the production to track the system demand and therefore keep the stock level close to zero. This problem will be formulated as a control problem. For this purpose, let us assume that we have complete access to the production level at each time t and let us design a state feedback controller so that the closed loop dynamics will be asymptotically stable. Let us make the following change of variable:

$$v_1(t) = u_1(t) - d_1.$$

Based on this change of variables, the system becomes:

$$\begin{cases} \dot{x}_1(t) = [\rho_{r1} + \Delta\rho_{r1}] x_1(t) + [-\rho_{d1} + \Delta\rho_{d1}] x_1(t - \tau) + v_1(t), \\ x(0) = 0. \end{cases} \quad (4.3)$$

The controller we will design is assumed to have the following form:

$$v_1(t) = -K_1 x_1(t),$$

which corresponds to the following real production rate:

$$u_1(t) = v_1(t) + d.$$

The controller constraints become:

$$-d_1 \leq v_1(t) \leq \bar{u}_1 - d_1.$$

Let us now generalize our model to a production system that produces n items. Let $x(t) \in \mathbb{R}^n$ and $u(t) \in \mathbb{R}^n$ denote respectively the stock level vector, and the production rate vector at time t. Following the same steps as for the case of one part type, the dynamics of the production system can be described by the following system of differential equations:

$$\begin{cases} \dot{x}(t) = [A_r + \Delta A_r] x(t) + [A_d + \Delta A_d] x(t - \tau) + Bv(t), \\ x(0) = 0 \end{cases} \quad (4.4)$$

where

$$x(t) = \begin{bmatrix} x_1(t) \\ \vdots \\ x_n(t) \end{bmatrix}, v(t) = \begin{bmatrix} v_1(t) \\ \vdots \\ v_n(t) \end{bmatrix},$$

$$A_r = \text{diag}\,[\rho_{r1}, \cdots, \rho_{rn}], \Delta A_r = \text{diag}\,[\Delta\rho_{r1}, \cdots, \Delta\rho_{rn}],$$
$$A_d = \text{diag}\,[-\rho_{d1}, \cdots, -\rho_{dn}], \Delta A_d = \text{diag}\,[\Delta\rho_{d1}, \cdots, \Delta\rho_{dn}],$$
$$B = \mathbb{I}.$$

The control input $v(t)$ is constrained component by component, i.e.: $-d_i \leq v_i(k) \leq \bar{u}_i - d_i, i = 1, \cdots, n$, with $\bar{u}_i$ being a real known scalar. The uncertainties ΔA_r and ΔA_d satisfy the following:

$$\begin{cases} \|\Delta A_r\| \leq \alpha, \\ \|\Delta A_d\| \leq \beta. \end{cases} \quad (4.5)$$

REMARK 2.2 *The uncertainties that satisfy the conditions (4.5) are referred to as admissible uncertainties and may be considered as time varying.*

The goal of this work is to develop a control law that satisfies all the constraints and and at the same time tracks precisely the demand, d at each time t. To solve this problem, we formulate it as a control problem and synthesize a state feedback controller that assures the asymptotic stability of the closed-loop dynamics.

Before closing this section, notice that from optimal control theory (Kirk (1970)) we know that for given matrices R (symmetric and positive-definite) and Q (symmetric and positive-semidefinite) if the following Riccati equation:

$$A^\top P + PA - PBR^{-1}B^\top P + Q = 0 \qquad (4.6)$$

has a solution P (symmetric and positive-definite matrix), then we can construct a state feedback controller of the following form:

$$v(t) = -Kx(t) \qquad (4.7)$$

with $K = R^{-1}B^\top P$.

Due to the presence of the time-delay, the uncertainties and the bounds on the control, extra conditions should be added to guarantee that the controller (4.7) will assure asymptotic stability of the closed-loop dynamics.

3.   Controller design

To solve the production planning problem we formulated in the previous section we will firstly relax the constraints on the control input and design a state feedback controller that guarantees the asymptotic stability of the closed-loop dynamics. To design such controller, a decomposition of the matrix A_d is used and a Riccati equation based on this decomposition is employed to compute the controller gain. Secondly, the results are extended to take of the control constraints.

Combining the system dynamics (4.4) and the controller expression (4.7), we get:

$$\dot{x}(t) = [A_{cl} + \Delta A_r] x(t) + [A_d + \Delta A_d] x(t - \tau)$$

with $A_{cl} = A_r - BR^{-1}B^\top P$.

The matrix P used here is the solution of an equivalent Riccati equation to be provided later on.

¿From the other side notice that:

$$x(t - \tau) = x(t) - \int_{t-\tau}^{t} \dot{x}(s)ds$$

$$= x(t) - \int_{t-\tau}^{t} \left[[A_{cl} + \Delta A_r] x(s) + [A_d + \Delta A_d] x(s - \tau) \right] ds.$$

If we decompose A_d as follows:

$$A_d = \widetilde{A}_d + \widehat{A}_d$$

then the previous closed-loop system becomes:

$$\dot{x}(t) = \left[A_{cl} + \widetilde{A}_d\right] x(t) + \widehat{A}_d x(t - \tau) + \Delta A_r x(t) + \Delta A_d x(t - \tau)$$
$$- \widetilde{A}_d \left[x(t) - x(t - \tau)\right].$$

Using the expression $x(t) - x(t - \tau)$, we get:

$$\dot{x}(t) = \left[A_{cl} + \widetilde{A}_d\right] x(t) + \widehat{A}_d x(t - \tau) + \Delta A_r x(t) + \Delta A_d x(t - \tau)$$
$$- \widetilde{A}_d \int_{t-\tau}^{t} \left[[A_{cl} + \Delta A_r] x(s) + [A_d + \Delta A_d] x(s - \tau)\right] ds. \quad (4.8)$$

Now if we assume that the pair $(A_r + \widetilde{A}_d, B)$ is stabilizable, then it is possible to construct a state feedback control using the optimal control theory and the gain can be computed from the following Riccati equation:

$$\left[A_r + \widetilde{A}_d\right]^\top P + P \left[A_r + \widetilde{A}_d\right] - PBR^{-1}B^\top P + Q = 0. \quad (4.9)$$

If this equation has a solution P (symmetric and positive-definite matrix), then we can construct a state feedback controller of the following form:

$$v(t) = -Kx(t)$$

with $K = R^{-1}B^\top P$.

Notice the asymptotic stability will be guaranteed when the initial condition is in the domain $\mathscr{D}$. In fact the system will be stable if $V(x(t)) \leq V(x_0)$ for all $x_0 \in \mathscr{D}$ with $V(.)$ a Lyapunov candidate function. The domain is defined as follows:

$$\mathscr{D} = \{x \in \mathbb{R}^n | V(x) = x^\top Px \leq 1\}$$

which is an ellipsoid that contains all the state for the closed-loop.

Notice that this can be stated as a linear matrix inequality and solved using the interior point method. For more details on this refer the reader to Boukas (2005) and Rodrigues and Boukas (2005).

Our first result is stated by the following theorem.

THEOREM 3.1 *If there exists a positive scalar $q > 1$ such that the following holds:*

$$0 \leq \tau < \frac{\mu - \left[\alpha + q\nu \left[\|\widehat{A}_d\| + \beta\right]\right]}{q\nu \left[\|\widetilde{A}_d A_{cl}\| + \|\widetilde{A}_d A_d\| + \|\widetilde{A}_d\| \left[\alpha + \beta\right]\right]} \tag{4.10}$$

where $\mu = \frac{0.5\lambda_{min}(PBR^{-1}B^\top P + Q)}{\lambda_{max}(P)}$ and $\nu = \sqrt{\frac{\lambda_{min}(P)}{\lambda_{max}(P)}}$, then system (4.8) is asymptotically stable for all initial condition in the domain $\mathscr{D}$ and all admissible uncertainties.

Sketch of the proof. First of all let us consider a Lyapunov functional of the following form:

$$V(\mathbf{x}(t)) = x^\top(t)Px(t)$$

The derivative with respect to time t along any trajectory is given by:

$$
\begin{aligned}
\dot{V}(\mathbf{x}(t)) = {}& \dot{x}^\top(t)Px(t) + x^\top(t)P\dot{x}(t) \\
= {}& \Big[\big[A_{cl} + \widetilde{A}_d \big] x(t) + \widehat{A}_d x(t-\tau) + \Delta A_r x(t) + \Delta A_d x(t-\tau) \\
& - \widetilde{A}_d \int_{t-\tau}^t \left[[A_{cl} + \Delta A_r]x(s) + [A_d + \Delta A_d]x(s-\tau) \right] ds \Big]^\top Px \\
& + x^\top P \Big[\big[A_{cl} + \widetilde{A}_d \big] x(t) + \widehat{A}_d x(t-\tau) + \Delta A_r x(t) + \Delta A_d x(t-\tau) \\
& - \widetilde{A}_d \int_{t-\tau}^t \left[[A_{cl} + \Delta A_r]x(s) + [A_d + \Delta A_d]x(s-\tau) \right] ds \Big]
\end{aligned}
$$

which can be rewritten as follows:

$$
\begin{aligned}
\dot{V}(\mathbf{x}(t)) = {}& x^\top(t) \left[\big[A_{cl} + \widetilde{A}_d \big]^\top P + P \big[A_{cl} + \widetilde{A}_d \big] \right] x(t) \\
& + x^\top(t) \left[\Delta A_r^\top P + P\Delta A_r \right] x(t) \\
& + x^\top(t-\tau)\widehat{A}_d^\top Px(t) + x^\top(t)P\widehat{A}_d x(t-\tau) \\
& + x(t-\tau)\Delta A_d^\top Px(t) + x^\top P\Delta A_d x(t-\tau) \\
& - 2x^\top(t)P\widetilde{A}_d \int_{t-\tau}^t \left[[A_{cl} + \Delta A_r]x(s) + [A_d + \Delta A_d]x(s-\tau) \right] ds
\end{aligned}
$$

which implies in turn after using (4.9):

$$\dot{V}(\mathbf{x}(t)) \leq -x^\top(t)\left[PBR^{-1}B^\top P + Q\right]x(t)$$
$$+\left\|x^\top(t)\left[\Delta A_r^\top P + P\Delta A_r\right]x(t)\right\| + \left\|x^\top(t-\tau)\widehat{A}_d^\top Px(t)\right\|$$
$$+\left\|x^\top(t)P\widehat{A}_dx(t-\tau)\right\| + \left\|x(t-\tau)\Delta A_d^\top Px(t)\right\|$$
$$+\left\|x^\top P\Delta A_dx(t-\tau)\right\|$$
$$+2\left\|x^\top(t)P\widetilde{A}_d\int_{t-\tau}^t \left[[A_{cl}+\Delta A_r]\,x(s)+[A_d+\Delta A_d]\,x(s-\tau)\right]ds\right\|.$$

Based on Razumikhin-type theorem, if there exists a positive scalar $q > 1$ such that:

$$V(x(t-\tau)) < q^2 V(x(t))$$

then we have:

$$\|x(t-\tau)\| < q\nu\|x(t)\|$$

with $\nu = \sqrt{\frac{\lambda_{min}(P)}{\lambda_{max}(P)}}$.

Using now this fact, we get:

$$\dot{V}(\mathbf{x}(t)) \leq -\gamma\|x(t)\|^2$$

with

$$\gamma = \lambda_{min}(PBR^{-1}B^\top P + Q) - 2\left[\left[\alpha + q\nu\left[\|\widehat{A}_d\| + \beta\right]\right]\right.$$
$$+q\nu\tau\left[\|\widetilde{A}_dA_{cl}\| + \|\widetilde{A}_dA_d\| + \|\widetilde{A}_d\|\,[\alpha + \beta]\right]\right]\lambda_{max}(P).$$

Based on (3.1), we conclude that $\gamma > 0$ and therefore $\dot{V}(\mathbf{x}(t)) < 0$ which concludes that the closed-loop system is asymptotically stable for all admissible uncertainties. $\square$

Notice that if the returned rate and the inspection rate are known without uncertainties, our previous results become:

COROLLARY 4.1 *If there exists a positive scalar $q > 1$ such that the following holds:*

$$0 \leq \tau < \frac{\mu - q\nu\|\widehat{A}_d\|}{q\nu\left[\|\widetilde{A}_dA_{cl}\| + \|\widetilde{A}_dA_d\|\right]} \tag{4.11}$$

then system (4.8) is asymptotically stable for all initial condition in $\mathcal{D}$.

The controller we designed in Theorem 3.1 does not take care of the imposed constraints on control and therefore the control may violate the bounds which makes it infeasible. To overcome this we should include these constraints in our design. For this purpose, notice that if we make the analogy with the results on saturating actuators (Boukas (2003)or see Liu (2005)) we can solve our problem. The dynamics of the system can be rewritten as follows:

$$\begin{cases} \dot{x}(t) = [A_r + \Delta A_r] x(t) + [A_d + \Delta A_d] x(t - \tau) + B v_m(t) \\ x(0) = 0 \end{cases} \quad (4.12)$$

with $v_m(t) = Sat(v(t)) = [Sat(v_1(t)), \cdots, Sat(v_n(t))]$ where

$$Sat(v_i(t)) = \begin{cases} -d_i & \text{if } v_i < -d_i, \\ v_i & \text{if } -d_i \leq v_i \leq \bar{u}_i - d_i, \\ \bar{u}_i - d_i & \text{if } v_i > \bar{u}_i - d_i. \end{cases}$$

Notice also that for any saturating actuator, the function $Sat(v(t))$ which saturates at $-d$ and $\bar{u} - d$, the following inequality holds (see Su et al. (1991))

$$\left\| Sat(v(t)) - \frac{v(t)}{2} \right\| \leq \frac{v(t)}{2}. \quad (4.13)$$

Combining now the system dynamics (4.12) and the controller expression (4.7) and after adding and subtracting the term $B\frac{v(t)}{2}$, with $v(t) = -R^{-1}B^\top P x(t)$, we have:

$$\dot{x}(t) = [A_{cl} + \Delta A_r] x(t) + [A_d + \Delta A_d] x(t - \tau) + B \left[v_m(t) - \frac{v(t)}{2} \right]$$

where $A_{cl} = A_r - \frac{BR^{-1}B^\top P}{2}$.

Again, following the same steps as we did before by decomposing A_d, we get the following dynamics:

$$\dot{x}(t) = \left[A_{cl} + \tilde{A}_d \right] x(t) + \hat{A}_d x(t - \tau) + \Delta A_r x(t) + \Delta A_d x(t - \tau)$$
$$- \tilde{A}_d \int_{t-\tau}^{t} [[A_{cl} + \Delta A_r] x(s) + [A_d + \Delta A_d] x(s - \tau)]$$
$$+ B \left[v_m(s) - \frac{v(s)}{2} \right] ds + B \left[v_m(t) - \frac{v(t)}{2} \right]. \quad (4.14)$$

Now if we assume that the pair $(A_r + \tilde{A}_d, B)$ is stabilizable, then it is possible to construct a state feedback control using the optimal control theory and the gain can be computed from the following Riccati equation:

$$\left[A_r + \widetilde{A}_d\right]^\top P + P\left[A_r + \widetilde{A}_d\right] - PBR^{-1}B^\top P + Q = 0. \qquad (4.15)$$

If this equation has a solution P (symmetric and positive-definite matrix), then we can construct a state feedback controller of the following form:

$$v(t) = -Kx(t)$$

with $K = R^{-1}B^\top P$.

The following theorem gives the results that determines the controller which asymptotically stabilizes the class of systems.

THEOREM 4.2 *If there exists a positive scalar $q > 1$ such that the following holds:*

$$0 \le \tau < \frac{\mu - \left[\alpha + 0.5\|B\|\|K\| + qv\left[\|\widehat{A}_d\| + \beta\right]\right]}{qv\left[\|\widetilde{A}_d A_{cl}\| + \|\widetilde{A}_d A_d\| + \|\widetilde{A}_d\|\left[\alpha + 0.5\|B\|\|K\| + \beta\right]\right]} \qquad (4.16)$$

then system (4.8) is asymptotically stable for for all initial condition in $\mathscr{D}$ all admissible uncertainties.

Proof: Let us again choose the same Lyapunov function as before. In this case, the derivative with respect to time t gives:

$$\dot{V}(\mathbf{x}(t)) = \dot{x}^\top(t)Px(t) + x^\top(t)P\dot{x}(t)$$

$$= \left[\left[A_{cl} + \widetilde{A}_d\right]x(t) + \widehat{A}_d x(t-\tau) + \Delta A_r x(t)\right.$$

$$+ \Delta A_d x(t-\tau) + B\left[v_m(t) - \frac{v(t)}{2}\right]$$

$$- \widetilde{A}_d \int_{t-\tau}^t \left[[A_{cl} + \Delta A_r]x(s) + \left[A_d + \Delta A_d\right]x(s-\tau)\right.$$

$$\left.\left. + B\left[v_m(s) - \frac{v(s)}{2}\right]\right]ds\right]^\top Px + x^\top P\left[\left[A_{cl} + \widetilde{A}_d\right]x(t)\right.$$

$$+ \widehat{A}_d x(t-\tau) + \Delta A_r x(t) + \Delta A_d x(t-\tau) + B\left[v_m(t) - \frac{v(t)}{2}\right]$$

$$- \widetilde{A}_d \int_{t-\tau}^t \left[[A_{cl} + \Delta A_r]x(s) + \left[A_d + \Delta A_d\right]x(s-\tau)\right.$$

$$\left.\left. + B\left[v_m(s) - \frac{v(s)}{2}\right]\right]ds\right]$$

which can be rewritten as follows:

$$\dot{V}(\mathbf{x}(t)) = x^\top(t)\left[\left[A_{cl} + \tilde{A}_d\right]^\top P + P\left[A_{cl} + \tilde{A}_d\right]\right]x(t)$$
$$+ x^\top(t)\left[\Delta A_r^\top P + P\Delta A_r\right]x(t) + x^\top(t-\tau)\widehat{A}_d^\top Px(t)$$
$$+ x^\top(t)P\widehat{A}_d x(t-\tau) + x(t-\tau)\Delta A_d^\top Px(t) + x^\top P\Delta A_d x(t-\tau)$$
$$+ \left[B\left[v_m(t) - \frac{v(t)}{2}\right]\right]^\top Px(t) + x^\top(t)PB\left[v_m(t) - \frac{v(t)}{2}\right]$$
$$- 2x^\top(t)P\tilde{A}_d\int_{t-\tau}^t\left[\left[A_{cl} + \Delta A_r\right]x(s) + \left[A_d + \Delta A_d\right]x(s-\tau)\right.$$
$$\left. + B\left[v_m(s) - \frac{v(s)}{2}\right]\right]ds$$

which implies in turn after using (4.15):

$$\dot{V}(\mathbf{x}(t)) \le -x^\top(t)\left[PBR^{-1}B^\top P + Q\right]x(t)$$
$$+ \|x^\top(t)\left[\Delta A_r^\top P + P\Delta A_r\right]x(t)\| + \|x^\top(t-\tau)\widehat{A}_d^\top Px(t)\|$$
$$+ \|x^\top(t)P\widehat{A}_d x(t-\tau)\| + \|x(t-\tau)\Delta A_d^\top Px(t)\|$$
$$+ \|x^\top P\Delta A_d x(t-\tau)\| + \left\|\left[B\left[v_m(t) - \frac{v(t)}{2}\right]\right]^\top Px(t)\right\|$$
$$+ \left\|x^\top(t)PB\left[v_m(t) - \frac{v(t)}{2}\right]\right\|$$
$$+ 2\left\|x^\top(t)P\tilde{A}_d\int_{t-\tau}^t\left[\left[A_{cl} + \Delta A_r\right]x(s) + \left[A_d + \Delta A_d\right]x(s-\tau)\right.\right.$$
$$\left.\left. + B\left[v_m(s) - \frac{v(s)}{2}\right]\right]ds\right\|.$$

Based on Razumikhin-type theorem, if there exists a positive scalar $q > 1$ such that:

$$V(x(t-\tau)) < q^2 V(x(t))$$

then we have:

$$\|x(t-\tau)\| < q\nu\|x(t)\|$$

with $\nu = \sqrt{\frac{\lambda_{min}(P)}{\lambda_{max}(P)}}$.

Using now this fact, the inequality (4.13) and the expression of the control $v(t) = -Kx(t)$, we get:

$$\dot{V}(\mathbf{x}(t)) \leq -\gamma \|x(t)\|^2$$

with

$$\gamma = \lambda_{min}(PBR^{-1}B^\top P + Q) - 2\left[\left[\alpha + 0.5\|B\|\|K\| + q\nu\left[\|\widehat{A}_d\| + \beta\right]\right]\right.$$

$$\left. + q\nu\tau\left[\|\widetilde{A}_d A_{cl}\| + \|\widetilde{A}_d A_d\| + \|\widetilde{A}_d\|\left[\alpha + \beta + 0.5\|B\|\|K\|\right]\right]\right]\lambda_{max}(P).$$

Based on (4.2), we conclude that $\gamma > 0$ and therefore $\dot{V}(\mathbf{x}(t)) < 0$ which concludes that the closed-loop dynamics is asymptotically stable for all admissible uncertainties. ☐

Similarly as we did for the previous theorem, when the uncertainties are equal to zero we get the following results.

COROLLARY 4.3 *If there exists a positive scalar $q > 1$ such that the following holds:*

$$0 \leq \tau < \frac{\mu - \left[0.5\|B\|\|K\| + q\nu\|\widehat{A}_d\|\right]}{q\nu\left[\|\widetilde{A}_d A_{cl}\| + \|\widetilde{A}_d A_d\| + 0.5\|\widetilde{A}_d\|\|B\|\|K\|\right]} \quad (4.17)$$

then system (4.8) is asymptotically stable for all initial condition in $\mathcal{D}$.

All the results we gave in this chapter are only sufficient conditions and therefore, if we can not find a solution, this does not imply that the system is not asymptotically stable. It is also clair that our results may by restrictive in some cases.

4. Numerical example

To show the effectiveness of our results let us consider a manufacturing system producing three part types. The data of the system is as follows:

$$A_r = \begin{bmatrix} 0.01 & 0.0 & 0.0 \\ 0.0 & 0.011 & 0.0 \\ 0.0 & 0.0 & 0.012 \end{bmatrix}, \quad A_d = \begin{bmatrix} -0.001 & 0.0 & 0.0 \\ 0.0 & -0.001 & 0.0 \\ 0.0 & 0.0 & -0.001 \end{bmatrix},$$

$$B = \begin{bmatrix} 1.0 & 0.0 & 0.0 \\ 0.0 & 1.0 & 0.0 \\ 0.0 & 0.0 & 1.0 \end{bmatrix},$$

Assume that $d_i = 1$, $\bar{u}_i = 2$, $i = 1, \cdots, 3$, $\|\Delta A_r\| \leq 0.2$ and $\|\Delta A_d\| \leq 0.2$, and the matrices Q and R are chosen as follows:

$$Q = \begin{bmatrix} 1.0 & 0.0 & 0.0 \\ 0.0 & 1.0 & 0.0 \\ 0.0 & 0.0 & 1.0 \end{bmatrix}, \quad R = \begin{bmatrix} 1.0 & 0.0 & 0.0 \\ 0.0 & 1.0 & 0.0 \\ 0.0 & 0.0 & 1.0 \end{bmatrix}.$$

If we decompose A_d into two matrices as it was described previously with:

$$\tilde{A}_d = \begin{bmatrix} -0.0011 & 0.0 & 0.0 \\ 0.0 & -0.0011 & 0.0 \\ 0.0 & 0.0 & -0.0011 \end{bmatrix}, \hat{A}_d = A_d - \tilde{A}_d$$

and solving the following algebraic Riccati equation:

$$\left[A_r + \tilde{A}_d \right]^\top P + P \left[A_r + \tilde{A}_d \right] - PBR^{-1}B^\top P + Q = 0,$$

we get:

$$P = \begin{bmatrix} 1.0089 & 0.0 & 0.0 \\ 0.0 & 1.0099 & 0.0 \\ 0.0 & 0.0 & 1.0110 \end{bmatrix}, K = \begin{bmatrix} 1.0089 & 0.0 & 0.0 \\ 0.0 & 1.0099 & 0.0 \\ 0.0 & 0.0 & 1.0110 \end{bmatrix}.$$

Choosing $q = 1.1$ we can check that the condition of our theorem is satisfied and therefore, the state feedback controller with this gain stabilizes the closed-loop asymptotically.

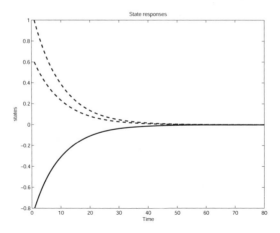

Figure 4.1. The behaviors of the states $x_1(t)$, $x_2(t)$ and $x_3(t)$ in function of time t

We have simulated this manufacturing systems with the controller we designed with the initial conditions $(x_1(0) = -0.8, x_2(0) = 1$, and $x_3(0) = 0.6$ and the results are illustrated by Fig. 4.1 and Fig. 4.2. These figures show that the closed-loop system is asymptotically stable under the computed state feedback controller and the constraints on the control are all satisfied. As it can be seen from the figures that all the stock levels goes to zero at the steady state and also the production rates converge to the neighborhood of the demand rates at the steady state as expected.

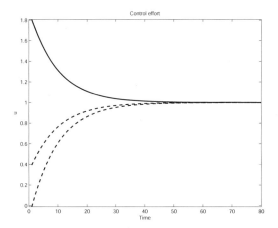

Figure 4.2. The behaviors of the control law $u(t)$ in function of time t

5. Conclusions

This chapter dealt with the production planning for manufacturing systems with inspection of produced parts. The model we proposed here considered the rejection of a certain rate after the inspection is performed. It also considered the possibility that a ceratin rate of delivered parts may be returned back to the stock level. The control theory is used to synthesize the state feedback controller that guarantees that the closed-loop dynamics will be asymptotically stable and at the same time respect the imposed constraints of the control input despite the uncertainties of the system.

References

Grubbstrom, R. W. and Wikner, J., *Inventory Trigger Control Policies Developed in Terms of Control Theory*, *International Journal of Production Economics*, vol. 154, pp. 397-406, 1996.

Hennet, J. C., *A Bimodal Scheme for Multi-Stage Production and Inventory Control*, *Automatica*, vol. 39, pp. 793–805, 2003.

Towill, D. R., *Dynamic Analysis of an Inventory and Order Based Production Control System*, *International Journal of Production Research*, vol. 63, no. 4, pp.671-687, 1982.

Boukas, E. K. and Liu, Z. K., *Deterministic and Stochastic Systems with Time-Delay*, *Birkhauser*, Boston, 2002.

John, S., Naim, M. M., and Towill, D. R., *Dynamic Analysis of a WIP Compensated Decision Support System*, *International Journal of Manufacturing Systems Design*, vol. 1, no. 4, 1994.

Wiendahl, H. and Breithaupt, J. W., *Automatic Production Control Applying Control Theory*, *International Journal of Production Economics*, vol. , no. 63, pp. 33-46, 2000.

Towill, D. R., Evans, G. N., and Cheema, P., *Analysis and Design of an Adaptive Minimum Reasonable Inventory Control System*, *Production Planning and Control*, vol. 8, No. 6, pp. 545-557, 1997.

Axsater, S., *Control Theory Concepts in Production and Inventory Control*, *International Journal of Systems Science*, vol. 16, no. 2, pp. 161-169, 1985.

Ridalls, C. E. and Bennett, S., *Production Inventory System Controller Design And Suplly Chain Dynamics*, *International Journal of Systems Science*, vol. 33, no. 3, pp. 181-195, 2002.

Simon, H. A., *On the Application of Servomechanism Theory in the Study of Production Control*, *Economitra*, vol. 20, pp. 247-268, 1952.

Gavish, B., and Graves, S., *A One-Product Production/Inventory Problem Under Continuous Review Policy*, *Operations Research*, vol 28, pp. 1228-1236, 1980.

Disney, S. M., Naim, M. M., and Towill, D. R., *Genetic Algorithm Optimization of a Class of Inventory Control Systems*, *International Journal of Production Economics*, Vol 68, no. 3, pp. 259-278, 2000 (Special Issue on Design and Implementaion of Agile Manufacturing Systems.

Sethi, S.P. and Zhang, Q., *Hierarchical Decision Making in Stochastic Manufacturing Systems*, *Birkhuser* Boston, Cambridge, MA, 1994.

Sethi, S.P., Zhang, H., and Zhang, Q., *Average-Cost Control of Stochastic Manufacturing Systems*, Springer, 2005.

Su, T. J., Liu, P. L., and Tsay, J. T., *Stabilization of Delay-dependence for Saturating Actuator Systems*, *Proceedings of the 30th IEEE Conference Decision and Control, Brighton, UK*, pp. 2991-2892, 1991.

Gershwin, S. B., *Manufacturing Systems Engineering*, *Prentice Hall*, New York, 1994.

Boukas, E. K. and Z. K. Liu, *Production and Maintenance Control for Manufacturing Systems*, *IEEE Transactions on Automatic Control*, vol. 46, no. 9, pp. 1455–1460, 2001.

Rodrigues, L. and Boukas, E. K., *Piecewise-Linear H_∞ Controller Synthesis wih Applications to Inventory Control of Switched Production Systems To appear in Automatica*, 2006.

D. E. Kirk, *Optimal Control Theory: An Introduction Englewood Cliffs, N.J., Prentice Hall, 1970*.

E. K. Boukas, *Stochastic Output Feedback of Uncertain Time-Delay Systems with Saturating Actuators*, *Journal of Optimization Theory and Applications*, Vol. 118, No. 2, pp. 255–273, 2003.

P. L. Liu, *Delay-Dependent Asymptotic Stabilization for Uncertain Time-Delay Systems with Saturating Actuators*, *International Journal of Mathematical Computing Science*, Vol. 15, No. 1, pp. 45-51, 2005.

Chapter 5

ADMISSION CONTROL IN THE PRESENCE OF PRIORITIES: A SAMPLE PATH APPROACH

Feng Chen

Department of Statistics and Operations Research
University of North Carolina at Chapel Hill

chenf@email.unc.edu

Vidyadhar G. Kulkarni

Department of Statistics and Operations Research
University of North Carolina at Chapel Hill

vkulkarn@email.unc.edu

Abstract   We consider the admission control problem for a two-class-priority $M/M/1$ queueing system. Two classes of customers arrive to the system according to Poisson processes. Class 1 customers have preemptive priority in service over class 2 customers. Each customer can be either accepted or rejected. An accepted customer stays in the system and incurs holding cost at a class-dependent rate until the service is finished, at which time a reward is generated. The objective is to minimize the expected total discounted net cost. We analyze the optimal control policies under three criteria: individual optimization, class optimization, and social optimization. Using sample path analysis, we prove that the optimal policy is of threshold-type under each optimization criterion. We also compare policies under different criteria numerically.

Keywords: M/M/1 queue with priorities, Admission control, Coupling method, Markov Decision Processes.

Introduction

This paper considers an $M/M/1$ queueing system serving two classes of customers. Class 1 customers have preemptive resume priority over class 2 customers. Within each class, the service is provided on a first-come, first-served basis. Class k customers arrive according to a Poisson process with parameter λ_k and require an i.i.d. $\exp(\mu_k)$ service time, $k = 1, 2$. A decision to accept or reject a customer needs to be made upon each arrival. There is no cost associated with rejecting a customer. When there are i class 1 customers and j class 2 customers in the system, the holding cost is incurred at rate $h(i, j) = h_1 i + h_2 j$. An expected reward of r_k is generated each time a class k customer finishes service. All rewards and costs are continuously discounted with rate $\alpha > 0$. The objective is to admit customers in an optimal way, i.e., minimize the expected total discounted net cost.

Priority issue arises in various queueing systems. For example, internet traffic protocols assign higher priority to data packages that require real-time transmission (e.g. on-line live audio and video) and lower priority to delay-insensitive packages (e.g. e-mails and file transmission). Service queues may give VIP customers higher priority over ordinary customers. In hospitals, patients in critical conditions receive higher priority in treatment over non-critical patients. Admission control problem in these kinds of multi-priority queues can be modeled by the framework presented here. This paper is originally motivated by the problem of outsourcing warranty repairs to outside vendors. Consider a manufacturer offering various types of warranties for its product. When different types of warranties specify different repair turnaround times, it is desirable to give higher priority to repairs with shorter turnaround time. Warranty repairs are outsourced to a number of outside vendors. When an item fails and is under warranty, it is sent to one of the vendors for repair. The manufacturer pays a fixed fee for each repair and incurs holding costs (good will cost) while items are at the vendor. The objective of the manufacturer is to assign warranty repairs in such a way that the expected long-run average cost is minimized. Analyzing the admission control problem studied here can serve as a starting point for solving this complicated routing problem. The results of this single-vendor admission control problem can be used to derive index-based dynamic routing policies for multi-vendor problems (See Opp, Kulkarni and Glazebrook (2005)).

Admission control for single class queueing systems is widely studied. See Stidham (1985) for a survey. Naor (1969) proposes the first quantitative model. He studies an $M/M/1$ system with a single class of customers, undiscounted reward and cost. The objective is to maximize the long-run average net reward. Naor considers only critical-number policies and shows that $n_S \leq n_I$, where n_S and n_I are the critical numbers for social optimization and individual optimization, respectively.

Many authors have generalized Naor's model. Among others, Yechiali (1971, 1972) shows that for $GI/M/s$ systems the socially optimal policy has critical-number form. Knudsen (1972) considers an $M/M/s$ queue with state-dependent net benefit. Lippman and Stidham (1977) study a birth-death process with general departure rate, random reward, with or without discounting and for a finite or infinite time horizon. Stidham (1978) considers a $GI/M/1$ queue with random reward and general holding cost, with or without discounting. Other models of admission control problem for single-class queues include Adiri and Yechiali (1974), Stidham and Weber (1989), and Rykov (2001).

Multi-class admission control problem has also been studied extensively. Papers in the area can be classified into two categories based on whether or not service is prioritized. For models without service priorities, see Miller (1969), Blanc et al. (1992), Kulkarni and Tedijanto (1998), and Nair and Bapma (2001). In these models, different classes are distinguished by different arrival rates, service rates, rewards, holding costs, etc. Among papers that consider service priorities, Mendelson and Whang (1990) study a priority pricing problem for a multi-class $M/M/1$ queueing system, where each customer decides by himself whether to join the system or not, and, if join, at what priority level. Hassin (1995) studies a bidding mechanism for a $GI/M/1$ queue without balking. Ha (1997) considers the production control problem in a make-to-stock production system with two prioritized customer classes.

To the best of our knowledge, Chen and Kulkarni (2005) are the first to consider the admission control problem for a multi-priority queue with the objective of minimizing expected total discounted cost. This paper differs from Chen and Kulkarni's paper mainly in two aspects: (i) We assume the rewards are generated at the time of service completion instead of the time of joining the repair queue as assumed by Chen and Kulkarni (2005). This shift of reward times changes the nature of the problem in some critical ways, e.g. the optimal value function is no longer non-decreasing in the number of customers of each type in starting state, and the cases where every customer is accepted do not exist anymore. (ii) We prove the structural results using sample path analysis (specifically, the coupling method) (Lindvall (1992), Wu et al. (2005)), while Chen and Kulkarni (2005) use standard value iteration method. The sample path approach provides more concise proofs.

Following Chen and Kulkarni (2005), we analyze the optimal control policies under 3 criteria: individual optimization, class optimization, and social optimization. Under individual optimization, each customer decides whether to join the system or not in order to minimizes his own expected total discounted cost. Under class optimization, there is a controller for each class. The controller of class k decides whether to accept an arriving class k customer or not in order to minimize the expected total discounted cost incurred by all class k customers, $k = 1, 2$. Under social optimization, there is a single controller for the whole system. The system controller decides whether to accept an arriving customer or not in order to minimize the expected total discounted net cost incurred by all customers. Using sample path argument, we obtain the same structural results as in Chen and Kulkarni (2005), i.e., the optimal control policy under each optimization criterion is of threshold type. We also compare different policies numerically. The numerical results suggest the same relationships as shown in Chen and Kulkarni (2005), i.e., the socially optimal policy accepts more low priority customers and less high priority customers than the class-optimal policy; the individually optimal policy accepts the most high priority customers while, depending on the parameters, it can accept either more or less low priority customers than the other two policies.

The remainder of the paper is organized as follows. Section 1, 2, 3 are dedicated to the structural properties of the optimal policies under individual optimization, social optimization, and class optimization, respectively. Section 4 compares different policies numerically. We end with the summary in Section 5.

1. Individual Optimization

Following the same approach as in Chen and Kulkarni (2005), one can easily derive the following results for individually optimal policies.

THEOREM 5.1 *Under the individual optimization criterion, an arriving class 1 customer who sees the system in state (i, j) joins the queue if and only if $i < L_1^I$, where*

$$L_1^I = \lfloor \log \frac{h_1}{h_1 + \alpha r_1} / \log \frac{\mu}{\mu + \alpha} \rfloor. \tag{5.1}$$

An arriving class 2 customer who sees the system in state (i, j) joins the queue if and only if $j < L_2^I(i)$, where

$$L_2^I(i) = \begin{cases} \lfloor \log \frac{h_2}{(h_2 + \alpha r_2)\phi_i(\alpha)} / \log \beta \rfloor, & \text{if } i \leq L_1^I \\ \lfloor (\log \frac{h_2}{(h_2 + \alpha r_2)\phi_{L_1^I}(\alpha)} + (i - L_1^I)(\log \frac{\mu + \alpha}{\mu})) / \log \beta \rfloor, & \text{if } i > L_1^I \end{cases} \tag{5.2}$$

where $\phi_i(\alpha)$ is the LST of the busy period initiated by i class 1 customers and $\beta = \frac{\mu}{\alpha + \mu + \lambda_1(1 - \phi_1(\alpha))}$. $\lfloor x \rfloor$ is the largest integer less than or equal to x. Furthermore, $L_2^I(i)$ is decreasing in i.

Note that shifting the reward time (from the moment a customer joins the queue to the moment a customer finishes service) not only changes the form of the threshold functions but also eliminates the cases where everyone is accepted.

2. Social Optimization

We consider socially optimal policies in this section. The objective of a socially optimal policy is to minimize the expected total discounted net cost generated by all customers. Let $v(i, j)$ be the expected total discounted net cost generated by a socially optimal policy over an infinite horizon starting from state (i, j). Following Lippman (1975), we uniformize the process by defining the uniform rate $\Lambda = \lambda_1 + \lambda_2 + \mu$. Rescaling time so that $\Lambda + \alpha = 1$, we have the following optimality equations

$$
\begin{aligned}
v(i, j) = \ &h_1 i + h_2 j + \lambda_1 \min\{v(i, j), v(i + 1, j)\} \\
&+ \lambda_2 \min\{v(i, j), v(i, j + 1)\} \\
&+ \mu \begin{cases} v(i - 1, j) - r_1, & \text{if } i \geq 1 \\ v(0, j - 1) - r_2, & \text{if } i = 0, j \geq 1 \\ v(0, 0), & \text{if } i = 0, j = 0. \end{cases}
\end{aligned} \tag{5.3}
$$

LEMMA 5.2 $v(0, 1) - v(0, 0) + r_2 \geq 0$.

Proof. Define two processes on the same probability space so that they see the same arrivals and potential services. Process 1 starts in state $(0, 1)$ and follows optimal policy. Process 2 starts in state $(0, 0)$ and follows policy ϕ which is described below. Let τ be the first time Process 1 reaches state $(0, 0)$. Let Process 2 take the same action as Process 1 upon each arrival until time τ, then follow the optimal policy afterwards. If a new class 2 customer is accepted while Process 1 is serving the last class 2 customer, we resample the remaining service time of the class 2 customer currently under service in Process 1 so that he finishes service at the same time as the new class 2 customer in Process 2. (This resampling argument can be applied to similar situations in the rest of this paper.) Therefore, Process 1 and 2 have identical customers except for one extra class 2 customer in Process 1 until time τ. Two processes become identical from then on. Using $v^\phi(i, j)$ to denote the expected total discounted net cost generated by policy ϕ starting from state (i, j), we get

$$
\begin{aligned}
v(0, 1) - v(0, 0) \ &\geq \ v(0, 1) - v^\phi(0, 0) \\
&= \ E \int_0^\tau e^{-\alpha t} h_2 dt + E e^{-\alpha \tau}(-r_2 + v(0, 0) - v(0, 0))
\end{aligned}
$$

78

$$\geq \ -r_2 E e^{-\alpha \tau} \geq -r_2. \quad \square$$

LEMMA 5.3 v *is supermodular, i.e.,*

$$v(i+1, j+1) - v(i+1, j) - v(i, j+1) + v(i, j) \geq 0. \qquad (5.4)$$

Proof. Fix i and j. Define four processes on the same probability space so that they see the same arrivals and potential services. Process 1 and 4 follow optimal policies and start in states $(i+1, j+1)$ and (i, j), respectively. Process 2 and 3 start in states $(i+1, j)$ and $(i, j+1)$, respectively, and use policies ϕ_2 and ϕ_3 which are described below. Denote the state of Process k at time t by (X_t^k, Y_t^k), $k = 1, 2, 3, 4$.

Let τ_1 be the first time Process 2 and 3 have 0 customers entirely. Note that if Process 2 and 3 take the same action upon each arrival they will reach state (0,0) at the same time, since service rates are the same for both classes. Let τ_2 be the first time Process 1 and 4 take different actions. Define $\tau = \min\{\tau_1, \tau_2\}$. Let Process 2 and 3 take the same action as Process 1 and 4 until time τ, then follow the optimal policy afterwards. Thus

$$\begin{aligned}
&v(i+1, j+1) - v(i+1, j) - v(i, j+1) + v(i, j) \\
\geq\ &v(i+1, j+1) - v^{\phi_2}(i+1, j) - v^{\phi_3}(i, j+1) + v(i, j) \\
=\ &E \int_0^\tau e^{-\alpha t} [h(X_t^4 + 1, Y_t^4 + 1) - h(X_t^4 + 1, Y_t^4) \\
&\qquad - h(X_t^4, Y_t^4 + 1) + h(X_t^4, Y_t^4)] dt \\
&+ E e^{-\alpha \tau}(-R_1 + R_2 + R_3 - R_4) \\
&+ E e^{-\alpha \tau}(v(X_\tau^1, Y_\tau^1) - v(X_\tau^2, Y_\tau^2) - v(X_\tau^3, Y_\tau^3) + v(X_\tau^4, Y_\tau^4)),
\end{aligned}$$

where R_i is the potential reward generated in Process i at time τ. It can be easily seen that the first term is 0 because of the linear holding cost rate.

To simplify notation, define

$$D = v(i+1, j+1) - v(i+1, j) - v(i, j+1) + v(i, j) \qquad (5.5)$$

$$A = -R_1 + R_2 + R_3 - R_4, \qquad (5.6)$$

$$B = v(X_\tau^1, Y_\tau^1) - v(X_\tau^2, Y_\tau^2) - v(X_\tau^3, Y_\tau^3) + v(X_\tau^4, Y_\tau^4). \qquad (5.7)$$

Case 1: $\tau = \tau_1$. Then, at τ, the four processes are in states $(0,1)$, $(0,0)$, $(0,0)$, and $(0,0)$, respectively. The two distinct paths by which this state is reached are: (i) $\{(1,2)\,(1,1)\,(0,2)\,(0,1)\} \rightarrow \{(0,2)\,(0,1)\,(0,1)\,(0,0)\} \rightarrow \{(0,1)\,(0,0)\,(0,0)\,(0,0)\}$; (ii) $\{(2,1)\,(2,0)\,(1,1)\,(1,0)\} \rightarrow \{(1,1)\,(1,0)\,(0,1)\,(0,0)\} \rightarrow \{(0,1)\,(0,0)\,(0,0)\,(0,0)\}$. In the former case, $R_1 = R_2 = R_3 = r_2$, and $R_4 = 0$. In the latter case, $R_1 = R_2 = r_1$, $R_3 = r_2$, and $R_4 = 0$. In both cases, we have

$$D \geq Ee^{-\alpha\tau}(r_2 + v(0,1) - v(0,0)) \geq 0,$$

where the last inequality follows from Lemma 5.2.

Case 2: $\tau = \tau_2$. Then $A = 0$. We have the following possibilities.

Case 2.1: A class 1 arrival is accepted by Process 1 and rejected by Process 4. Let Process 2 accept the arrival and Process 3 reject it. Then after this event the states in four processes are $(X_\tau^4 + 2, Y_\tau^4 + 1)$, $(X_\tau^4 + 2, Y_\tau^4)$, $(X_\tau^4, Y_\tau^4 + 1)$, and (X_τ^4, Y_τ^4), respectively. Adding and subtracting $v(X_\tau^4 + 1, Y_\tau^4 + 1) + v(X_\tau^4 + 1, Y_\tau^4)$, we have

$$\begin{aligned}
B &= v(X_\tau^4 + 2, Y_\tau^4 + 1) - v(X_\tau^4 + 1, Y_\tau^4 + 1) - v(X_\tau^4 + 2, Y_\tau^4) \\
&\quad + v(X_\tau^4 + 1, Y_\tau^4) + v(X_\tau^4 + 1, Y_\tau^4 + 1) - v(X_\tau^4, Y_\tau^4 + 1) \\
&\quad - v(X_\tau^4 + 1, Y_\tau^4) + v(X_\tau^4, Y_\tau^4).
\end{aligned}$$

Note that the first four terms and the last four terms above are inequality (5.4) evaluated at $(X_\tau^4 + 1, Y_\tau^4)$ and (X_τ^4, Y_τ^4), respectively. Thus the above argument can be repeated until either Case 1 or Case 2.2 or Case 2.4 happens.

Case 2.2: A class 1 arrival is rejected by Process 1 and accepted by Process 4. Let Process 2 reject the arrival and Process 3 accept it. Then after this event the states in four processes are $(X_\tau^4 + 1, Y_\tau^4 + 1)$, $(X_\tau^4 + 1, Y_\tau^4)$, $(X_\tau^4 + 1, Y_\tau^4 + 1)$, and $(X_\tau^4 + 1, Y_\tau^4)$, respectively. Note that Process 1 and 3 couple, so do Process 2 and 4. Therefore $B = 0$ and (5.4) holds.

Case 2.3: A class 2 arrival is accepted by Process 1 and rejected by Process 4. Let Process 2 reject the arrival and Process 3 accept it. Then after this event the states in four processes are $(X_\tau^4 + 1, Y_\tau^4 + 2)$, $(X_\tau^4 + 1, Y_\tau^4)$, $(X_\tau^4, Y_\tau^4 + 2)$, and (X_τ^4, Y_τ^4), respectively. Adding and subtracting $v(X_\tau^4 + 1, Y_\tau^4 + 1) + v(X_\tau^4, Y_\tau^4 + 1)$, we have

$$\begin{aligned}
B &= v(X_\tau^4 + 1, Y_\tau^4 + 2) - v(X_\tau^4 + 1, Y_\tau^4 + 1) - v(X_\tau^4, Y_\tau^4 + 2) \\
&\quad + v(X_\tau^4, Y_\tau^4 + 1) + v(X_\tau^4 + 1, Y_\tau^4 + 1) - v(X_\tau^4 + 1, Y_\tau^4) \\
&\quad - v(X_\tau^4, Y_\tau^4 + 1) + v(X_\tau^4, Y_\tau^4).
\end{aligned}$$

Note that the first four terms and the last four terms are inequality (5.4) evaluated at $(X_\tau^4, Y_\tau^4 + 1)$ and (X_τ^4, Y_τ^4), respectively. Thus the above argument can be repeated until either Case 1 or Case 2.2 or Case 2.4 happens.

Case 2.4: A class 2 arrival is rejected by Process 1 and accepted by Process 4. Let Process 2 accept the arrival and Process 3 reject it. Then after this event the states in four processes are $(X_\tau^4 + 1, Y_\tau^4 + 1)$, $(X_\tau^4 + 1, Y_\tau^4 + 1)$, $(X_\tau^4, Y_\tau^4 + 1)$, and $(X_\tau^4, Y_\tau^4 + 1)$, respectively. Note that Process 1 and 2 couple, so do Process 3 and 4. Therefore $B = 0$ and (5.4) holds. $\square$

LEMMA 5.4 $v(i,j)$ *is a unimodal function in* i, *i.e., if* $v(i+1,j) - v(i,j) \geq 0$, *then* $v(i+2,j) - v(i+1,j) \geq 0$.

Proof. Define two processes on the same probability space so that they see the same arrivals and potential services. Process 1 follows the optimal policy and starts in state $(i+2,j)$. Process 2 starts in state $(i+1,j)$ and follows policy ϕ that is described below.

Let τ be the first time Process 1 has $i+1$ class 1 customers. Process 2 takes the same action as Process 1 upon arrivals until τ then follow the optimal policy afterwards. Thus, at time τ Process 2 has i class 1 customers and the same number of class 2 customers, say j', as in Process 1. We have $j' \geq j$, since no class 2 customers have started service yet. Hence

$$
\begin{aligned}
v(i+2,j) - v(i+1,j) &\geq v(i+2,j) - v^\phi(i+1,j) \\
&= E\int_0^\tau e^{-\alpha t}h_1 dt + Ee^{-\alpha\tau}(v(i+1,j') - v(i,j')),
\end{aligned}
$$

where $j' \geq j$. From supermodularity, we have

$$
v(i+1,j') - v(i,j') \geq v(i+1,j) - v(i,j) \geq 0.
$$

Therefore $v(i+2,j) - v(i+1,j) \geq 0$. $\square$

THEOREM 5.5 *The socially optimal policy for admitting class 1 customers is characterized by a monotonically decreasing switching curve, i.e., for each* $j \geq 0$, *there exists a threshold* $L_1^s(j)$, *such that a class 1 arrival in state* (i,j) *is accepted if and only if* $i < L_1^s(j)$. *Furthermore,* $L_1^s(j)$ *is monotonically decreasing in* j.

Proof. We follow the convention that a customer is accepted when accepting or rejecting that customer makes no difference in terms of cost. Then a class 1 arrival in state (i,j) is accepted if and only if $v(i+1,j) \leq v(i,j)$. For any fixed j, let

$$
L_1^s(j) = \min\{i : v(i+1,j) > v(i,j)\}.
$$

Using Lemma 5.4, one can easily show that a class 1 arrival is accepted if and only if $i < L_1^s(j)$.

For $j_1 \leq j_2$, we have $v(i+1, j_2) - v(i, j_2) \geq v(i+1, j_1) - v(i, j_1)$, which follows from supermodularity. By definition of $L_1^s(j_1)$, we have $v(L_1^s(j_1)+1, j_1) > v(L_1(j_1), j_1)$, so $v(L_1^s(j_1)+1, j_2) > v(L_1^s(j_1), j_2)$. By definition of $L_1^s(j_2)$, we have $L_1^s(j_1) \geq L_1^s(j_2)$. Thus, $L_1^s(j)$ is decreasing in j. $\square$

LEMMA 5.6 *If $h_1 \geq h_2$ and $r_1 \geq r_2$, then v is diagonally dominant in both i and j, i.e.,*

$$v(i, j+2) - v(i, j+1) - v(i+1, j+1) + v(i+1, j) \geq 0, \qquad (5.8)$$

$$v(i, j+1) - v(i+1, j) - v(i+1, j+1) + v(i+2, j) \geq 0. \qquad (5.9)$$

Proof. (a). Consider (5.8) first.

Define four processes on the same probability space so that they see the same arrivals and potential services. Process 1 and 4 follow optimal policies and start in state $(i, j+2)$ and $(i+1, j)$, respectively. Process 2 and 3 start in state $(i, j+1)$ and $(i+1, j+1)$, respectively, and use policies ϕ_2 and ϕ_3 which are described below. Denote the state of Process k at time t by (X_t^k, Y_t^k), $k = 1, 2, 3, 4$.

Let τ_1 be the first time Process 3 and 4 have 0 class 1 customers. Since service rates are the same for both classes, Process 1 and 2 finish serving the first class 2 customer at τ_1. Let τ_2 be the first time Process 1 and 4 take different actions. Define $\tau = \min\{\tau_1, \tau_2\}$. Let Process 2 and 3 take the same action as Process 1 and 4 upon each arrival until time τ, then follow the optimal policy afterwards. Thus

$$\begin{aligned}
&v(i, j+2) - v(i, j+1) - v(i+1, j+1) + v(i+1, j) \\
\geq\ &v(i, j+2) - v^{\phi_2}(i, j+1) - v^{\phi_3}(i+1, j+1) + v(i+1, j) \\
=\ &E \int_0^\tau e^{-\alpha t}[h(X_t^4 - 1, Y_t^4 + 2) - h(X_t^4 - 1, Y_t^4 + 1) \\
&\qquad - h(X_t^4, Y_t^4 + 1) + h(X_t^4, Y_t^4)]dt \\
&+ Ee^{-\alpha\tau}(-R_1 + R_2 + R_3 - R_4) \\
&+ Ee^{-\alpha\tau}(v(X_\tau^1, Y_\tau^1) - v(X_\tau^2, Y_\tau^2) - v(X_\tau^3, Y_\tau^3) + v(X_\tau^4, Y_\tau^4)),
\end{aligned}$$

where R_i is the potential reward generated in Process i at time τ. It can be easily seen that the first term is 0 because of the linear holding cost rate.

Define A, B as in (5.6), (5.7).

Case a.1: $\tau = \tau_1$. Then the states in four processes at time τ are $(0, Y_\tau^4 + 1)$, $(0, Y_\tau^4)$, $(0, Y_\tau^4 + 1)$, and $(0, Y_\tau^4)$, respectively. Note that Process 1 and 3 couple, so do Process 2 and 4. Therefore $B = 0$. Also, $R_1 = R_2 = r_2$ and $R_3 = R_4 = r_1$, so $A = 0$. Thus (5.8) holds.

Case a.2: $\tau = \tau_2$. Then $A = 0$. We have the following possibilities.

Case a.2.1: A class 1 arrival is accepted by Process 1 and rejected by Process 4. Let Process 2 accept the arrival and Process 3 reject it. Then the states in four processes at time τ are $(X_\tau^4, Y_\tau^4 + 2)$, $(X_\tau^4, Y_\tau^4 + 1)$, $(X_\tau^4, Y_\tau^4 + 1)$, and (X_τ^4, Y_τ^4), respectively. Adding and subtracting $v(X_\tau^4 + 1, Y_\tau^4 + 1) + v(X_\tau^4 + 1, Y_\tau^4)$, we have

$$
\begin{aligned}
B = {} & v(X_\tau^4, Y_\tau^4 + 2) - v(X_\tau^4 + 1, Y_\tau^4 + 1) - v(X_\tau^4, Y_\tau^4 + 1) \\
& + v(X_\tau^4 + 1, Y_\tau^4) + v(X_\tau^4 + 1, Y_\tau^4 + 1) - v(X_\tau^4, Y_\tau^4 + 1) \\
& - v(X_\tau^4 + 1, Y_\tau^4) + v(X_\tau^4, Y_\tau^4).
\end{aligned}
$$

Note that the first four terms are inequality (5.8) evaluated at (X_τ^4, Y_τ^4), so the above argument can be repeated until Case a.1 or Case a.2.4 happens. The last four terms are inequality (5.4) evaluated at (X_τ^4, Y_τ^4), which is non-negative by Lemma 5.3.

Case a.2.2: A class 1 arrival is rejected by Process 1 and accepted by Process 4. Let Process 2 accept the arrival and Process 3 reject it. Then the states in four processes at time τ are $(X_\tau^4 - 2, Y_\tau^4 + 2)$, $(X_\tau^4 - 1, Y_\tau^4 + 1)$, $(X_\tau^4 - 1, Y_\tau^4 + 1)$, and (X_τ^4, Y_τ^4), respectively. Adding and subtracting $v(X_\tau^4 - 2, Y_\tau^4 + 1) + v(X_\tau^4 - 1, Y_\tau^4)$, we have

$$
\begin{aligned}
B = {} & v(X_\tau^4 - 2, Y_\tau^4 + 2) - v(X_\tau^4 - 2, Y_\tau^4 + 1) - v(X_\tau^4 - 1, Y_\tau^4 + 1) \\
& + v(X_\tau^4 - 1, Y_\tau^4) + v(X_\tau^4 - 2, Y_\tau^4 + 1) - v(X_\tau^4 - 1, Y_\tau^4 + 1) \\
& - v(X_\tau^4 - 1, Y_\tau^4) + v(X_\tau^4, Y_\tau^4).
\end{aligned}
$$

Note that the first four terms are inequality (5.8) evaluated at $(X_\tau^4 - 2, Y_\tau^4)$, so the above argument can be repeated until Case a.1 or Case a.2.4 happens. The last four terms are inequality (5.9) evaluated at $(X_\tau^4 - 2, Y_\tau^4)$, so the argument in part (b) can be repeated until Case b.1 or Case b.2.1 happens.

Case a.2.3: A class 2 arrival is accepted by Process 1 and rejected by Process 4. Let Process 2 accept the arrival and Process 3 reject it. Then the states in four processes at time τ are $(X_\tau^4 - 1, Y_\tau^4 + 3)$, $(X_\tau^4 - 1, Y_\tau^4 + 2)$, $(X_\tau^4, Y_\tau^4 + 1)$, and (X_τ^4, Y_τ^4), respectively. Adding and subtracting $v(X_\tau^4 - 1, Y_\tau^4 + 2) + v(X_\tau^4, Y_\tau^4 + 2) + v(X_\tau^4, Y_\tau^4 + 1)$, we have

$$
\begin{aligned}
B = {} & v(X_\tau^4 - 1, Y_\tau^4 + 3) - v(X_\tau^4 - 1, Y_\tau^4 + 2) - v(X_\tau^4, Y_\tau^4 + 2) \\
& + v(X_\tau^4, Y_\tau^4 + 1) + v(X_\tau^4 - 1, Y_\tau^4 + 2) - v(X_\tau^4 - 1, Y_\tau^4 + 1)
\end{aligned}
$$

$$-v(X_\tau^4, Y_\tau^4 + 1) + v(X_\tau^4, Y_\tau^4) + v(X_\tau^4 - 1, Y_\tau^4 + 1)$$
$$-v(X_\tau^4 - 1, Y_\tau^4 + 2) - v(X_\tau^4, Y_\tau^4 + 1) + v(X_\tau^4, Y_\tau^4 + 2).$$

Note that the first four terms and the second four terms are inequality (5.8) evaluated at $(X_\tau^4 - 1, Y_\tau^4 + 1)$ and $(X_\tau^4 - 1, Y_\tau^4)$, respectively. So the above argument can be repeated until Case a.1 or Case a.2.4 happens. The last four terms are inequality (5.4) evaluated at $(X_\tau^4 - 1, Y_\tau^4 + 1)$, which is non-negative by Lemma 5.3.

Case a.2.4: A class 2 arrival is rejected by Process 1 and accepted by Process 4. Let Process 2 accept the arrival and Process 3 reject it. Then the states in four processes at time τ are $(X_\tau^4 - 1, Y_\tau^4 + 1)$, $(X_\tau^4 - 1, Y_\tau^4 + 1)$, (X_τ^4, Y_τ^4), (X_τ^4, Y_τ^4), respectively. Note that Process 1 and 2 couple, so do Process 3 and 4. Therefore $B = 0$ and hence (5.8) holds.

(b). Consider (5.9) next.

Define four processes on the same probability space so that they see the same arrivals and potential services. Process 1 and 4 follow optimal policies and start in state $(i, j+1)$ and $(i+2, j)$, respectively. Process 2 and 3 start in state $(i+1, j)$ and $(i+1, j+1)$, respectively, and use policies ϕ_2 and ϕ_3 which are described below. Denote the state of Process k at time t by (X_t^k, Y_t^k), $k = 1, 2, 3, 4$.

Let β be the first time Process 2 and 3 have 0 class 1 customers. Let τ_1 be the first time Process 4 has 0 class 1 customers. Since service rates are the same for both classes, Process 1 finishes serving the first class 2 customer at β and the second class 2 customer (if any) at τ_1. Process 2 and 3 finish serving the first class 2 customer (if any) at τ_1. So between β and τ_1, Process 1 and 2 have identical customers, and Process 3 has one more class 2 customer but one less class 1 customer than Process 4. While Process 4 is serving the last class 1 customer, the servers in Process 1 and 2 are either serving class 2 customers or idle. In the former case, the rewards generated in four processes at τ_1 are respectively r_2, r_2, r_2, and r_1. In the latter case, the rewards are respectively $0, 0, r_2$, and r_1. Let τ_2 be the first time Process 1 and 4 take different actions. Define $\tau = \min\{\tau_1, \tau_2\}$. Let Process 2 and 3 take the same action as Process 1 and 4 upon each arrival until time τ, then follow the optimal policy afterwards.

Case b.1: $\tau = \tau_1$. Then

$$
\begin{aligned}
& v(i, j+1) - v(i+1, j) - v(i+1, j+1) + v(i+2, j) \\
\geq\ & v(i, j+1) - v^{\phi_2}(i+1, j) - v^{\phi_3}(i+1, j+1) + v(i+2, j) \\
=\ & E \int_0^\beta e^{-\alpha t} [h(X_t^4 - 2, Y_t^4 + 1) - h(X_t^4 - 1, Y_t^4) \\
& \qquad - h(X_t^4 - 1, Y_t^4 + 1) + h(X_t^4, Y_t^4)] dt
\end{aligned}
$$

$$+ Ee^{-\alpha\beta}(-r_2 + r_1 + r_1 - r_1)$$

$$+ E \int_\beta^\tau e^{-\alpha t}(h_1 - h_2)dt + Ee^{-\alpha\tau}(r_2 - r_1)$$

$$+ Ee^{-\alpha\tau} \begin{cases} [v(0,0) - v(0,0) - v(0,0) + v(0,0)], \text{if } Y^1_{\tau-} = Y^2_{\tau-} = 0 \\ \\ [v(0, Y^4_\tau - 1) - v(0, Y^4_\tau - 1) - v(0, Y^4_\tau) + v(0, Y^4_\tau)], \text{o.w.} \end{cases}$$

The first term is 0 because of the linear holding cost rate. Using the fact that $h_1 \geq h_2$ and $\tau \geq \beta$, one can show that (5.9) holds.

Case b.2: $\tau = \tau_2$. Then

$$v(i, j+1) - v(i+1, j) - v(i+1, j+1) + v(i+2, j)$$
$$\geq v(i, j+1) - v^{\phi_2}(i+1, j) - v^{\phi_3}(i+1, j+1) + v(i+2, j)$$
$$= E \int_0^\tau e^{-\alpha t}[h(X^4_t - 2, Y^4_t + 1) - h(X^4_t - 1, Y^4_t)$$
$$- h(X^4_t - 1, Y^4_t + 1) + h(X^4_t, Y^4_t)]dt$$
$$+ Ee^{-\alpha\tau}(v(X^1_\tau, Y^1_\tau) - v(X^2_\tau, Y^2_\tau) - v(X^3_\tau, Y^3_\tau) + v(X^4_\tau, Y^4_\tau)).$$

We have the following possibilities.

Case b.2.1: A class 1 arrival is accepted by Process 1 and rejected by Process 4. Let Process 2 accept the arrival and Process 3 reject. Then the states in four processes at τ are $(X^4_\tau - 1, Y^4_\tau + 1)$, (X^4_τ, Y^4_τ), $(X^4_\tau - 1, Y^4_\tau + 1)$, and (X^4_τ, Y^4_τ), respectively. Note that Process 1 and 3 couple, so do Process 2 and 4. So (5.9) holds.

Case b.2.2: A class 1 arrival is rejected by Process 1 and accepted by Process 4. Let Process 2 accept the arrival and Process 3 reject. Then the states in four processes at τ are $(X^4_\tau - 3, Y^4_\tau + 1)$, $(X^4_\tau - 1, Y^4_\tau)$, $(X^4_\tau - 2, Y^4_\tau + 1)$, and (X^4_τ, Y^4_τ), respectively. Adding and subtracting $v(X^4_\tau - 2, Y^4_\tau) + v(X^4_\tau - 2, Y^4_\tau + 1) + v(X^4_\tau - 1, Y^4_\tau + 1)$, we have

$$\begin{aligned} B &= v(X^4_\tau - 3, Y^4_\tau + 1) - v(X^4_\tau - 2, Y^4_\tau) - v(X^4_\tau - 2, Y^4_\tau + 1) \\ &\quad + v(X^4_\tau - 1, Y^4_\tau) + v(X^4_\tau - 2, Y^4_\tau + 1) - v(X^4_\tau - 1, Y^4_\tau) \\ &\quad - v(X^4_\tau - 1, Y^4_\tau + 1) + v(X^4_\tau, Y^4_\tau) + v(X^4_\tau - 2, Y^4_\tau) \\ &\quad - v(X^4_\tau - 2, Y^4_\tau + 1) - v(X^4_\tau - 1, Y^4_\tau) + v(X^4_\tau - 1, Y^4_\tau + 1). \end{aligned}$$

Note that the first four terms and the second four terms are inequality (5.9) evaluated at $(X^4_\tau - 3, Y^4_\tau)$ and $(X^4_\tau - 2, Y^4_\tau)$, respectively, so the above argument can be repeated until Case b.1 or Case b.2.1 happens. The last four terms are inequality (5.4) evaluated at $(X^4_\tau - 2, Y^4_\tau)$, which is non-negative by Lemma 5.3.

Case b.2.3: A class 2 arrival is accepted by Process 1 and rejected by

Process 4. Let Process 2 accept the arrival and Process 3 reject. Then the states in four processes at τ are $(X_\tau^4 - 2, Y_\tau^4 + 2)$, $(X_\tau^4 - 1, Y_\tau^4 + 1)$, $(X_\tau^4 - 1, Y_\tau^4 + 1)$, and (X_τ^4, Y_τ^4), respectively. Adding and subtracting $v(X_\tau^4 - 2, Y_\tau^4 + 1) + v(X_\tau^4 - 1, Y_\tau^4)$, we have

$$
\begin{aligned}
B = {} & v(X_\tau^4 - 2, Y_\tau^4 + 2) - v(X_\tau^4 - 1, Y_\tau^4 + 1) - v(X_\tau^4 - 2, Y_\tau^4 + 1) \\
& + v(X_\tau^4 - 1, Y_\tau^4) + v(X_\tau^4 - 2, Y_\tau^4 + 1) - v(X_\tau^4 - 1, Y_\tau^4) \\
& - v(X_\tau^4 - 1, Y_\tau^4 + 1) + v(X_\tau^4, Y_\tau^4).
\end{aligned}
$$

Note that the first four terms are inequality (5.8) evaluated at $(X_\tau^4 - 2, Y_\tau^4)$, so the argument in part (a) can be repeated until Case a.1 or Case a.2.4 happens. The last four terms are inequality (5.9) evaluated at $(X_\tau^4 - 2, Y_\tau^4)$, so the above argument can be repeated until Case b.1 or Case b.2.1 happens.

Case b.2.4: A class 2 arrival is rejected by Process 1 and accepted by Process 4. Let Process 2 accept the arrival and Process 3 reject. Then the states in four processes at τ are $(X_\tau^4 - 2, Y_\tau^4)$, $(X_\tau^4 - 1, Y_\tau^4)$, $(X_\tau^4 - 1, Y_\tau^4)$, and (X_τ^4, Y_τ^4), respectively. Adding and subtracting $v(X_\tau^4 - 2, Y_\tau^4 + 1) + v(X_\tau^4 - 1, Y_\tau^4 + 1)$, we have

$$
\begin{aligned}
B = {} & v(X_\tau^4 - 2, Y_\tau^4) - v(X_\tau^4 - 2, Y_\tau^4 + 1) - v(X_\tau^4 - 1, Y_\tau^4) \\
& + v(X_\tau^4 - 1, Y_\tau^4 + 1) + v(X_\tau^4 - 2, Y_\tau^4 + 1) - v(X_\tau^4 - 1, Y_\tau^4) \\
& - v(X_\tau^4 - 1, Y_\tau^4 + 1) + v(X_\tau^4, Y_\tau^4).
\end{aligned}
$$

Note that the first four terms are inequality (5.4) evaluated at $(X_\tau^4 - 2, Y_\tau^4)$, which is non-negative by Lemma 5.3. The last four terms are inequality (5.9) evaluated at $(X_\tau^4 - 2, Y_\tau^4)$, so the above argument can be repeated until Case b.1 or Case b.2.1 happens. $\square$

COROLLARY 5.7 *If $h_1 \geq h_2$ and $r_1 \geq r_2$, then v is convex in both i and j, i.e.,*

$$
v(i + 2, j) - v(i + 1, j) \geq v(i + 1, j) - v(i, j), \tag{5.10}
$$

$$
v(i, j + 2) - v(i, j + 1) \geq v(i, j + 1) - v(i, j). \tag{5.11}
$$

Proof. (5.10) is implied by (5.4) and (5.9), and (5.11) is implied by (5.4) and (5.8). $\square$

THEOREM 5.8 *The socially optimal policy for admitting class 2 customers is characterized by a monotonically decreasing switching curve, i.e., for each $i \geq 0$, there exists a threshold $L_2^s(i)$, such that a class 2* *arrival in state (i, j) is accepted if and only if $j < L_2^s(i)$. Furthermore, $L_2^s(i)$ is monotonically decreasing in i.*

Proof. Using supermodularity and convexity in j, one can prove this theorem by following similar argument as in the proof for Theorem 5.5. □

3.   Class Optimization

We consider class-optimal policies in this section. The objective of a class-optimal policy for class k, $k = 1, 2$, is to minimize the expected total discounted net cost generated by all customers in class k.

Optimal Policies for Class 1

We consider optimal admission control policies for class 1 customers first. Denote $v(i)$ the expected total discounted net cost generated by a class-optimal policy for class 1 over an infinite horizon starting from state i, where i is the number of class 1 customers in the system. Note that class 1 customers don't see class 2 customers under class optimization because of their higher priority. Thus, after uniformizing, the optimality equation can be written as

$$v(i) = h_1 i + \lambda_1 \min\{v(i), v(i+1)\} + \mu v((i-1)^+ - r_1 I_{\{i \geq 1\}}). \quad (5.12)$$

LEMMA 5.9 $v(1) - v(0) + r_1 \geq 0$.

Proof. Define two processes on the same probability space so that they see the same arrivals and potential services. Process 1 starts with 1 class 1 customer and follows optimal policy. Process 2 starts with 0 class 1 customers and follows policy ϕ which is described below. Let τ be the first time Process 1 has 0 class 1 customers. Let Process 2 take the same action as Process 1 upon each arrival until time τ, then follow the optimal policy afterwards. Therefore, Process 1 has one more class 1 customer than Process 2 until time τ. Two processes become identical from then on. Thus,

$$
\begin{aligned}
v(1) - v(0) &\geq v(1) - v^\phi(0) \\
&= E \int_0^\tau e^{-\alpha t} h_1 dt + E e^{-\alpha \tau}(-r_1 + v(0) - v(0)) \\
&\geq -r_1 E e^{-\alpha \tau} \geq -r_1. \quad \square
\end{aligned}
$$

LEMMA 5.10 *v is convex, i.e.,*

$$v(i+2) - v(i+1) - v(i+1) + v(i) \geq 0. \tag{5.13}$$

Proof. Define four processes on the same probability space so that they see the same arrivals and potential services. Process 1 and 4 follow optimal policies and start in state $i+2$ and i, respectively. Process 2 and 3 start in state $i+1$ and use policies ϕ_2 and ϕ_3, respectively, which are described below. Denote the state of Process k at time t by (X_t^k, Y_t^k), $k = 1, 2, 3, 4$.

Let τ_1 be the first time Process 2 and 3 have 0 class 1 customers. Let τ_2 be the first time Process 1 and 4 take different actions. Define $\tau = \min\{\tau_1, \tau_2\}$. Let Process 2 and 3 take the same action as Process 1 and 4 upon each arrival until time τ, then follow the optimal policy afterwards. Thus

$$
\begin{aligned}
& v(i+2) - v(i+1) - v(i+1) + v(i) \\
\geq\ & v(i+2) - v^{\phi_2}(i+1) - v^{\phi_3}(i+1) + v(i) \\
=\ & E \int_0^\tau e^{-\alpha t}[h(X_t^4 + 2) - h(X_t^4 + 1) - h(X_t^4 + 1) + h(X_t^4)]dt \\
& + Ee^{-\alpha\tau}(-R_1 + R_2 + R_3 - R_4) \\
& + Ee^{-\alpha\tau}(v(X_\tau^1) - v(X_\tau^2) - v(X_\tau^3) + v(X_\tau^4)),
\end{aligned}
$$

where R_i is the potential reward generated in Process i at time τ. It can be easily seen that the first term is 0 because of the linear holding cost rate.

To simplify notation, define

$$\bar{D} = v(i+2) - v(i+1) - v(i+1) + v(i), \tag{5.14}$$

$$\bar{B} = v(X_\tau^1) - v(X_\tau^2) - v(X_\tau^3) + v(X_\tau^4). \tag{5.15}$$

Also define A as in (5.6).

Case 1: $\tau = \tau_1$. Then the states in four processes at τ are 1, 0, 0, 0, respectively. The rewards generated at τ are $R_1 = R_2 = R_3 = r_1$, and $R_4 = 0$. Therefore,

$$\bar{D} \geq Ee^{-\alpha\tau}(v(1) - v(0) + r_1) \geq 0,$$

where the last inequality follows from Lemma 5.9.

Case 2: $\tau = \tau_2$. Then $A = 0$. We have the following possibilities.

Case 2.1: A class 1 arrival is accepted by Process 1 and rejected by Process 4. Let Process 2 accept and Process 3 reject the arrival. Then the states in four processes at τ are $X_\tau^4 + 3, X_\tau^4 + 2, X_\tau^4 + 1, X_\tau^4$, respectively. Adding and subtracting $v(X_\tau^4 + 1) + v(X_\tau^4 + 2)$, we have

$$
\begin{aligned}
\bar{B} = \; & v(X_\tau^4 + 3) - v(X_\tau^4 + 2) - v(X_\tau^4 + 2) + v(X_\tau^4 + 1) \\
& + v(X_\tau^4 + 2) - v(X_\tau^4 + 1) - v(X_\tau^4 + 1) + v(X_\tau^4).
\end{aligned}
$$

Note that the first four terms and the last four terms are inequality (5.13) evaluated at $X_\tau^4 + 1$ and X_τ^4, respectively. So the above argument can be repeated until Case 1 or Case 2.2 happens.

Case 2.2: A class 1 arrival is rejected by Process 1 and accepted by Process 4. Let Process 2 accept and Process 3 reject the arrival. Then the states in four processes at τ are $X_\tau^4 + 2, X_\tau^4 + 2, X_\tau^4 + 1, X_\tau^4 + 1$, respectively. Notice that Process 1 and 2 couple, so do Process 3 and 4. So $\bar{B} = 0$ and hence (5.13) holds. □

THEOREM 5.11 *The class-optimal policy for admitting class 1 customers is characterized by a critical number, i.e., there exists a threshold L_1^c, such that a class 1 arrival in state i is accepted if and only if $i < L_1^c$.*

Proof. Define

$$
L_1^c = \min\{i : v(i+1) > v(i)\}.
$$

Using Lemma 5.10 one can easily show that a class 1 arrival is accepted if and only if $i < L_1^c$. □

Optimal Policies for Class 2

We consider optimal admission control policies for class 2 customers next. Denote $v(i, j)$ the expected total discounted net cost generated by a class-optimal policy for class 2 over an infinite horizon starting from state (i, j). Assuming class 1 customers are admitted according to the class-optimal policy for class 1, the optimality equation can be written as

$$
\begin{aligned}
v(i, j) = \; & h_2 j + \lambda_1 \begin{cases} v(i+1, j), & \text{if } i < L_1^c \\ v(i, j), & \text{if } i >= L_1^c \end{cases} \\
& + \lambda_2 \min\{v(i, j+1), v(i, j)\} \\
& + \mu \begin{cases} v(i-1, j), & \text{if } i \geq 1 \\ v(0, j-1) - r_2, & \text{if } i = 0, j \geq 1 \\ v(0, 0), & \text{if } i = 0, j = 0. \end{cases}
\end{aligned} \qquad (5.16)
$$

LEMMA 5.12 $v(0, 1) - v(0, 0) + r_2 \geq 0.$

Proof. Same argument as in the proof for Lemma 5.2 applies. □

LEMMA 5.13 *v is convex in j, i.e.,*

$$v(i, j+2) - v(i, j+1) - v(i, j+1) + v(i, j) \geq 0. \tag{5.17}$$

Proof. Same argument as in the proof for Lemma 5.10 applies after the following changes. Replace class 1 by class 2. Replace $v(i)$ by $v(i, j)$, $v(i+1)$ by $v(i, j+1)$, etc. Replace r_1 by r_2. □

LEMMA 5.14 *v is supermodular, i.e.,*

$$v(i+1, j+1) - v(i, j+1) - v(i+1, j) + v(i, j) \geq 0. \tag{5.18}$$

Proof. Same argument as in the proof for Lemma 5.3 applies after the following changes. No reward is generated when a class 1 customer finishes service, i.e., $r_1 = 0$. Case 2.1 does not exist, since a class 1 arrival is always accepted in state (i, j) if it is accepted in state $(i+1, j)$. Case 2.2 is the same as in Lemma 5.3 except that it only happens when $i = L_1^c - 1$. □

THEOREM 5.15 *The class-optimal policy for admitting class 2 customers is characterized by a monotonically decreasing switching curve, i.e., for each $i \geq 0$, there exists a threshold $L_2^c(i)$, such that a class 2 arrival in state (i, j) is accepted if and only if $j < L_2^c(i)$. Furthermore, $L_2^c(i)$ is monotonically decreasing in i.*

Proof. Using supermodularity and convexity in j, one can prove this theorem by following similar argument as in the proof for Theorem 5.5. □

4. Numerical Comparison

We compare policies under different criteria numerically in this section. The numerical examples are computed by using standard value iteration algorithm. We approximate the infinite state space by assuming that no customers arrive when the total number of customers in the system reaches an upper bound B, which is much larger than the expected queue length. Thus the state space is $S = \{(i, j) : 0 \leq i, j \leq B\}$. The stopping criterion is $\max\{|v_{n+1}(i, j) - v_n(i, j)| : (i, j) \in S\} \leq 10^{-5}$, where $v_n(i, j)$ is the value function at the nth iteration.

Figure 5.1 plots the cost of class-optimal policy for class 1 customers against i, the number of class 1 customers in starting state. Figure 5.2 plots the cost of class-optimal policy for class 2 customers against j, the number of class 2 customers in starting state, for different i. Figure 5.1 and 5.2 use the following parameters $\alpha = 0.2, \mu = 0.5, \lambda_1 = 0.15, \lambda_2 = 0.15, h_1 = 0.3, h_2 = 0.2, r_1 = 25, r_2 = 18$. Note that the class-optimal v function is not non-decreasing in i or j, which is different from the results in Chen and Kulkarni (2005).

Figure 5.3 and 5.4 plot the cost of socially optimal policy against i and j for fixed j and i, respectively. They use the same parameters as in Figure 1 and 2 except that $r_1 = 15, r_2 = 10$. Again the socially optimal v function is not non-decreasing in i or j, which is different from the results in Chen and Kulkarni (2005). So moving the reward from when the customer join the system to when the service is completed changes the nature of the problem.

Figure 5.5 and 5.6 plot the switching curves under three optimization criteria for class 1 and class 2, respectively. Figure 5.5 uses the following parameters $\alpha = 0.1, \mu = 0.5, \lambda_1 = 0.1, \lambda_2 = 0.3, h_1 = 25, h_2 = 20, r_1 = 450, r_2 = 300$. Figure 5.6 uses the following parameters $\alpha = 0.1, \mu = 0.5, \lambda_1 = 0.39, \lambda_2 = 0.01, h_1 = 2, h_2 = 0.3, r_1 = 550, r_2 = 500$.

Note that for class 1 (higher priority) customer, individually optimal policy accepts the most and socially optimal policy accepts the least number of customers, which is consistent with the existing results in the literature. For class 2 (lower priority) customer, socially optimal policy accepts more customers than class optimal policy, which is the exact opposite to the comparison result for class 1. Depending on the parameters, individually optimal policy can accept either more or less customers than either of the other two policies. The above observations are the same as in Chen and Kulkarni (2005). We restate the explanation for the counter intuitive behavior of class 2 customers provided by Chen and Kulkarni (2005) in the following. The total cost incurred by each accepted customer can be divided into two parts: internal cost, i.e., individual holding cost, and external cost, i.e., effect on other customers. The internal cost of a class 2 customer is the highest under individual optimization and the lowest under social optimization, since the more class 1 customers accepted, the longer a class 2 customer needs to wait. The external cost of a class 2 customer is zero under individual optimization and the same positive amount under class and social optimization, which is the delay caused on later class 2 customers. Therefore, the total cost of a class 2 customer is higher under class optimization than under social optimization. It can be either higher or lower under individual optimization than under the other two criteria depending on which effect dominates. This behavior of class 2 customers also shows an interesting socio-economic fact: the lower priority customers fare better under centralized control than under class-based control.

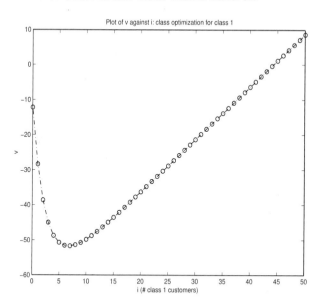

Figure 5.1. Plot of v against i: class optimization for class 1

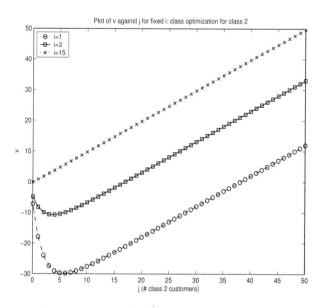

Figure 5.2. Plot of v against j for fixed i: class optimization for class 2

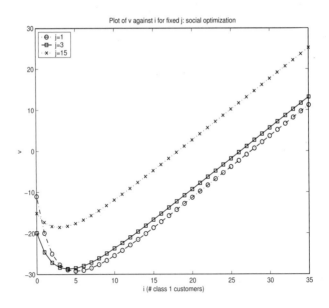

Figure 5.3.   Plot of v against i for fixed j: social optimization

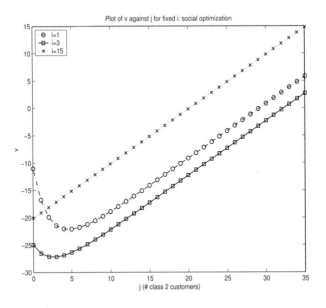

Figure 5.4.   Plot of v against j for fixed i: social optimization

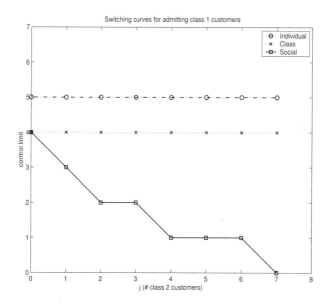

Figure 5.5. Switching curves for admitting class 1 customers

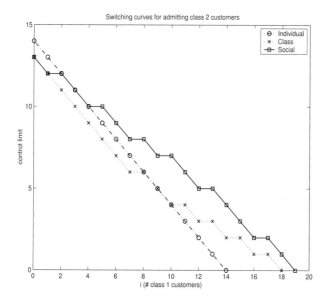

Figure 5.6. Switching curves for admitting class 2 customers

94

5. Conclusion

We have studied the admission control problem to a two-priority $M/M/1$ queueing system. We consider the optimal policy from three perspectives: individual optimization, class-optimization, and social optimization. Using sample path argument, we show that the optimal policy for each priority class from each perspective is of threshold type. We also compare optimal policies from different perspectives numerically.

References

Adiri, I. and U. Yechiali. (1974). "Optimal priority-purchasing and price decisions in nonmonopoly and monopoly queues." *Opns. Res.* 22, 1051-1066.

Blanc, J. P. C., P. R. de Waal, P. Nain, and D. Towsley. (1992). "Optimal control of admission to a multiserver queue with two arrival streams." *IEEE Trans. Automat. Control* 37, 785-797.

Chen, F. and V. G. Kulkarni. (2005). "Individual, class-based, and social optimal admission policies in two-priority queues." Technical.Report, Department of Statistics and Operations Research, University of North Carolina at Chapel Hill. (submitted to *Stochastic Models*)

Ha, A. (1997). "Stock-rationing policy for a make-to-stock production system with two priority classes and backordering." *Naval Res. Logist.* 44, 457-472.

Hassin, R. (1995). "Decentralized regulation of a queue." *Mgmt. Sci.* 41, 163-173.

Knudsen, N. C. (1972). "Individual and social optimization in a multi-server queue with a general cost-benefit structure." *Econometrica* 40, 515-528.

Kulkarni, V. G. and T. E. Tedijanto. (1998). "Optimal admission control of markov-modulated batch arrivals to a finite-capacity buffer." *Stochastic Models* 14, 95-122.

Lindvall, T. (1992) *Lectures on the coupling method.* John Wiley & Sons, New York.

Lippman, S. (1975). "Applying a new device in the optimization of exponential queueing systems." *Opns. Res.* 23, 687-710.

Lippman, S. and S. Stidham. (1977). "Individual versus social optimization in exponential congestion systems." *Opns. Res.* 25, 233-247.

Mendelson, H. and S. Whang. (1990). "Optimal incentive-compatible priority pricing for the $M/M/1$ queue." *Opns. Res.* 38, 870-883.

Miller, B. (1969). "A queueing reward system with several customer classes." *Mgmt. Sci.* 16, 234-245.

Nair, S. K. and R. Bapna. (2001). "An application of yield management for internet service providers." *Nav. Res. Log.* 48, 348-362.

Naor, P. (1969). "The regulation of queue size by levying tolls." *Econometrica* 37, 15-24.

Opp, M., V. G. Kulkarni, and K. Glazebrook. (2005). "Outsourcing warranty repairs: dynamic allocation." *Nav. Res. Log.* 52, 381-398.

Puterman, M. L. (1994). *Markov decision processes.* Wiley, New York.

Rykov, V. V. (2001). "Monotone control of queueing systems with heterogeneous servers." *Queueing Syst.* 37, 391-403.

Stidham, S. (1978). "Socially and individually optimal control of arrivals to a $GI/M/1$ queue." *Mgmt. Sci.* 24, 1598-1610.

Stidham, S. (1985). "Optimal control of admission to a queueing system." *IEEE Trans. Automat. Control* 30,705-713.

Stidham, S. and R. R. Weber. (1989). "Monotonic and insensitive optimal policies for control of queues with undiscounted costs." *Opns. Res.* 87, 611-625.

Wu, C. H., M. E. Lewis, and M. Veatch. (2005). "Dynamic allocation of reconfigurable resources in a two-stage tandem queueing system with reliability considerations." to appear in *IEEE Trans. Automat. Control*

Yechiali, U. (1971). "On optimal balking rules and toll charges in a $GI/M/1$ queuing process." *Opns. Res.* 19, 349-370.

Yechiali, U. (1972). "Customers' optimal joining rules for the $GI/M/s$ queue." *Mgmt. Sci.* 18, 434-443.

Chapter 6

SOME BILINEAR STOCHASTIC EQUATIONS WITH A FRACTIONAL BROWNIAN MOTION

T.E. Duncan

Department of Mathematics
University of Kansas, Lawrence, KS 66045

duncan@math.ku.edu

Abstract   An explicit solution is given for a bilinear stochastic differential equation with a fractional Brownian motion that is described by noncommuting linear operators and that has the Hurst parameter H in the interval $(\frac{1}{2}, 1)$. It is shown that the expression for this family of solutions for H in $(\frac{1}{2}, 1)$ extends to the solution for a Brownian motion, $H = \frac{1}{2}$. Some examples are given to contrast the solutions for commuting and noncommuting linear operators, in particular, the asymptotic behavior of the solutions can be significantly different for commuting and noncommuting operators. The methods to obtain the explicit solutions use a stochastic calculus for a fractional Brownian motion and some Lie theory.

Keywords: Fractional Brownian motions, bilinear stochastic differential equations, explicit solutions of stochastic equations

1.   Introduction

Fractional Brownian motion is a family of Gaussian processes that is indexed by the Hurst parameter H in $(0, 1)$. For $H = \frac{1}{2}$, this process is a Brownian motion. These processes have a self-similarity and for $(\frac{1}{2}, 1)$ they have a long range dependence. Since for $H \neq \frac{1}{2}$ these processes are

---

*Research supported in part by NSF Grants DMS 0204669, DMS 0505706 and ANI 0125410

not semimartingales, the usual stochastic calculus cannot be used so a different calculus is required. In recent years, a stochastic calculus has been developed for fractional Brownian motions with $H \in (\frac{1}{2}, 1)$ (e.g. [1,3,4]) that has some properties of the stochastic calculus for Brownian motion. The fractional Brownian motions were defined by Kolmogorov [9] and some properties were given by Mandelbrot and Van Ness [10]. Hurst [8] noted the initial potential application of these processes and subsequently their potential applicability has been noted in a wide range of physical phenomena, especially for $H \in (\frac{1}{2}, 1)$.

Since there are no general conditions for the existence or the uniqueness of the solutions of a significant family of nonlinear stochastic differential equations with a fractional Brownian motion having $H \neq \frac{1}{2}$, it is important to consider special families of equations that should be particularly useful as models of physical phenomena.

One important family for its wide potential applicability is bilinear stochastic differential equations which are also called multiplicative noise equations. While these equations have been solved explicitly for commuting linear operators in a finite dimensional space [2] and in a Hilbert space [5], apparently no results are available for explicit solutions for noncommuting linear operators.

In this paper, explicit solutions are given for bilinear stochastic differential equations with a fractional Brownian motion with $H \in (\frac{1}{2}, 1)$, where the linear operators are noncommutative but they commute with their commutator. The method of solution combines some stochastic analysis for a fractional Brownian motion (e.g. [1,3,4]) and some Lie theory for these equations, especially the Baker-Campbell-Hausdorff formula (e.g. [11]). These solutions are shown to extend in H to $H = \frac{1}{2}$, that is, Brownian motion.

With explicit solutions, the asymptotic behavior of these solutions can be determined and compared for commuting and noncommuting linear operators. Using some simple examples it is shown that the asymptotic behavior of solutions can be quite different in these two cases.

Bilinear stochastic differential equations have been used with Brownian motions to model diverse physical phenomena, e.g. the Merton-Samuelson model of a stock price, as well as to describe mathematical objects e.g. Radon-Nikodym derivatives. In studies of stock prices, it has been noted that there is often a long range dependence which indicates the potential use of a fractional Brownian motion in some aspects of financial modeling.

2. Preliminaries

The family of fractional Brownian motions is a family of Gaussian processes that is indexed by the Hurst parameter H in the interval $(0,1)$.

For some fixed $H \in (0,1)$, a standard fractional Brownian motion $(B(t), t \geq 0)$ is a process with continuous sample paths such that $\mathbb{E}[B(t)] = 0$ and

$$\mathbb{E}[B(s)B(t)] = \frac{1}{2}[s^{2H} + t^{2H} - |s-t|^{2H}] \tag{1}$$

for all $s, t \in \mathbb{R}_+$.

Let $(\Omega, \mathcal{F}, \mathbb{P})$ be a complete probability space for $(B(t), t \geq 0)$. It suffices to choose $\Omega = C(\mathbb{R}_+, \mathbb{R})$ with the topology of local uniform convergence, $\mathcal{F}$ is the completion of the Borel σ-algebra and $\mathbb{P}$ is the Gaussian measure for the fractional Brownian motion. A fractional Brownian motion is self-similar, that is, if $a > 0$ then $(B(at), t \geq 0)$ and $(a^H B(t), t \geq 0)$ have the same probability law. If $H > \frac{1}{2}$ then $(B(t), t \geq 0)$ has a long range dependence, that is, $\Sigma r(n) = \infty$, where

$$r(n) = \mathbb{E}[B(1)(B(n+1) - B(n))].$$

For $H = \frac{1}{2}$, the process is a standard Brownian motion. For $H \neq \frac{1}{2}$, these processes are not semimartingales. So the stochastic calclus for semimartingales is not applicable. A stochastic calculus is available (e.g. [1,3,4]) which uses either the Wick product or some methods from Malliavin calculus for defining a stochastic integral. This stochastic integral for a suitable family of integrands has zero mean and an explicitly computable second moment. However, there is no general theory for the solutions of stochastic differential equations with a fractional Brownian motion if $H \neq \frac{1}{2}$. Nevertheless, especially for $H \in (\frac{1}{2}, 1)$, solutions can be given for various families of stochastic differential equations such as linear and bilinear . Fractional calculus plays an important role for fractional Brownian motion because for $s, t \in [0, T]$

$$\mathbb{E}[B(s)B(t)] = \rho(H) \int_0^T u_{\frac{1}{2}-H}^2(r) \tag{2}$$
$$(I_{T-}^{H-\frac{1}{2}} u_{H-\frac{1}{2}} 1_{[0,s]})(r)(I_{T-}^{H-\frac{1}{2}} u_{H-\frac{1}{2}} 1_{[0,t]})(r) dr$$

where $u_a(r) = r^a$ for $a \in \mathbb{R}$,

$$(I_{T-}^\alpha f)(s) = \frac{1}{\Gamma(\alpha)} \int_s^T f(r)(r-s)^{\alpha-1} dr$$

for $\alpha > 0$,

$$(I_{T-}^{-\alpha} f)(s) = \frac{-1}{\Gamma(1-\alpha)} \frac{d}{ds} \int_s^T f(r)(r-s)^{-\alpha} dr$$

for $\alpha \in (0,1)$, and

$$\rho(H) = \frac{2H\Gamma(H+\frac{1}{2})\Gamma(\frac{3}{2}-H)}{\Gamma(2-2H)}.$$

While solutions for bilinear stochastic differential equations with a fractional Brownian motion for $H \in (\frac{1}{2}, 1)$ have been given explicitly ([1,5]), it was assumed that the linear operators in the equations are commutative. Since this commutivity condition is quite restrictive, in this paper, explicit solutions are given for some noncommuting operators.

3. Main Result

Consider the following bilinear stochastic differential equation

$$dX(t) = X(t)(Adt + CdB(t)), \qquad (3)$$
$$X(0) = I,$$

where $X(t) \in GL(n) \subset L(\mathbb{R}^n, \mathbb{R}^n)$, $A, C \in L(\mathbb{R}^n, \mathbb{R}^n)$. $(B(t), t \geq 0)$ is a real-valued standard fractional Brownian motion with $H \in (\frac{1}{2}, 1)$. It is assumed that A and C may not commute but they commute with their commutator, that is,

$$[A, [A, C]] = [C, [A, C]] = 0 \qquad (4)$$

where $[\cdot, \cdot]$ is the Lie bracket in $gl(n) = L(\mathbb{R}^n, \mathbb{R}^n)$, that is, $[E, F] = EF - FE$.

The following result gives an explicit solution to the stochastic differential equation (3).

Theorem A solution to the equation (3) with the Lie bracket assumption (4) is the process $(\hat{X}(t), t \geq 0)$ where

$$\hat{X}(t) = \prod_{j=1}^6 e^{Z_j(t)}, \qquad (5)$$

$$Z_1(t) = tA, \qquad (6)$$

$$Z_2(t) = CB(t), \qquad (7)$$

$$Z_3(t) = -[A, C] \int_0^t B(s)ds, \tag{8}$$

$$Z_4(t) = \frac{1}{2}C[A, C]t^{2H+1}, \tag{9}$$

$$Z_5(t) = \frac{-1}{2(2H+2)}[A, C]^2 t^{2H+2}, \tag{10}$$

$$Z_6(t) = -\frac{1}{2}C^2 t^{2H}. \tag{11}$$

Proof. Initially some commutation properties of A and the other terms in the solution (5) are proved that arise in the verification by substitution that $\hat{X}$ is a solution of (3). Consider the product of A and $e^{Z_2(t)}$,

$$
\begin{aligned}
Ae^{Z_2(t)} &= e^{Z_2(t)}e^{-Z_2(t)}Ae^{Z_2(t)} \\
&= e^{CB(t)}Ad_{e^{CB(t)}}(A) \\
&= e^{CB(t)}e^{-B(t)ad_C}(A).
\end{aligned} \tag{12}
$$

The second and third equality in (12) follow from the definitions of Ad and ad, the adjoint representation of the group and the algebra, respectively, and naturally occur in such computations (e.g. p.65, [6]). These equalities are valid here because the real-valued random variable $B(t)$ may be considered as a parameter and the (measurable) computation is done for almost all $\omega \in \Omega$. It follows from the assumption (4) that

$$
\begin{aligned}
e^{-B(t)ad_C}(A) &= A - B(t)[C, A] \\
&= A + B(t)[A, C].
\end{aligned} \tag{13}
$$

Since A commutes with $[A, C]$, it also commutes with $e^{Z_3(t)}$ and $e^{Z_5(t)}$ for each $t \geq 0$. It remains to investigate the commutivity property of A with $e^{Z_4(t)}$ and $e^{Z_6(t)}$. For the product A and $e^{Z_4(t)}$ compute the bracket $[A, C[A, C]]$,

$$
\begin{aligned}
[A, C[A, C]] &= AC(AC - CA) - C(AC - CA)A \\
&= AC(AC - CA) - CA(AC - CA) \\
&= [A, C]^2.
\end{aligned} \tag{14}
$$

Making computations analogous to (12) and (13), the following equalities are satisfied

$$Ae^{Z_4(t)} = e^{Z_4(t)}e^{-\alpha ad_{C[A,C]}}(A).$$

$$e^{-\alpha ad_{C[A,C]}}(A) = A - \alpha[C[A,C],A]$$
$$= A + \alpha[A,C]^2 \qquad (15)$$

where $\alpha = \frac{1}{2} t^{2H+1}$.

Now consider A and $e^{Z_6(t)}$ and initially compute $[C^2, A]$. By the assumption (4), it follows that

$$[C,[C,A]] = C^2A - CAC - CAC + AC^2 = 0.$$

So
$$C^2A = 2CAC - AC^2$$

and

$$[C^2, A] = C^2A - AC^2 = 2CAC - AC^2 - AC^2$$
$$= 2[C,A]C = 2C[C,A]. \qquad (16)$$

It is clear that $[C^2, [C^2, A]] = 0$ and

$$Ae^{Z_6(t)} = e^{Z_6(t)}e^{-Z_6(t)}Ae^{Z_6(t)}$$
$$= e^{Z_6(t)}e^{\frac{1}{2}t^{2H}ad_{C^2}}(A). \qquad (17)$$

So

$$e^{\frac{1}{2}t^{2H}ad_{C^2}}(A) = A + \frac{1}{2}t^{2H}[C^2, A]$$
$$= A + t^{2H}C[C, A]. \qquad (18)$$

The terms $[A, C]$ and $C[C, A]$ that have arisen from the commutation computations for A above, commute with $e^{Z_i(t)}$ for $i = 2, \cdots, 6$.

Collecting the computations in (13), (15), (17) and (18), it follows that

$$A\hat{X}(t) = \hat{X}(A + B(t)[A, C] + \frac{1}{2} t^{2H+1}[A, C]^2 - t^{2H}C[A, C]). \qquad (19)$$

Since C commutes with $[C, A]$, $C[A, C]$, $[A, C]^2$ and C^2, it commutes with $e^{Z_i(t)}$ for $i = 3, \cdots, 6$. Since $[A, C]$ commutes with $C[A, C]$, $[A, C]^2$

and C^2, it commutes with $e^{Z_i(t)}$ for $i = 4, 5, 6$. Since $C[A, C]$ commutes with $[A, C]^2$ and C^2, it commutes with $e^{Z_i(t)}$ for $i = 5, 6$ and since $[A, C]^2$ commutes with C^2, it commutes with $e^{Z_6(t)}$. Summarizing, the following equalities are satisfied

$$C \prod_{j=2}^{6} e^{Z_j(t)} = \prod_{j=2}^{6} e^{Z_j(t)} C, \tag{20}$$

$$[A, C] \prod_{j=3}^{6} e^{Z_j(t)} = \prod_{j=3}^{6} e^{Z_j(t)} [A, C], \tag{21}$$

$$C[A, C] \prod_{j=4}^{6} e^{Z_j(t)} = \prod_{j=4}^{6} e^{Z_j(t)} C[A, C], \tag{22}$$

$$[A, C]^2 e^{Z_6(t)} = e^{Z_6(t)} [A, C]^2. \tag{23}$$

To determine the additional terms that arise from the stochastic calculus for a fractional Brownian motion, some terms that occur in an application of Taylor's formula applied at the points of a partition of $[0, t]$ are considered because no Itô formula for fractional Brownian motion is available for noncommutting linear operators.

Let $P = \{t_0, \cdots, t_n\}$ be a partition of $[0, t]$ and let $t_i, t_{i+1} \in P$ for $i \in \{0, \cdots, n-1\}$. It follows from the stochastic calculus for a fractional Brownian motion that

$$\prod_{j=1}^{6} e^{Z_j(t_i)} \int_{t_i}^{t_{i+1}} C dB(s) = \int_{t_i}^{t_{i+1}} \prod_{j=1}^{6} e^{Z_j(t_i)} C dB(s)$$

$$+ \int_{t_i}^{t_{i+1}} \int_{0}^{t_i} \prod_{j=1}^{6} e^{Z_j(t_i)} C^2 \phi_H(r - q) dr dq$$

$$+ \int_{t_i}^{t_{i+1}} \int_{0}^{t_i} \int_{0}^{t_i} \prod_{j=1}^{6} e^{Z_j(t_i)} [C, A] C 1_{[0,q]}(r) \phi_H(s - r) dq dr ds. \tag{24}$$

The third term on the RHS of (24) can be rewritten to eliminate the indicator function by interchanging the order of integration for q and r as

$$\int_{t_i}^{t_{i+1}} \int_{0}^{t_i} \int_{0}^{q} \prod_{j=1}^{6} e^{Z_j(t_i)} [C, A] C \phi_H(s - r) dr dq ds.$$

The equality (24) is satisfied by a basic property of the stochastic calculus for $H \in (\frac{1}{2}, 1)$. (e.g. [1, 4])

Choosing a sequence of partitions $(P_n, n \in \mathbb{N})$ of $[0, t]$ such that $|P_n| \to 0$ as $n \to \infty$, $P_n \subset P_{n+1}$ for each $n \in \mathbb{N}$, and $P_n = \{t_0^{(n)}, \cdots, t_n^{(n)}\}$ where $0 = t_0^{(n)} < t_1^{(n)} < \cdots < t_n^{(n)} = t$, the limit in probability of the sum of terms of the form (24) from P_n as $n \to \infty$ is

$$
\int_0^t \hat{X}(s)C dB(s) + \int_0^t \hat{X}(s)C^2 H s^{2H-1} ds
$$
$$
+ \int_0^t \hat{X}(s)[C, A]C(H - \frac{1}{2})s^{2H} ds. \tag{25}
$$

A brief explanation is given for the occurrence of the six terms $Z_1, \cdots, Z_6$ that describe the solution. The term Z_1 in (6) arises from the linear operator $A dt$ in (3) and the term Z_2 in (7) arises from the linear operator $C dB(t)$. The term Z_3 in (8) arises from the noncommutivity of A and C. The term Z_4 in (9) arises from the stochastic calculus that follows from the equality (24) and from the noncommutivity of A and C^2. The term Z_5 in (10) arises from the noncommutivity of A and $C[A, C]$ and the term Z_6 in (11) arises from the stochastic calculus.

Combining (19), (25) and the first order differentials of $e^{Z_i(t)}$ for $i = 3, \cdots, 6$, it follows that

$$
\int_0^t d\hat{X}(s) = \int_0^t \hat{X}(s)(A + B(s)[A, C]
$$
$$
+ \frac{1}{2} s^{2H+1}[A, C]^2 - s^{2H}[A, C]C + H s^{2H-1}C^2
$$
$$
- (H - \frac{1}{2})s^{2H}[A, C]C - B(s)[A, C] + (H + \frac{1}{2})s^{2H}C[A, C] \tag{26}
$$
$$
- \frac{1}{2} s^{2H+1}[A, C]^2 - H s^{2H-1}C^2)ds + \int_0^t \hat{X}(s)C dB(s)
$$
$$
= \int_0^t \hat{X}(s)(A ds + C dB(s)).
$$

This equality verifies that $(\hat{X}(t), t \geq 0)$ is a solution of (3). $\qquad \square$

It is useful to compare the solution (5) for $H \in (\frac{1}{2}, 1)$ with the solution for $H = \frac{1}{2}$, that is, where the process $(B(t), t \geq 0)$ in (3) is a standard Brownian motion.

Corollary If $(B(t), t \geq 0)$ in (3) is a standard Brownian motion, that is, $H = \frac{1}{2}$, then a solution of (3) is $(\hat{X}(t), t \geq 0)$ where

$$\hat{X}(t) = \prod_{j=1}^{6} e^{Z_j(t)} \tag{27}$$

and

$$Z_1(t) = tA, \tag{28}$$

$$Z_2(t) = CB(t), \tag{29}$$

$$Z_3(t) = -[A, C] \int_0^t B(s)ds, \tag{30}$$

$$Z_4(t) = \frac{1}{2}C[A, C]t^2, \tag{31}$$

$$Z_5(t) = \frac{1}{6}[A, C]^2 t^3, \tag{32}$$

$$Z_6(t) = -\frac{1}{2}C^2 t. \tag{33}$$

Proof. It suffices to note that Itô formula implies that the same terms arise except the third term in (25) which is zero for $H = \frac{1}{2}$. □

It follows immediately that the solution (5) is valid for $H \in [\frac{1}{2}, 1)$.

4. Some Examples

A few examples are considered to show some different asymptotic behavior for the solution (5) depending on the commutation properties of A and C in (3).

Let $n = 3$ so that a solution of (3), $X(t) \in GL(3) \subset L(\mathbb{R}^3, \mathbb{R}^3)$. Let $E_{ij} \in L(\mathbb{R}^3, \mathbb{R}^3)$ be the elementary matrix with 1 in the (i, j) position and 0 elsewhere, and let I be the identity and $H \in (\frac{1}{2}, 1)$. Let

$$A_0 = C_0 = I, \tag{34}$$

$$A_1 = E_{12} + E_{13}, \tag{35}$$

$$C_1 = E_{23}. \tag{36}$$

A solution of

$$dY(t) = Y(t)(A_0 dt + C_0 dB(t)),$$
$$Y(0) = I, \tag{37}$$

is the process $(Y_0(t), t \geq 0)$ that follows from the scalar case [2] as well as [5], that is,

$$Y_0(t) = e^{tI} e^{B(t)I} e^{-\frac{1}{2}t^{2H}I}. \tag{38}$$

¿From the Law of the Iterated Logarithm for a fractional Brownian motion [7], it is elementary to verify that

$$\lim_{t \to \infty} Y_0(t) = 0, \quad a.s. \tag{39}$$

Now consider the equation (37) where A_0 and C_0 are replaced by $A_0 + A_1$ and $C_0 + C_1$ respectively, that is,

$$dY(t) = Y(t)((A_0 + A_1)dt + (C_0 + C_1)dB(t)), \\ Y(0) = I. \tag{40}$$

The following equalities are easily verified

$$[A_0 + A_1, C_0 + C_1] = [A_1, C_1] = E_{13},$$
$$[A_0 + A_1, [A_1, C_1]] = 0,$$
$$[C_0 + C_1, [A_1, C_1]] = 0,$$
$$[A_1, C_1]^2 = 0,$$
$$(C_0 + C_1)[A_1, C_1] = 0,$$
$$(C_0 + C_1)^2 = I + 2E_{23}.$$

A solution for (40) is given by

$$Y_1(t) = e^{(tA_1)} e^{(tA_0 + \Sigma_{j=2}^6 Z_j(t))}$$
$$= (I + t(E_{12} + E_{13})) e^{tI} e^{B(t)I} e^{B(t)E_{23}}$$
$$(I + (- \int_0^t B(s)ds + \alpha t^{2H+1})E_{13}) e^{-\frac{t^{2H}}{2}(I + 2E_{23})} \tag{41}$$
$$= (I + t(E_{12} + E_{13}))(e^{t + B(t) - \frac{t^{2H}}{2}} I)(I + B(t)E_{23})$$
$$(I - (\int_0^t B(s)ds + \alpha t^{2H+1})E_{13})(I - t^{2H} E_{23})$$

where $\alpha = \frac{1}{2}$.

The following limit follows directly

$$\lim_{t \to \infty} Y_1(t) = 0, \quad a.s. \tag{42}$$

Now consider the equation (37) where A_0 and C_0 are replaced by A_1 and C_1, respectively, that is,

$$dY(t) = Y(t)(A_1 dt + C_1 dB(t)),$$
$$Y(0) = I. \tag{43}$$

It easily follows that

$$[A_1, C_1]^2 = 0,$$

$$C_1[A_1, C_1] = 0,$$

$$C_1^2 = 0.$$

A solution of (43) is

$$Y_2(t) = e^{(tA_1)} e^{C_1 B(t)} e^{[A_1, C_1] \int_0^t B(s)ds}$$

$$= (I + t(E_{12} + E_{13}))(I + B(t)E_{23})(I - E_{13} \int_0^t B(s)ds) \tag{44}$$

$$= I + B(t)E_{23} + tE_{12} + (t + tB(t) - \int_0^t B(s)ds)E_{13}.$$

Clearly $(Y_2(t), t \geq 0)$ does not converge as $t \to \infty$, e.g.

$$\limsup_{t \to \infty} (Y_2(t))_{23} = +\infty, \quad a.s. \tag{45}$$

$$\liminf_{t \to \infty} (Y_2(t))_{23} = -\infty, \quad a.s. \tag{46}$$

These examples provide some indication that the asymptotic behavior of the solutions of these bilinear equations with noncommuting operators can deviate significantly from the commutative case.

References

[1] E. Alos and D. Nualart, *Stochastic integration with respect to fractional Brownian motion*, Stoch. Stoch. Rep., 75 (2003), 129-152.

[2] T.E. Duncan, *Some applications of fractional Brownian motion to linear systems*, Systems Theory: Modeling, Analysis and Control, eds. T.J. Djaferis and I.C. Schick, Kluwer, 2000, 97-105.

[3] T.E. Duncan, Y. Hu and B. Pasik-Duncan, *Stochastic calculus for fractional Brownian motion. I Theory*, SIAM J. Contr. Optim., 38 (2000), 582-612.

[4] T.E. Duncan, J. Jakubowski and B. Pasik-Duncan, *Stochastic integration for fractional Brownian motion in a Hilbert space*, Stoc. Dynamics, to appear.

[5] T.E. Duncan, B. Maslowski and B. Pasik-Duncan, *Stochastic equations in Hilbert space with a multiplicative fractional Gaussian noise*, Stoc. Proc. Appl., 115 (2005), 1357-1383.

[6] B.C. Hall, *Lie Groups, Lie Algebras and Representations*, Graduate Texts in Math., 222, Springer 2003.

[7] G.A. Hunt, *Random Fourier transform*, Trans. Amer. Math. Soc., 71, (1951), 38-69.

[8] H.E. Hurst, *Long-term storage capacity in reservoirs*, Trans. Amer. Soc. Civil Eng. 116 (1951), 400-410.

[9] A.N. Kolmogorov, *Wienersche Spiralen und einige andere interessante Kurven im Hilbertschen Raum*, C. R. (Doklady) Acad. URSS (N. S.) 26 (1940), 115-118.

[10] B.B. Mandelbrot and J.W. Van Ness, *Fractional Brownian motion, fractional noises and applications*, SIAM Rev., 10 (1968), 422-437.

[11] V.S. Varadarajan, *Lie Groups, Lie Algebras and Their Representation*, Graduate Texts in Math., 102 Springer 1984.

Chapter 7

TWO TYPES OF RISK

Jerzy A. Filar

Center for the Industrial and Applied Mathematics,
School of Mathematics & Statistics,
University of South Australia,
Mawson Lakes, SA 5095,Australia

jerzy.filar@unisa.edu.au

Boda Kang

Center for the Industrial and Applied Mathematics,
School of Mathematics & Statistics,
University of South Australia,
Mawson Lakes, SA 5095,Australia

boda.kang@unisa.edu.au

Abstract   The risk encountered in many environmental problems appears to exhibit special "two-sided" characteristics. For instance, in a given area and in a given period, farmers do not want to see too much or too little rainfall. They hope for rainfall that is in some given interval. We formulate and solve this problem with the help of a "two-sided loss function" that depends on the above range. Even in financial portfolio optimization a loss and a gain are "two sides of a coin", so it is desirable to deal with them in a manner that reflects an investor's relative concern. Consequently, in this paper, we define Type I risk: "the loss is too big" and Type II risk: "the gain is too small". Ideally, we would want to minimize the two risks simultaneously. However, this may be impossible and hence we try to balance these two kinds of risk. Namely, we tolerate certain amount of one risk when minimizing the other. The latter problem is formulated as a suitable optimization problem and illustrated with a numerical example.

Keywords: Two-sided risk, rainfall, temperature, value-at-risk, conditional value-at-risk, Type I risk, value-of-gain, conditional value-of-gain, Type II risk, assurance, scenarios, portfolio optimization.

Introduction

The risk encountered in many environmental problems appears to exhibit special "two-sided" characteristics. The "fundamental security" in an environmental problem may be a variable such as a rainfall or a temperature. For instance, in a given area and in a given period, farmers do not want to see too much or too little rainfall. They hope for rainfall that is in some given interval. Similarly, we often hope that the temperature is neither too high nor too low. We formulate and solve this problem with the help of a "two-sided loss function" that depends on the above range.

In financial mathematics, there is an extensive literature discussing the risk of a financial portfolio using the value-at-risk concept, see [3–5, 14–18]. However, these authors consider only the "one-sided" risk using the return of the portfolio. We argue that - even in financial context - a loss and a gain are "two sides of a coin", so it is desirable to differentiate between the loss and the gain of a portfolio and deal with them in a manner that reflects an investor's relative concern about loss and gain. This is because different people have different attitudes toward a loss and a gain. Thus, it might be useful to provide models that trade-off the aversion to these two types of risk. Trying to minimize these kinds of risks is somewhat different from minimizing conditional value-at-risk using the usual loss or gain function (as is done in, for instance, [14, 15]).

Based on the above discussion, we define Type I risk: "the loss is too big" and Type II risk: "the gain is too small". Ideally, we would want to minimize the two risks simultaneously. However, this may be impossible and hence we try to balance these two kinds of risk. Namely, we tolerate certain amount of one risk when minimizing the other. In the financial context, investors can then suitably choose parameters according to their own attitude towards the loss and gain risk.

The paper is organized as follows: we provide a new loss function for the two-sided problem such as rainfall or temperature. Using the new loss function together with conditional value-at-risk, we show how to formulate such a risk in Section 1. In Section 2 we introduce two types of risk associated with the new loss and gain function. We suggest a way to balance a loss and a gain in a more general case. We also provide a criterion for users to choose parameters in these problems. Finally, in Section 3, together with the portfolio problem, we put forward the concept and some properties of conditional value-at-risk and conditional value-of-gain with the new loss and gain functions respectively. All proofs are provided in Section 4.

1. Two-sided risk

Risks in environmental problems are different from financial market risk in some aspects. For example, in the rainfall problem, too much rain or too little rain are both undesirable. Too much rain will lead to a flood, whereas too little rain will lead to a drought. A similar problem arises with temperature. We do not want the temperature in a location to be too high or too low. In this section we introduce one natural formulation of this "two-sided risk" problem.

Let a random variable (r.v.,for short) X denote the rainfall in a location during some specific period. Let us suppose, for instance, that farmers in this location hope that the rainfall in this season is in the interval $[\nu_1, \nu_2]$. That is, exceeding ν_2 or being lower than ν_1 are both risky in some sense. We shall call $[\nu_1, \nu_2]$ the *riskless interval*.

Let $X_1 := \max\{\nu_1 - X, 0\}$ define the *lower risk* random variable. Obviously, as X falls below ν_1, X_1 increases above 0. Since, insufficient rain is undesirable, so are large values of X_1. Similarly, we define $X_2 := \max\{X - \nu_2, 0\}$ as the *upper risk* random variable. Again, as X raises above ν_2, the r.v. X_2 increases above 0. Therefore, the smaller are the values of both X_1 and X_2, the better is the result for our risk sensitive farmers. Figure 7.1 illustrates this situation.

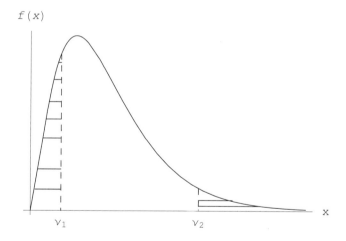

Figure 7.1. Two-sided risk.

It follows immediately that

$$P\{X \notin [\nu_1, \nu_2]\} = P\{X < \nu_1\} + P\{X > \nu_2\} = P\{X_1 > 0\} + P\{X_2 > 0\}.$$

While the above probability of an undesirable event is an entity that we would like to be "small", in practical situations, it could well be that the event $\{X_1 > 0\}$ is relatively more, or less, undesirable than the event $\{X_2 > 0\}$ (for instance, some crops may withstand drought better than excessive moisture). To capture this, unequal, concern about the lower and upper risks we now introduce a single *two-sided risk function* (also called loss function) parameterized by $\gamma \in [0, 1]$:

$$h(X, \gamma, \nu_1, \nu_2) = \gamma X_1 + (1-\gamma) X_2 = \gamma \max\{\nu_1 - X, 0\} + (1-\gamma) \max\{X - \nu_2, 0\}.$$

Here, γ captures the relative importance of lower risk versus the upper risk.

Note that, if we assume the distribution of X is $F(x)$, namely $F(x) = P(X \le x)$, then the distribution function of $h(X, \gamma, \nu_1, \nu_2)$ is:

$$H(x, \gamma, \nu_1, \nu_2) = P(h(X, \gamma, \nu_1, \nu_2) \le x) = F\left(\nu_2 + \frac{x}{1-\gamma}\right) - F\left(\nu_1 - \frac{x}{\gamma}\right).$$

Given γ, ν_1 and ν_2, it is now easy to see $h(X, \gamma, \nu_1, \nu_2)$ is a convex function in X. It is also clear that the function $h(X, \gamma, \nu_1, \nu_2)$ is nonnegative everywhere in its domain. Since here we use nonnegative numbers to describe the risk, we can now use the loss function $h(X, \gamma, \nu_1, \nu_2)$ in place of the "portfolio $f(x, y)$" in [15]. Similarly to [15] we now define *value-at-risk (VaR)* and *conditional value-at-risk (CVaR)* based on this two-sided risk function as follows.

For a given distribution or given data sample of X and the confidence level α, we can obtain the VaR $(\zeta_\alpha(X, \gamma, \nu_1, \nu_2))$ and CVaR $(\phi_\alpha(X, \gamma, \nu_1, \nu_2))$ of $h(X, \gamma, \nu_1, \nu_2)$ as follows:

$$\phi_\alpha(X, \gamma, \nu_1, \nu_2) = \min_\zeta F_\alpha(\zeta, X, \gamma, \nu_1, \nu_2),$$

where $F_\alpha(\zeta, X, \gamma, \nu_1, \nu_2) = \zeta + \frac{1}{1-\alpha} E[h(X, \gamma, \nu_1, \nu_2) - \zeta]^+$, and

$$\zeta_\alpha(X, \gamma, \nu_1, \nu_2) \in \operatorname{argmin}_\zeta F_\alpha(\zeta, X, \gamma, \nu_1, \nu_2).$$

In fact, value-at-risk is the maximal loss the farmer will face with the confidence level $\alpha \in [0, 1]$ and conditional value-at-risk is the mean loss in the $(1 - \alpha)$ worst case of the two-sided risk function $h(X, \gamma, \nu_1, \nu_2)$.

Figure 7.2 below portrays the essence of these concepts in the case where the distribution function of $h(X, \gamma, \nu_1, \nu_2)$ is continuous. Note that the shaded area is on the right (rather than left) tail of the distribution because large values of our two-sided risk function are undesirable.

Based on these definitions, some limit properties of $\zeta_\alpha(X, \gamma, \nu_1, \nu_2)$ and $\phi_\alpha(X, \gamma, \nu_1, \nu_2)$ follow immediately:

$$\lim_{\alpha \to 0} \zeta_\alpha(X, \gamma, \nu_1, \nu_2) = 0, \quad \lim_{\alpha \to 0} \phi_\alpha(X, \gamma, \nu_1, \nu_2) = E[h(X, \gamma, \nu_1, \nu_2)];$$

$$\lim_{\alpha \to 1} \zeta_\alpha(X, \gamma, \nu_1, \nu_2) = \lim_{\alpha \to 1} \phi_\alpha(X, \gamma, \nu_1, \nu_2) = \sup_X h(X, \gamma, \nu_1, \nu_2).$$

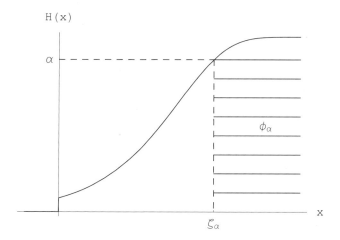

Figure 7.2. α–VaR (ζ_α) and α–CVaR (ϕ_α).

Two-sided risk as an optimization problem

Assume we obtain a sample of observations of X denoted by $x_1, x_2, \cdots, x_N$. After specifying γ, ν_1 and ν_2, using the method provided in [15] we can state the following mathematical programming problem.

$$\min_{\zeta} \; \zeta + \frac{1}{N(1-\alpha)} \sum_{k=1}^{N} u_k$$

subject to

$$m_k \geq 0, \; m_k - \gamma(\nu_1 - x_k) \geq 0, k = 1, \cdots, N;$$

$$n_k \geq 0, \; n_k - (1-\gamma)(x_k - \nu_2) \geq 0, k \doteq 1, \cdots, N;$$

$$u_k \geq 0, \; u_k - m_k - n_k + \zeta \geq 0, k = 1, \cdots, N,$$

where $\zeta; u_1, u_2, \cdots, u_N; m_1, m_2, \cdots, m_N; n_1, n_2, \cdots, n_N$ are the decision variables of this optimization problem.

The optimal objective value of this mathematical program constitutes an estimate - based on the sample - of the conditional value-at-risk of the two-sided risk function $h(X, \gamma, \nu_1, \nu_2)$. Furthermore, the ζ_α^* entry of an optimal solution is an estimate of value-at-risk corresponding to this CVaR.

To explain how the above mathematical program arises, we note that after obtaining the sample $x_1, x_2, \ldots, x_N$ from the r.v. X, the sample mean $\frac{1}{N} \sum_{k=1}^{N} [h(x_k, \gamma, \nu_1, \nu_2) - \zeta]^+$ approximates the nonnegative deviation of the loss from ζ, that is, $E[h(X, \gamma, \nu_1, \nu_2) - \zeta]^+$. Hence, we can use following function to approximate the function $F_\alpha(\zeta, X, \gamma, \nu_1, \nu_2)$ defined above:

$$\tilde{F}_\alpha(\zeta, X, \gamma, \nu_1, \nu_2) = \zeta + \frac{1}{N(1-\alpha)} \sum_{k=1}^{N} [h(x_k, \gamma, \nu_1, \nu_2) - \zeta]^+ =$$

$$\zeta + \frac{1}{N(1-\alpha)} \sum_{k=1}^{N} [\gamma \max\{\nu_1 - x_k, 0\} + (1-\gamma)\max\{x_k - \nu_2, 0\} - \zeta]^+.$$

Thus, instead of minimizing $F_\alpha(\zeta, X, \gamma, \nu_1, \nu_2)$, we try to minimize $\tilde{F}_\alpha(\zeta, X, \gamma, \nu_1, \nu_2)$:

$$\min_\zeta \ \zeta + \frac{1}{N(1-\alpha)} \sum_{k=1}^{N} [\gamma \max\{\nu_1 - x_k, 0\} + (1-\gamma)\max\{x_k - \nu_2, 0\} - \zeta]^+.$$

In terms of auxiliary real variables u_k, m_k and n_k, for $k = 1, \cdots, N$, after setting $u_k = [\gamma \max\{\nu_1 - x_k, 0\} + (1-\gamma)\max\{x_k - \nu_2, 0\} - \zeta]^+$, $m_k = \gamma \max\{\nu_1 - x_k, 0\}$ and $n_k = (1-\gamma)\max\{x_k - \nu_2, 0\}$, the preceding is equivalent to minimizing the linear expression

$$\min_\zeta \ \zeta + \frac{1}{N(1-\alpha)} \sum_{k=1}^{N} u_k$$

subject to the linear constrains as follows:

$$m_k \geq 0, \ m_k - \gamma(\nu_1 - x_k) \geq 0, k = 1, \cdots, N;$$

$$n_k \geq 0, \ n_k - (1-\gamma)(x_k - \nu_2) \geq 0, k = 1, \cdots, N;$$

$$u_k \geq 0, \ u_k - m_k - n_k + \zeta \geq 0, k = 1, \cdots, N.$$

Note that the above linear constraints can be obtained from properties of the function $[x]^+ = \max\{x, 0\}$.

Numerical examples

In the first example we generated 1000 observations with log normal distribution $N(1,2)$, namely, $\log X \sim N(1,2)$, and set $\nu_1 = 15, \nu_2 = 50, \gamma = 0.5$. We obtained $\zeta_{0.95} = 62.93$ and $\zeta_{0.99} = 225.50$. If we choose $\nu_1 = 0.3, \nu_2 = 250$, we will know that $\zeta_{0.95} = 0.28$.

For instance, this means that with probability 0.95, the two-sided risk associated with our hypothetical rainfall r.v. X satisfies

$$P\{h(X, .5, 15, 50) \leq 62.93)\} = P\{.5X_1 + .5X_2 \leq 62.93)\} \geq 0.95.$$

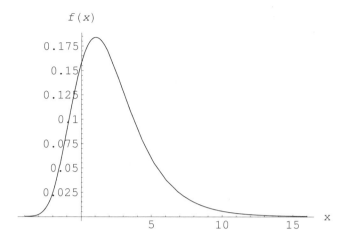

Figure 7.3. Density function of Extreme(1,2).

In the next example we will see how the parameter γ influences this problem. Here we use the asymmetric extreme distribution: $X \sim$ Extreme(1,2), see Figure 7.3.

Again, we obtained $N = 1,000$ observations for which $\max x_i = 16.33$, $\min x_i = -3.25$, range $= 19.59$. We chose $\alpha = 0.9$, then the following results were obtained from the optimization problem above:

$$\zeta_\alpha(X, 0.7, 10, 15) = 7.48, \ \zeta_\alpha(X, 0.3, 10, 15) = 3.21;$$

$$\zeta_\alpha(X, 0.7, 1, 15) = 1.18, \ \zeta_\alpha(X, 0.3, 1, 15) = 0.52.$$

Take $\zeta_\alpha(X, 0.7, 10, 15) = 7.48$ as an example to explain the meaning of value-at-risk here. As before, the following inequality holds:

$$P(h(X, 0.7, 10, 15) \leq 7.48) = P\{.7X_1 + .3X_2 \leq 7.48)\} \geq 0.9.$$

This means that our, unequally weighted, two-sided risk of missing the rainfall interval $[10, 15]$ is less than 7.48 with probability 0.9. Similarly,

$$P(h(X, 0.3, 10, 15) \leq 3.21) = P\{.3X_1 + .7X_2 \leq 3.21)\} \geq 0.9.$$

We can see that, in this instance, the VaR drops sharply as we place less weight on the lower risk X_1.

This shows that the weight γ plays an important role. However, its influence is interconnected with the size and location of the interval $[\nu_1, \nu_2]$ in the domain of the density function of this asymmetric distribution. From Figure 7.4 we see that VaR can both decrease or increase as γ

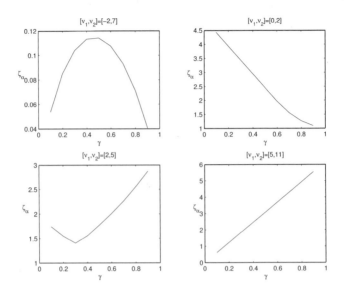

Figure 7.4. Relationship between ζ_α for fixed α and γ for different intervals.

increases, depending on the exact specification of the riskless interval. For instance, as a function of γ, VaR could be concave, convex, linear, or nonlinear.

Of course, with γ and the riskless interval held fixed, VaR and CVaR exhibit the usual dependence on the percentile parameter α. For instance, in the above, with fixed $\gamma = 0.3, \nu_1 = 10, \nu_2 = 15$, we observe the relationship between $\zeta_\alpha, \phi_\alpha$ and α that is displayed in Figure 7.5.

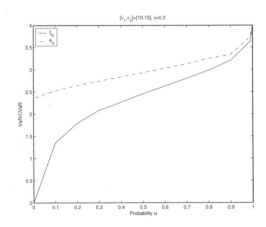

Figure 7.5. Relationship between ζ_α, ϕ_α and α.

2. Two types of risk

Recall that, X_1 was defined as the part of X that is lower than ν_1, and X_2 as the part of X that exceeds ν_2. In the environmental problems that motivated the preceding section it was natural to aim to minimize X_1 and X_2 simultaneously. Hence, a convex combination of X_1 and X_2 was a good choice for that purpose.

However, there are some applications (e.g., the standard financial return) where we have a different requirement with respect to X_1 and X_2. For example, we may want X_1 (a loss below ν_1 threshold) to be small and X_2 (a gain above ν_2 threshold) to be large. In such a case, we require a different analysis of the two tails of the underlying probability distribution. In what follows, we discuss this problem in the special case where $\nu_1 = \nu_2 = 0$. The analysis in the general, $\nu_1 \neq \nu_2$, case can be performed in an analogous manner.

Thus, as before, we begin by considering $X_1 = \max\{-X, 0\}$ and $X_2 = \max\{X, 0\}$, where X_1, X_2 are negative part and positive part of X respectively. In a typical financial market, X_1, X_2 will, respectively, represent the loss and the gain resulting from an investment. However, in this case, we clearly want X_1 to be small whereas X_2 to be large.

Unlike the discussion in Section 1, it will be convenient to deal with the above as two separate, yet interrelated, aspects of the underlying portfolio optimization problem. The essential observation is that in many (most?) situations, investments that increase a probability of a large gain may also increase a probability of large loss. In this sense, the problem is reminiscent of the classical problem of Type I and II errors in Statistics.

Type I and Type II risk

Following the above motivation we define the risk associated with large values of X_1 as *Type I risk*, and the risk associated with small values of X_2 as *Type II risk*.

We note that the above formulation of Type I risk is similar to already standard concepts (e.g., see [16]). In particular, we now briefly recall definitions of VaR and CVaR on X_1. More detailed discussion together with some financial applications will be given in Section 3.

Mathematically, we treat above random variables (r.v.'s) as functions $X : \Omega \to \mathbb{R}$ that belong to the linear space $\mathcal{L}^2 = \mathcal{L}^2(\Omega, \mathcal{F}, P)$, that is, (measurable) functions for which the mean and variance exist.

We denote by $\Psi_1(\cdot)$ on $\mathcal{R}$ the distribution function of X_1 as follows:

$$\Psi_1(\zeta) = P\{X_1 \leq \zeta\}.$$

DEFINITION 7.1 *The value-at-risk (VaR) of the loss X_1 associated with a confidence level α is the functional $\zeta_\alpha : \mathcal{L}^2 \to (-\infty, \infty)$:*

$$\zeta_\alpha(X) := \inf\{\zeta | P\{X_1 \leq \zeta\} \geq \alpha\} = \inf\{\zeta | \Psi_1(\zeta) \geq \alpha\},$$

which shows the maximal loss the investor will face with the confidence level α. That is, $\zeta_\alpha(X)$ is the maximal amount of loss that will be incurred with probability at least α. However, with probability $1 - \alpha$, the loss will be greater than $\zeta_\alpha(X)$, so we will define:

DEFINITION 7.2 *Conditional value-at-risk (CVaR) is the functional $\phi_\alpha : \mathcal{L}^2 \to (-\infty, \infty)$:*

$$\phi_\alpha(X) = \text{mean of the } \alpha - \text{tail distribution of } X_1,$$

where the distribution in question is the one with distribution function $\Psi_{1,\alpha}(\zeta)$ defined by

$$\Psi_{1,\alpha}(\zeta) = \begin{cases} 0, & \zeta < \zeta_\alpha(X), \\ (\Psi_1(\zeta) - \alpha)/(1 - \alpha), & \zeta \geq \zeta_\alpha(X). \end{cases}$$

Since $X_1 = \max\{-X, 0\}$ is a convex function of X, $\phi_\alpha(X)$ defined above is a convex function of X as well.

Similarly, let us denote by $\Psi_2(\cdot)$ on $\mathcal{R}$ the distribution function of X_2 as follows:

$$\Psi_2(\xi) = P\{X_2 \leq \xi\}.$$

DEFINITION 7.3 *The value-of-gain (VoG) of the gain X_2 associated with a assurance level β is the functional $\xi_\beta : \mathcal{L}^2 \to (-\infty, \infty)$:*

$$\xi_\beta(X) = \sup\{\xi | P\{X_2 > \xi\} \geq \beta\} = \sup\{\xi | 1 - \Psi_2(\xi) \geq \beta\},$$

which shows the minimum gain the investor can achieve with a specified assurance level β. That is, $\xi_\beta(X)$ is the minimal amount of gain that will be incurred with probability at least β. However, with probability $1 - \beta$, the gain will be less than $\xi_\beta(X)$, so we will define:

DEFINITION 7.4 *Conditional value-of-gain (CVoG) is the functional $\psi_\beta : \mathcal{L}^2 \to (-\infty, \infty)$:*

$$\psi_\beta(X) = \text{mean of the } \beta - \text{left tail distribution of } X_2,$$

where the distribution in question is the one with distribution function $\Psi_{2,\beta}(\cdot)$ defined by

$$\Psi_{2,\beta}(x) = \begin{cases} \Psi(x)/(1 - \beta), & x \leq \xi_\beta(X), \\ 1, & x > \xi_\beta(X). \end{cases}$$

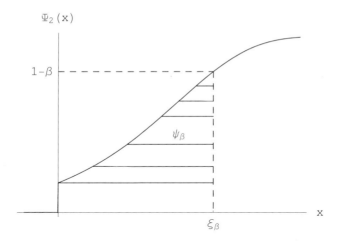

Figure 7.6. $\beta-\text{VoG}$ (ξ_β) and $\beta-\text{CVoG}$ (ψ_β).

Figure 7.6 portrays the essence of these concepts and you could see the differences between CVoG and CVaR.

One of the important properties of conditional value-of-gain, $\psi_\beta(X)$, concave in X, will be proved in Section 3 together with some financial applications.

Two problems and properties of parameters

Investors who want to minimize Type I risk will try to minimize CVaR, and those who want to minimize Type II risk will try to maximize CVoG. However, one of these types of risk will tend to stay high when the other one is minimized. In addition, some parties may want to minimize a combination of Type I and Type II risks. Our discussion below indicates one reasonable approach to these important and difficult problems.

Basically, we are assuming that by choosing a "portfolio" (defined formally in the next section) an investor can select the r.v. X from a family of r.v.'s with known probability distributions. Hence, an "optimal portfolio" may involve solving the following two problems:

Problem I

$$\min_X \phi_\alpha(X) \text{ (minimize the CVaR loss)}$$

Subject to:

$$\psi_\beta(X) \geq \tau. \text{ (guarantee a CVoG gain level of } \tau)$$

Since Conditional Value of Gain, $\psi_\beta(X)$ is a concave function of X, for any real number τ the set $\{X : \psi_\beta(X) \geq \tau\}$ is a convex set. Hence, the above optimization problem is a convex problem, that is, in principle, suitable for fast numerical solution.

Problem II

$$\max_{X} \ \psi_\beta(X), \quad \text{(maximize the CVoG gain)}$$

Subject to:

$$\phi_\alpha(X) \leq v. \quad \text{(tolerable CVaR risk level } v)$$

Since $-\psi_\beta(X)$ is a convex function of the decision variable x and the set $\{X : \phi_\alpha(X) \leq v\}$ is a convex set, above problem is also a convex problem.

In above problems, we have four parameters in total. Those are *confidence level* α, *assurance level* β, *Type I risk tolerance* v and *gain target* τ. Selection of values of these parameters constitutes a characterization of the investor's attitudes towards the "loss versus gain dilemma". However, an intelligent investor will want to select these values on the basis of their interrelationship that are, ultimately, influenced by the probability distribution function of the asset X. The analysis below, should enable such an investor to make an informed decision.

Firstly, we assume that we have chosen and fixed α and β, and in this case, we want to choose τ and v so that above two problems are meaningful and interesting. We shall require following notations:

$$\tau^*(\beta) := \max_{X} \psi_\beta(X), \quad v_*(\alpha) := \min_{X} \phi_\alpha(X),$$

and we shall denote the optimal objective function value of Problems I and II by

$$Z_1(\alpha, \beta, \tau), \quad Z_2(\alpha, \beta, v),$$

respectively. Select $X_*(\alpha) \in \operatorname{argmin}\phi_\alpha(X)$ and let $\psi_\beta(X_*(\alpha)) = \tau_*(\beta, \alpha)$, then we have following lemma.

Ideally, an investor would want a portfolio that is an optimal solution to both Problems I and II. However, in order to achieve this, some adjustments to the target τ (respectively, tolerance level v) maybe needed.

LEMMA 7.5 (1) *If* $(\alpha, \beta) \in \{(\alpha, \beta)|\tau_*(\beta, \alpha)\} = \tau^*(\beta)\}$, *then we can choose* $\tau \leq \tau^*(\beta)$, *and for any such* $\tau, Z_1(\alpha, \beta, \tau) = v_*(\alpha)$.

(2) *If* $(\alpha, \beta) \in \{(\alpha, \beta)|\tau_*(\beta, \alpha) < \tau^*(\beta)\}$, *then a choice of* $\tau \in (\tau_*(\beta, \alpha), \tau^*(\beta)]$ *yields* $Z_1(\alpha, \beta, \tau) > v_*(\alpha)$.

The first case in the Lemma corresponds to the ideal situation since we obtain the maximum gain while at the same time we minimize our risk of loss. However, when the second case occurs, namely the strict inequality holds, perhaps, the best we can do is to choose our gain target level τ in that kind of interval. Of course, we will face a greater risk of loss when we do this.

For Problem II, we can obtain similar conditions for choosing the risk tolerance v. As before, we define the notation: $X^*(\beta) \in \text{argmax}\psi_\beta(X)$, and let $v^*(\alpha, \beta) = \phi_\alpha(X^*(\beta))$, then the following lemma follows immediately:

LEMMA 7.6 (1) *If* $(\alpha, \beta) \in \{(\alpha, \beta)|v_*(\alpha) = v^*(\alpha, \beta)\}$, *then we can choose* $v \geq v_*(\alpha)$, *and for any such* v, $Z_2(\alpha, \beta, v) = \tau^*(\beta)$.

(2) *If* $(\alpha, \beta) \in \{(\alpha, \beta)|v_*(\alpha) < v^*(\alpha, \beta)\}$, *then a choice of* $v \in [v_*(\alpha), v^*(\alpha, \beta))$ *yields* $Z_2(\alpha, \beta, v) < \tau^*(\beta)$.

The first case of this lemma corresponds to the ideal situation where we attain minimum risk of loss while maximizing our gain. But, when the second case happens, that means the risk of loss has not been minimized. We can improve it while trying to maximize our gain but, of course, we will sacrifice part of the gain.

In fact, combining the analysis of Problem I with that of II, we obtain the following equation:

$$\{(\alpha, \beta)|\tau_*(\beta, \alpha) = \tau^*(\beta)\} =$$

$$\{(\alpha, \beta)| \{\text{argmin}_X \phi_\alpha(X)\} \bigcap \{\text{argmax}_X \psi_\beta(X)\} \neq \emptyset\} =$$

$$\{(\alpha, \beta)|v_*(\alpha) = v^*(\alpha, \beta)\}.$$

What we are interested in now is the set

$$\{(\alpha, \beta)| \{\text{argmin}_X \phi_\alpha(X)\} \bigcap \{\text{argmax}_X \psi_\beta(X)\} \neq \emptyset\}.$$

From an investor's point of view, the larger this set is, the better. We experimented with different distributions of X and obtained a range of results. For a symmetric distribution, e.g. normal distribution, it is easy to find parameters μ, σ that permit the above set to be large. However, for asymmetric distributions, it is harder to do so.

In fact, generally speaking, if we denote the distribution function of X by $F(x)$, then the distributions of X_1, X_2 are $(1 - F(-x))I_{x \geq 0}$ and $F(x)I_{x \geq 0}$, respectively. We observe their typical graphs in Figure 7.7.

The next lemma shows that the non-overlapping feature in the left panel of Figure 7.7 always occurs in the case of a symmetric underlying distribution of X.

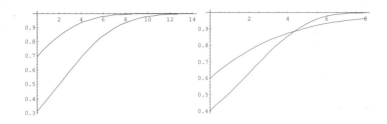

Figure 7.7. Distribution function of the loss and gain, when X is a normal distribution on the left and an extreme distribution on the right.

LEMMA 7.7 *(symmetric property) Assume X is a symmetric random variable with the distribution function $F(x)$. By symmetric with respect to μ, $F(\mu + x) + F(\mu - x) = 1$. Then if $\mu \neq 0$, $F(x) \neq 1 - F(-x)$, in fact, $F(x) < 1 - F(-x)$, when $\mu > 0$; $F(x) > 1 - F(-x)$, when $\mu < 0$ and $F(x) = 1 - F(-x)$, when $\mu = 0$.*

However, after calculating some examples, we found that for asymmetric distributions it is hard to find α, β such that $\{\mathrm{argmin}_X \phi_\alpha(X)\} \cap \{\mathrm{argmax}_X \psi_\beta(X)\} \neq \emptyset$.

Remark: We note that definitions of optimality for Problems I and II could be generalised to "ε-optimality", $\varepsilon > 0$ and (typically) very small. This is because, in practice, investors would be satisfied with portfolios that are only slightly sub-optimal. All of the previous analysis generalizes to this situation in a natural way. For details we refer the reader to Boda's thesis [8].

3. Financial interpretation

In this section, we explicitly apply the analysis of Type I and Type II risk to the portfolio optimization problem and interpret the results. Therefore we concentrate on financial analysis and related optimization algorithms.

Loss function, gain function and Type I risk

We now apply the general concepts of two types of risk to a specific portfolio optimization problem and derive methods to optimize and balance Type I and Type II risks.

Let vector $Y = (Y_1, \cdots, Y_m)^T$ be the random return on m stocks. We define a *portfolio* to be an m-vector $x = (x_1, \cdots, x_m)^T$ such that $x^T e = 1, x \geq 0$. We also define the random *loss function* and a *gain function*, induced by the portfolio $x = (x_1, \cdots, x_m)^T$ as follows:

$$l(x, Y) = \max\{-x^T Y, 0\}, \ g(x, Y) = \max\{x^T Y, 0\}.$$

Note that we are not assuming that the distribution of Y_j's is symmetric. For a portfolio, we believe it is a *loss* if it is negative, otherwise it is a *gain*. For the loss function $l(x, Y)$, we define *value-at-risk (VaR)* similarly to Definition 7.1 or, equivalently, [15].

Namely, for each x, we denote by $L(x, \cdot)$ on $\mathcal{R}$ the distribution function of $l(x, Y)$ as follows:

$$L(x, \varsigma) = P_Y\{l(x, Y) \le \varsigma\}.$$

Next, choose and fix a confidence level $\alpha \in [0, 1]$. The α-VaR of the loss associated with a portfolio x, and the loss function $l(x, Y)$ is the value:

$$\varsigma_\alpha(x) = \min\{\varsigma | L(x, \varsigma) \ge \alpha\},$$

which shows the maximal loss the investor will face with the confidence level α.

Further, we recall that the conditional value-at-risk (CVaR) was defined as:

$$\phi_\alpha(x) = \text{mean of the } \alpha - \text{tail distribution of } Z = l(x, Y),$$

where the distribution in question is the one with distribution function $L_\alpha(x, \cdot)$ defined by

$$L_\alpha(x, \varsigma) = \begin{cases} 0, & \varsigma < \varsigma_\alpha(x), \\ (L(x, \varsigma) - \alpha)/(1 - \alpha), & \varsigma \ge \varsigma_\alpha(x). \end{cases}$$

Analogously to the analysis in [15] we use $l(x, Y) = \max\{-x^T Y, 0\}$ in place of $f(x, Y)$ to define VaR and CVaR. Note that $l(x, Y)$ is convex with x, so if we let

$$F_\alpha(x, \varsigma) = \varsigma + \frac{1}{1 - \alpha} E\{[l(x, Y) - \varsigma]^+\},$$

then following conclusions will hold, by the same arguments as those given in [15].

THEOREM 7.8 *As a function of $\varsigma \in R$, $F_\alpha(x, \varsigma)$ is finite and convex (hence continuous), with*

$$\phi_\alpha(x) = \min_\varsigma F_\alpha(x, \varsigma),$$

and moreover,

$$\zeta_\alpha(x) \in argmin_\zeta F_\alpha(x, \zeta).$$

In particular, one always has:

$$\zeta_\alpha(x) \in argmin_\zeta F_\alpha(x, \zeta), \quad \phi_\alpha(x) = F_\alpha(x, \zeta_\alpha(x)),$$

COROLLARY 7.9 *The conditional value-at-risk, $\phi_\alpha(x)$, is convex with respect to x. Indeed, in this case $F_\alpha(x, \zeta)$ is jointly convex in (x, ζ).*

THEOREM 7.10 *Minimizing $\phi_a(x)$ with respect to $x \in X$ is equivalent to minimizing $F_\alpha(x, \zeta)$ over all $(x, \zeta) \in X \times R$, in the sense that*

$$\min_{x \in X} \phi_\alpha(x) = \min_{(x, \zeta) \in X \times R} F_\alpha(x, \zeta)$$

where moreover,

$$(x^*, \zeta^*) \in argmin_{(x, \zeta) \in X \times R} F_\alpha(x, \zeta) \Longleftrightarrow$$

$$x^* \in argmin_{x \in X} \phi_\alpha(x), \quad \zeta^* \in argmin_{\zeta \in R} F_\alpha(x^*, \zeta).$$

One kind of approximation of $F_\alpha(x, \zeta)$ obtained by sampling the probability distribution of Y. So a sample set $y_1, \cdots, y_N$ of observations of Y yields the approximation function

$$\tilde{F}_\alpha(x, \zeta) = \zeta + \frac{1}{N(1 - \alpha)} \sum_{k=1}^{N} \max\{\max\{-x^T y_k, 0\} - \zeta, 0\}.$$

Because here $l(x, y_k) = \max\{-x^T y_k, 0\}$ is a non-smooth function of x, the formulation of the problem $\min_{(x, \zeta)} \tilde{F}_\alpha(x, \zeta)$ in [15] should be changed to the following linear programming problem:

$$\min \, \zeta + \frac{1}{N(1 - \alpha)} \sum_{k=1}^{N} u_k$$

Subject to:

$$x \geq 0, x^T \mathbf{e} = 1;$$

$$l_k \geq 0, l_k + x^T y_k \geq 0, k = 1, \cdots, N;$$

$$u_k \geq 0, l_k - \zeta - u_k \leq 0, k = 1, \cdots, N.$$

Similarly to arguments in [14], VaR and CVaR corresponding to $l(x, Y)$ can be approximated by the optimizer and the optimal objective function value of the above linear programming problem. Of course, the quality of this approximation increases with the sample size N.

Type II risk

Whereas the preceding analysis of Type I risk was completely analogous to that in [14], when considering properties of Type II risk a few, natural, adjustments need to be made when considering the problem of minimizing the risk associated with the new gain function $g(x, Y) = \max\{x^T Y, 0\}$ failing to take sufficiently large values.

Concept of conditional value-of-gain. For each x, the distribution function of $g(x, Y)$ is defined by: $G(x, \xi) = P_Y\{g(x, Y) \leq \xi\}$. Choose and fix $\beta \in [0, 1]$, the investor's *assurance level*.

DEFINITION 7.11 *The value-of-gain (VoG) associated with a portfolio x and $g(x, Y)$ is the value:*

$$\xi_\beta(x) = \sup\{\xi | P_Y\{g(x, Y) > \xi\} \geq \beta\} = \sup\{\xi | 1 - G(x, \xi) \geq \beta\},$$

which shows the minimum gain the investor can achieve with a specified assurance level β.

However, with probability $1 - \beta$, the gain will be less than $\xi_\beta(x)$, so the following definition is now natural.

DEFINITION 7.12 *Conditional value-of-gain (CVoG):*

$$\psi_\beta(x) = \text{mean of the } \beta - \text{left tail distribution of } Z = g(x, Y),$$

where the distribution in question is the one with distribution function $G_\beta(x, \cdot)$ defined by

$$G_\beta(x, \xi) = \left\{ \begin{array}{ll} G(x, \xi)/(1 - \beta), & \xi \leq \xi_\beta(x), \\ 1, & \xi > \xi_\beta(x). \end{array} \right.$$

The fact that the distribution function of $Z = g(x, Y)$ need not be continuous necessitates the following two additional definitions.

DEFINITION 7.13 *The $\beta - CVoG^+$ ("upper" $\beta - CVoG$) of the gain associated with a decision x is the value:*

$$\psi_\beta^+(x) = E\{g(x, Y) | g(x, Y) \leq \xi_\beta(x)\},$$

whereas the $\beta - CVoG^-$ ("lower" $\beta - CVoG$) of the gain is the value:

$$\psi_\beta^-(x) = E\{g(x, Y) | g(x, Y) < \xi_\beta(x)\}.$$

It is important to differentiate between the cases where the upper and lower conditional values-of-gain coincide, or differ. This is done in the following proposition that is proved in Section 4.

PROPOSITION 7.14 *(Basic CVoG relations). If there is no probability atom at $\xi_\beta(x)$, one simply has:*

$$\psi_\beta^-(x) = \psi_\beta(x) = \psi_\beta^+(x).$$

If a probability atom does exist at $\xi_\beta(x)$, one has:

$$\psi_\beta^-(x) < \psi_\beta(x) = \psi_\beta^+(x), \text{ when } G(x, \xi_\beta(x)) = 1 - \beta,$$

or on the other hand,

$$\psi_\beta(x) = \psi_\beta^+(x), \text{ when } G(x, \xi_\beta(x)) = 0,$$

(with $\psi_\beta^-(x)$ then being ill defined). But in all the remaining cases, we have

$$0 < G(x, \xi_\beta(x)) < 1 - \beta,$$

and one has the strict inequality

$$\psi_\beta^-(x) < \psi_\beta(x) < \psi_\beta^+(x).$$

The next proposition (also proved in Section 4) shows that $\psi_\beta(x) =$ mean of the $\beta -$ left tail distribution of $Z = g(x, Y)$ can be expressed as convex combination of value-of-gain and the upper conditional value-of-gain.

PROPOSITION 7.15 *(CVoG as a weighted average). Let $\lambda_\beta(x)$ be the probability assigned to the gain amount $z = \xi_\beta(x)$ by the $\beta-$ left tail distribution, namely*

$$\lambda_\beta(x) = G(x, \xi_\beta(x))/(1 - \beta) \in [0, 1].$$

If $G(x, \xi_\beta(x)) > 0$, so there is a positive probability of a gain less than $\xi_\beta(x)$, then

$$\psi_\beta(x) = \lambda_\beta(x)\psi_\beta^+(x) + [1 - \lambda_\beta(x)]\xi_\beta(x),$$

with $\lambda_\beta(x) < 1$. However, if $G(x, \xi_\beta(x)) = 0$, $\xi_\beta(x)$ is the lowest gain that can occur (and thus $\lambda_\beta(x) = 0$ but $\psi_\beta^-(x)$ is ill defined), then

$$\psi_\beta(x) = \xi_\beta(x).$$

For a gain function in finance, following [15], we can easily derive some useful properties of CVoG as a measure of risk with significant advantages over VoG. For a discrete distribution and a "scenario case", the following results will illustrate a method to estimate VoG and CVoG from historical data.

PROPOSITION 7.16 *(CVoG for scenario models). Suppose the probability measure P is concentrated on finitely many points y_k of Y, so that for each $x \in X$ the distribution of the gain $Z = g(x, Y)$ is likewise concentrated on finitely many points, and $G(x, \cdot)$ is a step function with jumps at those points. Fixing x, let those corresponding gain points $z_k := g(x, y_k)$ be ordered as $z_1 < z_2 < \cdots < z_N$, with the probability of z_k being $p_k > 0$. For any fixed assurance level $\beta \in [0, 1]$, let k_β be the unique index such that*

$$\sum_{k=1}^{k_\beta} p_k \leq 1 - \beta < \sum_{k=1}^{k_\beta+1} p_k.$$

The $\beta-$VoG of the gain is given by

$$\xi_\beta(x) = z_{k_\beta},$$

whereas the $\beta-$CVoG of the gain is given by

$$\psi_\beta(x) = \frac{1}{1 - \beta} \left[\sum_{k=1}^{k_\beta} p_k z_k + \left(1 - \beta - \sum_{k=1}^{k_\beta} p_k \right) z_{k_\beta} \right].$$

Furthermore, in this situation the weight from Proposition 7.15 is given by

$$\lambda_\beta(x) = \frac{1}{1 - \beta} \sum_{k=1}^{k_\beta} p_k.$$

COROLLARY 7.17 *(Lowest gain). In the notation of Proposition 7.16, if z_1 is the lowest point with probability $p_1 > 1 - \beta$, then $\psi_\beta(x) = \xi_\beta(x) = z_1$.*

Maximization rule and coherence. We can define the function

$$H_\beta(x, \xi) := \xi - \frac{1}{1 - \beta} E\{[g(x, Y) - \xi]^-\}$$

that enables us to determine CVoG and VoG from solutions of an appropriate optimization problem formulated in the next theorem. The proof (given in Section 4) is inspired by the line of argument used to prove somewhat similar results in [14, 15].

THEOREM 7.18 *As a function of $\xi \in R$, $H_\beta(x, \xi)$ is finite and concave (hence continuous), with*

$$\psi_\beta(x) = \max_\xi H_\beta(x, \xi),$$

and moreover,

$$\xi_\beta(x) \in argmax_\xi H_\beta(x, \xi).$$

In particular, one always has:

$$\psi_\beta(x) = H_\beta(x, \xi_\beta(x)).$$

COROLLARY 7.19 *(Concavity of CVoG) If $g(x, y)$ is convex with respect to x, then $\psi_\beta(x)$ is concave with respect to x. Indeed, in this case $H_\beta(x, \xi)$ is jointly concave in (x, ξ).*

THEOREM 7.20 *Maximizing $\psi_\beta(x)$ with respect to $x \in X$ is equivalent to maximizing $H_\beta(x, \xi)$ over all $(x, \xi) \in X \times R$, in the sense that*

$$\max_{x \in X} \psi_\beta(x) = \max_{(x,\xi) \in X \times R} H_\beta(x, \xi).$$

Moreover,

$$(x^*, \xi^*) \in argmax_{(x,\xi) \in X \times R} H_\beta(x, \xi) \iff$$

$$x^* \in argmax_{x \in X} \psi_\beta(x), \ \zeta^* \in argmax_{\zeta \in R} H_\beta(x^*, \xi).$$

It is also possible, and interesting, to optimize an arbitrary portfolio performance function subject to a number of gain-assurance level constraints. This is summarized in the next theorem.

THEOREM 7.21 *(Gain-shaping with CVoG) Let g be any objective function chosen on X. For any selection of assurance levels β_i and corresponding target levels $\tau_i, i = 1, \cdots, l$, the problem*

minimize $g(x)$ over $x \in X$ satisfying $\psi_{\beta_i}(x) \geq \tau_i$ for $i = 1, \cdots, l$,

is equivalent to the problem

minimize $g(x)$ over $(x, \xi_1, \cdots, \xi_l) \in X \times \mathcal{R} \times \cdots \times \mathcal{R}$

satisfying $H_{\beta_i}(x, \xi_i) \geq \tau_i$, for $i = 1, \cdots, l$.

Indeed, $(x^*, \xi_1^*, \cdots, \xi_l^*)$ solves the second problem if and only if x^* solves the first problem and the inequalities $H_{\beta_i}(x^*, \xi_i^*) \geq \tau_i$, hold for $i = 1, \cdots, l$.

Moreover one then has $\psi_{\beta_i}(x^*) \geq \tau_i$ for every i, and actually $\psi_{\beta_i}(x^*) = \tau_i$ for each i such that $H_{\beta_i}(x^*, \xi_i^*) = \tau_i$ (i.e., those that correspond to active CVoG constraints).

Based on Theorem 7.18, we can construct the following approximating algorithm for maximizing the CVoG. We assume that Y_k's are i.i.d distributed according to $p(y)$, and a sample of observations from $p(y)$ is denoted by $y_1, y_2, \cdots, y_N$. Maximizing CVoG can then be approximated by the following mathematical programming problem:

$$\max \ \xi - \frac{1}{N(1-\beta)} \sum_{k=1}^{N} u_k$$

subject to

$$x \geq 0, \ x^T \mathbf{e} = 1;$$

$$u_k \geq 0, \ g(x, y_k) - \xi + u_k \geq 0, \ k = 1, \cdots, N.$$

However, since $g(x, y_k) = \max\{x^T y_k, 0\}$ is not a smooth function of x, the above mathematical programming problem is not in a directly tractable form. Hence, we introduce $0-1$ integer variables for each sample point to change the above mathematical programming problem to one that can be solved using mixed integer programming method as follows:

$$\max \ \xi - \frac{1}{N(1-\beta)} \sum_{k=1}^{N} u_k$$

subject to

$$x \geq 0, \ x^T \mathbf{e} = 1;$$

$$n_k \in \{0, 1\}, 0 \leq g_k \leq n_k M, 0 \leq l_k \leq (1 - n_k)M, \ k = 1, \cdots, N;$$

$$g_k - l_k = x^T y, \ k = 1, \cdots, N;$$

$$u_k \geq 0, \ g_k - \xi + u_k \geq 0, \ k = 1, \cdots, N,$$

where, M is a suitably chosen large number. Of course, when the sample size N is large, this means that the computational effort required to solve the above optimality can be prohibitively difficult. Nonetheless, there are now many heuristics for solving large scale integer programming problems.

Table 7.1. Portfolio mean return m

Instrument	Mean return
S&P	0.0101110
Gov.bond	0.0043532
Small cap	0.0137058

Table 7.2. Portfolio variance-covariance matrix V

	S&P	Gov.bond	Small cap
S&P	0.00324625	0.00022983	0.00420395
Gov.bond	0.00022983	0.00049937	0.00019247
Small cap	0.00420395	0.00019247	0.00764097

Examples. Here we use the data from [14] to calculate VaR, CVaR and VoG, CVoG using our new loss and gain function. The data are as follows:

We assume the return of three stocks satisfy a multivariate normal distribution $N(m, V)$. The mean vector and variance-covariance matrix are shown in Table 7.1 and Table 7.2 respectively. We can use these parameters to generate samples that satisfy multivariate normal distribution and then use the samples and the above constructions and mathematical programs to calculate VaR, VoG and optimize CVaR and CVoG.

The calculated VaR, optimized CVaR and the optimal portfolio corresponding to different confidence levels α and the sample size of $10,000$ are shown in Table 7.3. The calculated VoG, optimized CVoG and the optimal portfolio corresponding to different assurance levels β and the sample size of $1,000$ are shown in Table 7.4.

Table 7.3. Results of Minimizing Type I risk (minimizing CVaR) using the new loss function

α	Sample Size	S&P	Gov.bond	Small cap	VaR	CVaR
0.90	10000	0.0784	0.9215	0	0.0193	0.0233
0.95	10000	0.0741	0.9258	0	0.0321	0.0413
0.99	10000	0.0924	0.9075	0	0.0479	0.0532

The results are as might be expected. In particular, we note that when optimizing the portfolio with respect Type I risk, we find that zero weight is allocated to "Small cap", a small weight to "S&P" and a large weight

Table 7.4. Results of Minimizing Type II risk (maximizing CVoG) using the new gain function

β	Sample Size	S&P	Gov.bond	Small cap	VoG	CVoG
0.33	1000	0	0	1	0.0919	0.0210
0.4	1000	0	0	1	0.0575	0.0098
0.6	1000	0.0951	0	0.9048	0.0219	0.0024

to "Gov. bond". This merely reflects the fact that government bonds have a variance that is very close to zero, followed by "S&P", followed by "Small cap". Correspondingly, when optimizing the portfolio with respect to Type II risk, we find that zero weight is allocated to "Gov. bond", and very large weights are allocated to "Small cap". This reflects the fact that the mean return of "Small cap" is the highest while the government bonds have the lowest mean return.

Balancing two types of risks

Returning to the discussion of Section 2, we will continue to consider two problems associated with our loss and gain functions. Since we have seen that an investor who focuses on just Type I risk will obtain very different results to those that would be obtained if Type II risk were of the main concern. However, most investors will be sensitive - albeit, in varying degrees - to both types of risk. Hence, the challenge is to formulate a portfolio optimization problem that captures these dual concerns. Here, we will give two possible formulations of this "risk balancing problem" and illustrate them with an example based on the above data.

Two problems. Basically, we want to solve the following problems:
   Problem I

$$\min_x \phi_\alpha(x), \quad \text{(minimize the CVaR loss)}$$

Subject to:

$$\psi_\beta(x) \geq \tau. \quad \text{(guarantee a CVoG target level of } \tau \text{).}$$

Since conditional value-of-gain, $\psi_\beta(x)$ is a concave function of x, the set $\{x : \psi_\beta(x) \geq \tau\}$ for any real number τ is a convex set. Hence, the above Problem I is a convex programming problem that is, in principle, suitable for fast numerical solution.

Problem II

$$\max_{x} \; \psi_\beta(x), \quad \text{(maximize the CVoG gain)}$$

Subject to:

$$\phi_\alpha(x) \le v. \quad \text{(tolerable CVaR risk level } v\text{)}$$

Since $-\psi_\beta(x)$ is a convex function of the decision variable x and the set $\{x : \phi_\alpha(x) \le v\}$ is a convex set, the above problem is also a convex programming problem.

According to our Theorem 7.21 and Theorem 16 in [15], if we have observations of returns $y_k, k = 1, \cdots, N$ generated by the distribution $p(y)$, then we can change the above two problems to the following Mixed Integer Programming (MIP) problems:

Problem I'

$$\min \; \zeta + \frac{1}{N(1-\alpha)} \sum_{k=1}^{N} \eta_k$$

Subject to:

$$x \ge 0, \; x^T \mathbf{e} = 1;$$

$$n_k \in \{0,1\}, \; 0 \le g_k \le n_k M, \; 0 \le f_k \le (1-n_k)M, \; k = 1, \cdots, N;$$

$$g_k - f_k - x^T y_k = 0, \; k = 1, \cdots, N;$$

$$\eta_k \ge 0, \; f_k - \zeta - \eta_k \le 0, \; k = 1, \cdots, N;$$

$$u_k \ge 0, \; g_k - \xi + u_k \ge 0, \; k = 1, \cdots, N;$$

$$\xi - \frac{1}{N(1-\beta)} \sum_{k=1}^{N} u_k \ge \tau;$$

where, M is a suitably chosen large number.

Problem II'

$$\max \; \xi - \frac{1}{N(1-\beta)} \sum_{k=1}^{N} u_k$$

Subject to:

$$x \ge 0, \; x^T \mathbf{e} = 1;$$

$$n_k \in \{0,1\}, \; 0 \le g_k \le n_k M, \; 0 \le f_k \le (1-n_k)M, \; k = 1, \cdots, N;$$

$$g_k - f_k - x^T y_k = 0, \; k = 1, \cdots, N;$$

$$u_k \ge 0, \; g_k - \xi + u_k \ge 0, \; k = 1, \cdots, N;$$

$$\eta_k \ge 0, \; f_k - \zeta - \eta_k \le 0, \; k = 1, \cdots, N;$$

$$\zeta + \frac{1}{N(1-\alpha)} \sum_{k=1}^{N} \eta_k \leq v;$$

where, M is a suitably chosen large number.

From above, it is easy to see that if we chose α, β, τ and v, we can use the above two optimization problems to calculate two optimized portfolios that capture a given investor's attitude to Type I and Type II risks.

Examples. We still use the data in Table 7.1 and Table 7.2. We let $\alpha = 0.9, \beta = 0.6$ be fixed. Using a sample size of 1000, we first calculate the maximal conditional value-of-gain, $\tau^*(\beta) = \tau^*(0.6) = 0.0024$ and the minimal conditional value-at-risk $v_*(\alpha) = v_*(0.9) = 0.0233$. Then, we choose $\tau \leq \tau^*(0.6)$ and $v \geq v_*(0.9)$ and solve the preceding two mixed integer programming problems. We obtain the following results.

Table 7.5. Results of Problem I

τ	Sample Size	S&P	Gov.bond	Small cap	$Z_1(0.9, 0.6, \tau)$
0.001	1000	0.0809	0.9190	0	0.0249
0.0015	1000	0.0902	0.9097	0	0.0257
0.002	1000	0.2160	0.5468	0.2371	0.0626
0.0023	1000	0.3597	0.2655	0.3746	0.0955

From Table 7.5 we can see that, $Z_1(\alpha, \beta, \tau)$ is very close to $v_*(\alpha)$ when our target level τ is very small, that means low requirement for the gain will yield low risk and vice versa.

Table 7.6. Results of Problem II

v	Sample Size	S&P	Gov.bond	Small cap	$Z_2(0.9, 0.6, v)$
0.024	1000	0.0697	0.9302	0	0.0003
0.030	1000	0.0576	0.8566	0.0857	0.0016
0.060	1000	0.2046	0.5691	0.2262	0.0019
0.1	1000	0.3791	0.2276	0.3931	0.0023

We can see that $Z_2(\alpha, \beta, v)$ is close to $\tau^*(\beta)$ when our risk tolerance v is large from Table 7.6, that shows the relaxation of the risk requirement will lead to a larger gain and vice versa.

134

4.   Proofs

In this section, we will provide proofs of each lemma, proposition, theorem and corollary in above sections.

Proof of Lemma 7.5: If $(\alpha, \beta) \in \{(\alpha, \beta) | \tau_*(\beta, \alpha)\} = \tau^*(\beta)\}$, then $\psi_\beta(X_*(\alpha)) = \tau_*(\beta, \alpha) = \tau^*(\beta) = \max_X \psi_\beta(X)$. So $X_*(\alpha) \in \text{argmax}_X$ $\psi_\beta(X)$, that means for any $X \in \mathcal{L}^2$, we can't find one such that $\psi_\beta(X) > \tau^*(\beta)$, so we should choose $\tau \leq \tau^*(\beta)$, however based on the above relationship, there exists at least one X that will minimize $\phi_\alpha(X)$ and maximize $\psi_\beta(X)$ simultaneously, so in this case, $Z_1(\alpha, \beta, \tau) = v_*(\alpha)$.

In another case when $(\alpha, \beta) \in \{(\alpha, \beta) | \tau_*(\beta, \alpha) < \tau^*(\beta)\}$, that means we will obtain a lower gain $\tau_*(\beta, \alpha)$ when we try to minimize the risk $\phi_\alpha(X)$, but we can't obtain a gain that will exceed $\tau^*(\beta)$, reasonably, we will choose $\tau \in (\tau_*(\beta, \alpha), \tau^*(\beta)]$, however for this τ, there won't exist $X \in \text{argmin}_X \phi_\alpha(X)$ such that $\psi_\beta(X) \geq \tau$, this will yield a higher risk, that is $Z_1(\alpha, \beta, \tau) > v_*(\alpha)$. $\qquad \square$

Proof of Lemma 7.6: Similarly to the proof of Lemma 7.5, we can prove Lemma 7.6. $\qquad \square$

Proof of Lemma 7.7: Since $F(\mu + x) + F(\mu - x) = 1$, so $F(x) = F(\mu - \mu + x) = F(\mu - (\mu - x)) = 1 - F(\mu + (\mu - x)) = 1 - F(2\mu - x)$. It is easy to use the above relationship to prove the lemma together with the monotonicity of the distribution function $F(x)$. $\qquad \square$

Proof of Proposition 7.14: We define:

$$\beta^-(x) = G(x, \xi_\beta(x)^-), \ \beta^+(x) = G(x, \xi_\beta(x)^+).$$

In comparison with the definition of $\psi_\beta(x)$ in Definition 7.12, $\psi_\beta^-(x)$ is the mean of the gain distribution associated with

$$G_\beta^-(x, \xi) = \begin{cases} G(x, \xi)/(1 - \beta^-(x)), & \xi \leq \xi_\beta(x), \\ 1, & \xi > \xi_\beta(x), \end{cases}$$

whereas the $\psi_\beta^+(x)$ value is the mean of the gain distribution associated with

$$G_\beta^+(x, \xi) = \begin{cases} G(x, \xi)/(1 - \beta^+(x)), & \xi \leq \xi_\beta(x), \\ 1, & \xi > \xi_\beta(x). \end{cases}$$

It is easy to see that $\beta^-(x)$ and $\beta^+(x)$ mark the bottom and top of the vertical gap at $\xi_\beta(x)$ for the original distribution function $G(x, \cdot)$ (if a jump occurs there).

The case of there being no probability atom at $\xi_\beta(x)$ corresponds to having $\beta^-(x) = \beta^+(x) = \beta \in (0, 1)$. Then the first equation holds because the distribution functions $G_\beta^-(x, \xi), G_\beta(x, \xi)$ and $G_\beta^+(x, \xi)$ are identical.

When a probability atom exists but $\beta = \beta^+(x)$, we have: $\beta^-(x) < \beta^+(x) < 1$ and thus the second relations. If $\beta^+(x) = 0$, we can nevertheless get the third one since $\beta^-(x) < \beta^+(x) < 1$. Under the alternative of $0 < G(x, \xi_\beta(x)) < 1$, the strict inequalities in the fifth prevail. $\qquad \square$

Proof of Proposition 7.15: According to the definition of CVoG and when $G(x, \xi_\beta(x)) > 0$, we can calculate the mean in the definition directly as follows: for a fixed portfolio x,

$$
\begin{aligned}
\psi_\beta(x) &= \int_0^{\xi_\beta(x)} d(G(x,\xi)/(1-\beta)) + (1 - G(x,\xi_\beta(x))/(1-\beta))\xi_\beta(x) \\
&= \frac{G(x,\xi_\beta(x))}{1-\beta} \times \frac{\int_0^{\xi_\beta(x)} dG(x,\xi)}{G(x,\xi_\beta(x))} + \left(1 - \frac{G(x,\xi_\beta(x))}{1-\beta}\right)\xi_\beta(x) \\
&= \frac{G(x,\xi_\beta(x))}{1-\beta} E\{g(x,Y)|g(x,Y) \le \xi_\beta(x)\} + \left(1 - \frac{G(x,\xi_\beta(x))}{1-\beta}\right)\xi_\beta(x) \\
&= \frac{G(x,\xi_\beta(x))}{1-\beta} \psi_\beta^+(x) + \left(1 - \frac{G(x,\xi_\beta(x))}{1-\beta}\right)\xi_\beta(x),
\end{aligned}
$$

so we can obtain the following equation after defining $\lambda_\beta(x) = \frac{G(x,\xi_\beta(x))}{(1-\beta)}$,

$$
\psi_\beta(x) = \lambda_\beta(x)\psi_\beta^+(x) + [1 - \lambda_\beta(x)]\xi_\beta(x).
$$

We know $\lambda_\beta(x) \in [0,1]$ since $0 \le G_\beta(x, \xi_\beta(x)) \le 1 - \beta$. If $G(x,\xi_\beta(x)) = 0$, $\xi_\beta(x)$ is the lowest gain that can occur (and thus $\lambda_\beta(x) = 0$ but $\psi_\beta^-(x)$ is ill defined), then $\psi_\beta(x) = \xi_\beta(x)$. $\qquad \square$

Proof of Proposition 7.16: According to the following relationship:

$$
\sum_{k=1}^{k_\beta} p_k \le 1 - \beta < \sum_{k=1}^{k_\beta+1} p_k,
$$

we have

$$
G(x,\xi_\beta(x)) = \sum_{k=1}^{k_\beta} p_k, \quad G(x,\xi_\beta(x)^-) = \sum_{k=1}^{k_\beta-1} p_k,
$$

$$
G(x,\xi_\beta(x)) - G(x,\xi_\beta(x)^-) = p_{k_\beta}.
$$

The assertions then follow from Definition 7.12 and Proposition 7.15. $\square$

Proof of Corollary 7.17: This amounts to the special case in Proposition 7.16 with $k_\beta = 0$, then we know $\psi_\beta(x) = \xi_\beta(x) = z_1$. $\qquad \square$

Proof of Theorem 7.18: Firstly, we will prove the Theorem in the case that the distribution function $G(x,\xi)$ of the gain $g(x,Y)$ for fixed x is everywhere continuous with respect to ξ. We also assume the random return Y has desity function $p(y)$. Before proceeding main steps, we will give a lemma for preparation.

LEMMA 7.22 *With x fixed, let $Q(\xi) = \int_{y \in R^n} q(\xi, y) p(y) dy$, where $q(\xi, y) = [g(x, y) - \xi]^-$. Then Q is a convex continuously differentiable function with derivative*

$$Q'(\xi) = G(x, \xi).$$

Proof: This lemma follows from Proposition 2.1 of Shapiro and wardi (1994) in [20].

Now, let's prove the Theorem in this particular case. In view of the defining formula for $H_\beta(x, \xi)$,

$$H_\beta(x, \xi) = \xi - \frac{1}{1 - \beta} E\{[g(x, Y) - \xi]^-\},$$

it is immediate from Lemma 7.22 and the fact that linear function is a concave function that $H_\beta(x, \xi)$ is concave and continuously differentiable with derivative

$$\frac{\partial}{\partial \xi} H_\beta(x, \xi) = 1 - \frac{1}{1 - \beta} G(x, \xi).$$

Therefore, the values of ξ that furnish the maximum of $H_\beta(x, \xi)$ are precisely those for which $G(x, \xi) = 1 - \beta$. They form a nonempty closed interval, inasmuch as $G(x, \xi)$ is continuous and nondecreasing in ξ with limit 1 as $\xi \to \infty$ and limit 0 as $\xi \to -\infty$. This further yields the validity of the formula $\xi_\beta(x) \in \text{argmax}_\xi H_\beta(x, \xi)$. In particular, then, we have

$$\max_{\xi \in R} H_\beta(x, \xi) = H_\beta(x, \xi_\beta(x)) = \xi_\beta(x) - \frac{1}{1 - \beta} \int_{y \in R} [g(x, y) - \xi_\beta(x)]^- p(y) dy.$$

But the integral here equals

$$\int_{g(x,y) \leq \xi_\beta(x)} [\xi_\beta(x) - g(x, y)] p(y) dy = \xi_\beta(x) \int_{g(x,y) \leq \xi_\beta(x)} p(y) dy -$$

$$\int_{g(x,y) \leq \xi_\beta(x)} g(x, y) p(y) dy,$$

where the first integral on the right is by definition $G(x, \xi_\beta(x)) = 1 - \beta$ and the second is $(1 - \beta)\psi_\beta(x)$. Thus,

$$\max_{\xi \in R} H_\beta(x, \xi) = \xi_\beta(x) - \frac{1}{1 - \beta}[(1 - \beta)\xi_\beta(x) - (1 - \beta)\psi_\beta(x)] = \psi_\beta(x).$$

This confirms the formula for $\beta-$ CVoG, $\psi_\beta(x) = \max_\xi H_\beta(x, \xi)$, and completes the proof of the Theorem in this special case.

In the following, I'll prove it in a more general sense, including the discreteness of the distribution.

The finiteness of $H_\beta(x, \xi)$ is a consequence of our assumption that $E\{|g(x,y)|\} < \infty$ for each $x \in X$. It's concave follows at once from the convexity of $[g(x,y) - \xi]^-$ with respect to ξ. Similar to convex function, a finite concave function, $H_\beta(x, \xi)$ has finite right and left derivatives at any ξ. The following approach of proving the rest of the assertions in the theorem will rely on first establishing for these one-sided derivatives, the formulas,

$$\frac{\partial^+ H_\beta}{\partial \xi}(x, \xi) = \frac{1 - \beta - G(x, \xi)}{1 - \beta}, \quad \frac{\partial^- H_\beta}{\partial \xi}(x, \xi) = \frac{1 - \beta - G(x, \xi^-)}{1 - \beta}.$$
$$(7.1)$$

We start by observing that

$$\frac{H_\beta(x, \xi') - H_\beta(x, \xi)}{\xi' - \xi} = 1 - \frac{1}{1 - \beta} E\left\{ \frac{[g(x, Y) - \xi']^- - [g(x, Y) - \xi]^-}{\xi' - \xi} \right\}.$$

When $\xi' > \xi$, we have:

$$\frac{[g(x, Y) - \xi']^- - [g(x, Y) - \xi]^-}{\xi' - \xi} \begin{cases} = 0 & \text{if } g(x, Y) \geq \xi' \\ = 1 & \text{if } g(x, Y) \leq \xi \\ \in (0, 1) & \text{if } \xi < g(x, Y) < \xi' \end{cases}$$

Since $P_Y\{\xi < g(x, Y) \leq \xi'\} = G(x, \xi') - G(x, \xi)$, this yields the existence of a value $\rho(\xi, \xi') \in [0, 1]$ for which

$$E\left\{ \frac{[g(x, Y) - \xi']^- - [g(x, Y) - \xi]^-}{\xi' - \xi} \right\} = G(x, \xi) + \rho(\xi, \xi')[G(x, \xi') - G(x, \xi)].$$

Since furthermore $G(x, \xi') \searrow G(x, \xi)$ as $\xi' \searrow \xi$, it follows that

$$\lim_{\xi' \searrow \xi} E\left\{ \frac{[g(x, Y) - \xi']^- - [g(x, Y) - \xi]^-}{\xi' - \xi} \right\} = G(x, \xi).$$

So, we obtain

$$\lim_{\xi' \searrow \xi} \frac{H_\beta(x, \xi') - H_\beta(x, \xi)}{\xi' - \xi} = 1 - \frac{1}{1 - \beta} G(x, \xi) = \frac{1 - \beta - G(x, \xi)}{1 - \beta},$$

thereby verifying the first formula in (7.1). For the second formula in (7.1), we argue similarly that when $\xi' < \xi$ we have

$$\frac{[g(x, Y) - \xi']^- - [g(x, Y) - \xi]^-}{\xi' - \xi} \begin{cases} = 0 & \text{if } g(x, Y) \geq \xi \\ = 1 & \text{if } g(x, Y) \leq \xi' \\ \in (0, 1) & \text{if } \xi' < g(x, Y) < \xi \end{cases}$$

where $P_Y\{\xi' < g(x,Y) < \xi\} = G(x,\xi^-) - G(x,\xi')$. Since $G(x,\xi') \nearrow G(x,\xi^-)$ as $\xi' \nearrow \xi$ we obtain

$$\lim_{\xi' \nearrow \xi} E\left\{ \frac{[g(x,Y) - \xi']^- - [g(x,Y) - \xi]^-}{\xi' - \xi} \right\} = G(x,\xi^-),$$

and then

$$\lim_{\xi' \nearrow \xi} \frac{H_\beta(x,\xi') - H_\beta(x,\xi)}{\xi' - \xi} = 1 - \frac{1}{1 - \beta} G(x,\xi^-) = \frac{1 - \beta - G(x,\xi^-)}{1 - \beta}.$$

That gives the second formula in (7.1).

Because of concavity, the one-sided derivatives in (7.1) are non-increasing with respect to ξ, with the formulas assuring that

$$\lim_{\xi \to \infty} \frac{\partial^+ H_\beta}{\partial \xi}(x,\xi) = \lim_{\xi \to \infty} \frac{\partial^- H_\beta}{\partial \xi}(x,\xi) = -\frac{\beta}{1 - \beta}$$

and on the other hand,

$$\lim_{\xi \to -\infty} \frac{\partial^+ H_\beta}{\partial \xi}(x,\xi) = \lim_{\xi \to -\infty} \frac{\partial^- H_\beta}{\partial \xi}(x,\xi) = 1.$$

On the basis of these limits, we know that the level set $\{\xi | H_\beta(x,\xi) \geq c\}$ are bounded (for any choice of $c \in R$) and therefore that the maximum in the theorem is attained, with the argmax set being a closed, bounded interval. The values of ξ in that set are characterized as the ones such that

$$\frac{\partial^+ H_\beta}{\partial \xi}(x,\xi) \leq 0 \leq \frac{\partial^- H_\beta}{\partial \xi}(x,\xi).$$

According to the formulas in (7.1), they are the values of ξ satisfying $G(x,\xi^-) \leq 1 - \beta \leq G(x,\xi)$. The rest of the Theorem is a direct conclusion of the above results. □

Proof of Corollary 7.19: The joint concavity of $H_\beta(x,\xi)$ in (x,ξ) is an elementary consequence of the definition of this function, the relationship between convexity and concavity and the convexity of the function $(x,\xi) \to [g(x,y) - \xi]^-$ when $g(x,y)$ is convex in x. The concave of $\psi_\beta(x)$ in x follows immediately then from the maximization formula in Theorem 7.18. (In convex analysis, when a convex function of two vector variables is minimized with respect to one of them, the residual is a convex function of the other: see Rockafellar [13]. Here, we can obtain the result just simply applying the above theory to the convex function $-H_\beta(x,\xi)$.) □

Proof of Theorem 7.20: This rests on the principle in optimization that maximization with respect to (x,ξ) can be carried out by maximizing with respect to ξ for each x and then maximizing the residual with respect to x. In the situation at hand, we invoke Theorem 7.18 and in particular, in order to get the equivalence in the second formula in the Theorem, the fact there that the maximum of $H_\beta(x, \xi)$ in ξ (for fixed x) is always attained. $\square$

Proof of Theorem 7.21: This relies on the maximization formula in Theorem 7.18 and the assured attainment of the maximum there. The arguments are very much like that for Theorem 7.20. Because $\psi_\beta(x) = \max_\xi H_\beta(x, \xi)$, we have $\psi_{\beta_i}(x) \geq \tau_i$, if and only if there exists ξ_i such that $H_{\beta_i}(x, \xi_i) \geq \tau_i$. $\square$

References

[1] C. Acerbi, C. Nordio and C. Sirtori, Expected Shortfall as a Tool for Financial Risk Management, working paper, 2001.

[2] C. Acerbi and D. Tasche, On the coherence of Expected Shortfall, *Journal of Banking and Finance*, **26 (7)**, 1487–1503, 2002.

[3] P. Artzner, F. Delbaen, J. M. Eber and D. Heath, Coherent Measures of Risk, *Mathematical Finance*, **9**, 203–227, 1999.

[4] S. Basak and A. Shapiro, Value-at-Risk Based Risk Management: Optimal Policies and Asset Prices, *The Review of Financial Studies*, **14(2)**, 371–405, 2001.

[5] D. Bertsimas, G. J. Lauprete and A. Samarov, Shortfall as a Risk Measure: Properties Optimization and Application. *Journal of Economic Dynamics and Control*, **28**, 1227–1480, 2004.

[6] G. A. Holton, Value-at-risk: theory and practice. Amsterdam; Boston: Academic Press, 2003.

[7] P. Kall and S. W. Wallace, Stochastic Programming, Wiley, Chichester, 1994.

[8] Boda Kang, Measures of risk: issues of time consistency and surrogate processes, PhD thesis, University of South Australia, 2006.

[9] P. Krokhmal, J. Palmquist and S. Uryasev, Portfolio Optimization with Conditional Value-At-Risk Objective and Constraints, *The Journal of Risk*, **4 (2)**, 11-27, 2002.

[10] N. Larsen, H. Mausser and S. Uryasev, Algorithms for Optimization of value-at-risk, *P. Pardalos and V.K. Tsitsiringos, (Eds.) Financial Engineering, e-Commerce and Supply Chain*, Kluwer Academic Publishers, 129–157, 2002.

[11] H. Markowitz, Portfolio Selection: Efficient Diversification of Investment, J.Wiley and Sons., 1959.

[12] G. C. Pflug, Some Remarks on the Value-at-Risk and the Conditional Value-at-Risk, *Probabilistic Constrained Optimization: Methodology and Applications (S.P. Uryasev, ed.)*, Kluwer, 278–287, 2000.

[13] R. T. Rockafellar, Convex Analysis, Princeton University Press, available since 1997 in paperback in the series Princeton Landmarks in Mathematics and Physics, 1997.

[14] R. T. Rockafellar and S. Uryasev, Optimization of conditional value-at-risk, *Journal of Risk*, **2**, 21–42, 2000.

[15] R. T. Rockafellar and S. Uryasev, Conditional value-at-risk for general loss distributions, *Journal of Banking & Finance*, **26**, 1443–1471, 2002.

[16] R.T. Rockafellar and S. Uryasev, Deviation Measures in Risk Analysis and Optimization, *Research Report 2002-7*, Dept. of Industrial and Systems Engineering, University of Florida, 2002.

[17] R.T. Rockafellar, S. Uryasev and M. Zabarankin, Generalized Deviations in Risk Analysis. To appear on *Finance and Stochastics*, **9**, 2005.

[18] R.T. Rockafellar, S. Uryasev and M. Zabarankin, Optimality Conditions in Portfolio Analysis with Generalized Deviation Measures, *Mathematical Programming*, accepted for publication, 2005.

[19] R.T. Rockafellar, S. Uryasev and M. Zabarankin, Master Funds in Portfolio Analysis with General Deviation Measures, *The Journal of Banking and Finance*, Vol.29, 2005.

[20] A. Shapiro, Y. Wardi, Nondifferentiability of the steady-state function in discrete event dynamic systems, *IEEE Transactions on Automatic Control*, **39**, 1707-1711, 1994.

Chapter 8

OPTIMAL PRODUCTION POLICY IN A STOCHASTIC MANUFACTURING SYSTEM

Yongjiang Guo

Academy of Mathematics and System Sciences
The Chinese Academy of Sciences
Beijing 100080, P.R. China

yjguo@amss.ac.cn

Hanqin Zhang

Academy of Mathematics and System Sciences
The Chinese Academy of Sciences
Beijing 100080, P.R. China

hanqin@amt.ac.cn

Abstract   This paper is concerned with the optimal production planning in a dynamic stochastic manufacturing system consisting of a single or parallel machines that are failure prone and facing a constant demand. The objective is to choose the production rate over time to minimize the long-run average cost of production and surplus. The analysis is developed by the infinitesimal perturbation approach. The infinitesimal perturbation analysis and identification algorithms are used to estimate the optimal threshold value. The asymptotically optimal threshold value and the convergence rate of the identification algorithms are obtained. Furthermore, the central limit theorem of the identification algorithms is also established.

Keywords: Manufacturing system; perturbation analysis; stochastic approximation; truncated Robbins-Monro algorithm.

1.   Introduction

We consider the problem of a dynamic stochastic manufacturing system consisting of a single or parallel machines that must meet the demand for its products at a minimum long-run average cost of production and surplus. The stochastic nature of the system is due to the machine that are failure-prone. The machine capacity is assumed to be a finite-state Markov chain. There is a considerable literature devoted to the production planning for manufacturing systems with the long-run average cost criterion, see Bielecki and Kumar (1988) and Sethi et al. (2004). Bielecki and Kumar deal with a single machine (only with two sates: up and down), single product problem with linear surplus cost. Because of the simple structure of their problem, they are able to obtain the optimal production policy which is of threshold. Furthermore, the explicit threshold value is given. When their problem is generalized to convex costs, explicit solutions are no longer possible. As a result, some generalizations of the Bielecki-Kumar problem such as those by Sharifnia (1988), and Liberopoulos and Hu (1995) are only heuristic in nature. Sethi et al. develop appropriate dynamic programming equations to rigorously prove that the optimal production policy is in the class of threshold policies. However, they cannot get the explicit threshold values.

In this paper, the infinitesimal perturbation is used to analyze the threshold value. Formally, combining infinitesimal perturbation analysis (IPA) with the truncated Robbins-Monro algorithm gives the stochastic approximation to estimate the derivative of the long-run average cost function with respect to the threshold value. This estimation, in sequence, implies an approximation for the optimal threshold value. A similar procedure used in this paper appears in Tang and Boukas (1999), however, they are constitutionally different. In Tang and Boukas (1999), they are absorbed in the manufacturing system consisting of one machine with two states–up and down producing one part type. Based on the work in Hu and Xiang (1993) which proves the equivalence between the queueing and a manufacturing system with two states–up and down, the problem in Tang and Boukas (1999) is solved in the context of queueing theory. When the machine has more than two states, the equivalence given by Hu and Xiang (1993) does not hold. Using the asymptotic analysis for Markov chains (see Sethi et al. (2004), and Yin and Zhang (1997)), we directly prove the convergence of the truncated Robbins-Monro algorithm, and hence get the approximation for the optimal threshold value. Furthermore we obtain the asymptotical normality of the approximation.

The plan of the paper is as follows. In Section 2, we introduce the problem and specify required assumptions. Section 3 is devoted to study the approximation of the optimal threshold value including the strong consistence and convergence rate. Section 4 concludes the paper.

2.   Problem Formulation

We consider a product manufacturing system with stochastic production capacity and constant demand for its production over time. In order to specify the model, let $x(t)$, $u(t)$ and d denote, respectively, the surplus level, the production rate, and the positive constant demand rate at time $t \in \mathcal{R}_+ = [0, +\infty)$. Here, surplus refers to inventory when $x(t) \geq 0$ and backlog when $x(t) < 0$. We assume that $x(t) \in \mathcal{R} = (-\infty, +\infty)$ and $u(t) \in \mathcal{R}_+$. The system dynamic behavior is given by

$$\frac{\mathrm{d}x(t)}{\mathrm{d}t} = u(t) - d, \quad x(0) = x. \tag{8.1}$$

Let $M(t)$ represent the maximum production capacity of the system at time t. We assume that $M(\cdot)$ is a Markov process defined on a probability space $(\Omega, \mathcal{F}, P)$ with a finite state space $\mathcal{M} = \{0, 1, \cdots, m\}$. The representation for $\mathcal{M}$ stands usually, but not necessarily, for the case of m identical machines, each with a unit capacity and having two states–up and down. This is not an essential assumption. In general, $\mathcal{M}$ could be any finite set of nonnegative numbers representing production capacities in the various states of the system. Of course, the production rate $u(t)$ must satisfy the constraint $0 \leq u(t) \leq M(t)$. Just as what we mentioned in the introduction, we only consider the following hedging point control policy:

$$u(\theta, t) = \begin{cases} M(t), & x(t) < \theta, \\ d \wedge M(t), & x(t) = \theta, \\ 0, & \text{otherwise} \end{cases} \tag{8.2}$$

for all $t \geq 0$, where θ is the threshold value. Under control $u(\theta, t)$, we denote the inventory level process by $x(\theta, t)$. The problem is to find θ that minimizes the cost function

$$J(\theta) = \lim_{t \to \infty} \frac{1}{t} \mathsf{E} \int_0^t [H(x(\theta, s)) + C(u(\theta, s))] \, \mathrm{d}s, \tag{8.3}$$

where $H(\cdot)$ defines the cost of inventory/shortage, and $C(\cdot)$ is the production cost.

We impose the following assumptions on the Markov process $M(\cdot)$ and the cost functions $H(\cdot)$ and $C(\cdot)$ through the paper.

(A1) Let the $(m+1) \times (m+1)$ matrix $Q = (q_{ij})$ be the generator of Markov process $M(\cdot)$. We assume that Q is strongly irreducible in the following sense: The system of equations

$$(\nu_0, \nu_1, \cdots, \nu_m)Q = 0 \quad \text{and} \quad \sum_{i=0}^m \nu_i = 1$$

has a unique solution $(\nu_0, \nu_1, \cdots, \nu_m)$ with $\nu_i > 0, i = 0, 1, \cdots, m$. Furthermore, the average capacity level $\bar{m} = \sum_{i=0}^{m} i\nu_i > d$.

(A2) $H(\cdot)$ is a nonnegative continuously differentiable convex function with $H(0) = 0$. Let $h(z) = \frac{dH(z)}{dz}$. Furthermore, there exist positive constants C_h and K_h such that for any $z_1, z_2 \in \mathcal{R}$,

$$|h(z_1) - h(z_2)| \leq C_h \cdot (|z_1|^{K_h - 1} + |z_2|^{K_h - 1}) \cdot |z_1 - z_2|.$$

(A3) $C(\cdot)$ is a nonnegative continuously differentiable convex function defined on interval $[0, m]$. $C(0)$ is assumed to be zero. Let $c(z) = \frac{dC(z)}{dz}$.

For any $\theta \in \mathcal{R}$, if $x(\theta, 0) = \theta$, it follows from (8.2) that

$$x(\theta, t) = x(0, t) + \theta, \quad t \geq 0. \tag{8.4}$$

This simple relationship implies that $J(\theta)$ is a convex function of threshold value θ. In fact, for $r \in [0, 1]$ and $\theta_1, \theta_2 \in \mathcal{R}$, in view of Assumptions (A2) and (A3),

$$rJ(\theta_1) + (1 - r)J(\theta_2)$$
$$\geq \lim_{t \to \infty} \frac{1}{t} \mathsf{E} \int_0^\infty [rH(x(\theta_1, s)) + (1 - r)H(x(\theta_2, s)) + rC(u(\theta_1, s))$$
$$+ (1 - r)C(u(\theta_2, s))] \, ds$$
$$\geq \lim_{t \to \infty} \frac{1}{t} \mathsf{E} \int_0^\infty [H(rx(\theta_1, s) + (1 - r)x(\theta_2, s)) + C(ru(\theta_1, s)$$
$$+ (1 - r)u(\theta_2, s))] \, ds$$
$$= \lim_{t \to \infty} \frac{1}{t} \mathsf{E} \int_0^\infty [H(x(r\theta_1 + (1 - r)\theta_2, s)) + C(u(r\theta_1 +$$
$$(1 - r)\theta_2, s))] \, ds$$
$$= J(r\theta_1 + (1 - r)\theta_2),$$

this gives the convexity of $J(\cdot)$.

3. Approximation for the Optimal Threshold Value

Since $J(\theta)$ is convex, there is an optimal threshold value θ^0. In what follows, we need to estimate θ^0. The estimation is started with the derivative of the long-run average cost function $J(\theta)$ with respect to the threshold value θ. To get the derivative of $J(\theta)$, let

$$
\begin{aligned}
T_0^*(\theta) &= 0, \\
T_i^*(\theta) &= \{t > T_{i-1}^*(\theta) : x(\theta, t) = 0, M(t-) \geq d, M(t) < d\}, \\
\tau_i^*(\theta) &= T_i^*(\theta) - T_{i-1}^*(\theta), \quad i = 1, 2, \cdots.
\end{aligned}
$$

Since the cost function $J(\theta)$ is independent of the initial state, without loss of generality, we assume that $x(\theta, 0) = \theta$ and $M(0) < d$. It follows from (8.4) that $T_i^*(\theta)$ and $\tau_i^*(\theta)$ do not depend on the threshold value θ (in following, we simply write them as T_i^* and τ^* respectively). For a fixed θ, $\{(x(\theta, t), M(t)),\ t > 0\}$ is a regenerative process, where the regeneration points or regeneration times are $\{T_i^*, i \geq 1\}$. Moreover, $\{\tau_i^*, i = 1, 2, \cdots\}$ is an iid sequence. Define

$$J_t(\theta) = \frac{1}{t} \int_0^t [H(x(\theta, x)) + C(u(\theta, x))]\ \mathrm{d}s.$$

By Assumptions (A2) and (A3), it follows that with probability one

$$\frac{\mathrm{d}J_t(\theta)}{\mathrm{d}\theta} = \frac{1}{t} \int_0^t [h(x(\theta, s)) + c(u(\theta, s))]\ \mathrm{d}s. \qquad (8.5)$$

$\frac{\mathrm{d}J_t(\theta)}{\mathrm{d}\theta}$ is called the infinitesimal perturbation analysis (IPA) derivative of $J_t(\theta)$. In practice, we use $\frac{\mathrm{d}J_t(\theta)}{\mathrm{d}\theta}$ as an estimate for $\frac{\mathrm{d}J(\theta)}{\mathrm{d}\theta}$. One concern is whether $\frac{\mathrm{d}J(\theta)}{\mathrm{d}\theta} = \lim_{t \to \infty} \frac{\mathrm{d}J_t(\theta)}{\mathrm{d}\theta}$ a.s., which is called the strong consistency of the IPA derivative estimate, see Ho and Cao (1991), and Yan, Yin and Lou (1994). If $\mathsf{E}\tau_1^* < \infty$, along the same lines as in Glasserman (1991) and by Theorem 3.6.1 in Ross (1996), we can prove the following lemma. $\mathsf{E}\tau_1^* < \infty$ will be given by Lemma 3.2.

LEMMA 3.1 *If Assumptions (A1)–(A3) hold, then the IPA derivative estimate (8.5) is strongly consistent and*

$$\frac{\mathrm{d}J(\theta)}{\mathrm{d}\theta} = \frac{1}{\mathsf{E}[\tau_1^*]} \mathsf{E} \int_0^{\tau_1^*} [h(x(\theta, s)) + c(u(\theta, s))]\ \mathrm{d}s. \qquad (8.6)$$

In view of the convexity of $J(\cdot)$, we know that to find an optimal threshold value θ^0, it suffices to find the zero of $J(\theta)$. Using Lemma 3.1, we only need to find the zero of $\mathsf{E} \int_0^{\tau_1^*} [h(x(\theta, s)) + c(u(\theta, s))]\ \mathrm{d}s$. Next we design a sequence of hedging point policies to approximate this zero value. Let $T_0 = 0$, $\theta_0 = 0$ and $\{a_n, n \geq 0\}$ be a sequence of positive numbers with $a_n \geq a_{n+1}$ and $\lim_{n \to \infty} a_n = 0$. Define

$$x^{(1)}(t) = 0 + \int_{T_0}^t [u(\theta_0, s) - d]\ \mathrm{d}s, \qquad (8.7)$$

$$T_1 = \inf\{t > T_0 : x(\theta_0, t) = \theta_0, M(t-) \geq d, M(t) < d\}, \qquad (8.8)$$

$$f_1 = \int_{T_0}^{T_1} [h(x(\theta_0, s)) + c(u(\theta_0, s))]\ \mathrm{d}s, \qquad (8.9)$$

$$\hat{\theta}_1 = \theta_0 - a_0 f_1, \tag{8.10}$$

$$\theta_1 = \hat{\theta}_1 I_{\{|\hat{\theta}_1|<b\}} - b I_{\{\hat{\theta}_1 \leq -b\}} + b I_{\{\hat{\theta}_1 \geq b\}}, \tag{8.11}$$

$$\varepsilon_1 = f_1 - [\mathsf{E}\tau_1^*] \cdot f(\theta_0), \tag{8.12}$$

where $f(\theta) = \frac{dJ(\theta)}{d\theta}$, and $b > 0$ is an arbitrary constant with $\theta^0 \in [-b, b]$. Assume that $\{x^{(k)}(t), t \geq T_{k-1}, 1 \leq k \leq n\}$ and $\{(\theta_k, T_k), 0 \leq k \leq n\}$ have been defined, where θ_n is considered as the nth estimate for θ^0. For $t \geq T_n$, define

$$x^{(n+1)}(t) = x(T_n) + \int_{T_n}^t [u(\theta_n, s) - d] \, ds, \tag{8.13}$$

$$T_{n+1} = \inf\{t > T_n : x^{(n+1)}(t) = \theta_n,$$
$$M(t-) \geq d, M(t) < d\}, \tag{8.14}$$

$$f_{n+1} = \int_{T_n}^{T_{n+1}} \left[h(x^{(n+1)}(s)) + c(u(\theta_n, s)) \right] \, ds, \tag{8.15}$$

$$\hat{\theta}_{n+1} = \theta_n - a_n f_{n+1}, \tag{8.16}$$

$$\theta_{n+1} = \hat{\theta}_{n+1} I_{[|\hat{\theta}_{n+1}|<b]} - b I_{[\hat{\theta}_{n+1} \leq -b]} + b I_{[\hat{\theta}_{n+1} \geq b]}, \tag{8.17}$$

$$\varepsilon_{n+1} = f_{n+1} - [\mathsf{E}\tau_1^*] \cdot f(\theta_n). \tag{8.18}$$

θ_{n+1} is the $(n+1)$th estimate for θ_0, f_{n+1} is the $(n+1)$th IPA derivative estimate for $[\mathsf{E}\tau_1^*] \cdot f(\theta_n)$, and ε_{n+1} is the estimate noise. The above algorithm to approximate the optimal threshold value θ^0 is a truncated Robbins-Monro algorithm, a_n is the step-size of the Robbins-Monro algorithm, see Robbins and Monro (1951). Furthermore, $\{T_n, n \geq 1\}$ is a sequence of stopping times with respect to $M(\cdot)$. Based on $\{x^{(n)}(t), t \geq T_{n-1}, n \geq 1\}$, define $\hat{x}(t)$ as

$$\hat{x}(t) = x^{(n)}(t) \quad t \in [T_{n-1}, T_n), \quad n = 1, 2, \cdots. \tag{8.19}$$

As a standard procedure in the Robbins-Monro algorithm, we introduce conditions:

(A4) $a_n > 0$, $\lim_{n \to \infty} n a_n = \alpha \neq 0$, where α is a constant.

(A5) $0 \leq a_{n+1}^{-1} - a_n^{-1} \to \beta > 0$, $as\ n \to \infty$.

(A6) As $\theta \to \theta^0$, $f(\theta)$ can be expressed as $f(\theta) = \gamma \cdot (\theta - \theta^0) + O(|\theta - \theta^0|^2)$ for some constant $\gamma < -\beta/(2[\mathsf{E}\tau_1^*])$, where β is defined in Assumption (A5).

LEMMA 3.2 *Let Assumptions* (A1) *and* (A4)–(A5) *hold, then for any* $r > 0$ *and* $n \geq 0$, $\mathsf{E}[(T_{n+1} - T_n)^r] \leq \gamma_r$ *where* $\gamma_r < \infty$ *is a constant only depending on* r. *In particular*, $\mathsf{E}[\tau_1^*]^r < \gamma_r$.

Proof. The proof consists of two steps.

Case 1. $\theta_n \geq \theta_{n-1}$. For any $\triangle \geq 0$, define

$$\hat{T}_{n+1}(\triangle) = \inf\{t > T_n : \int_{T_n}^t [M(s) - d]\, ds = \triangle\}, \qquad (8.20)$$

$$\tilde{T}_{n+1} = \inf\{t \geq \hat{T}_{n+1}(\theta_n - \theta_{n-1}) : M(t) < d\}. \qquad (8.21)$$

It follows from the definition of T_n that

$$T_{n+1} - T_n = \tilde{T}_{n+1} - T_n = [\tilde{T}_{n+1} - \hat{T}_{n+1}(\theta_n - \theta_{n-1})] + [\hat{T}_{n+1}(\theta_n - \theta_{n-1}) - T_n].$$

Noting that $\theta_n - \theta_{n-1} < 2b$ and Using Lemma 3.1 and Corollary 3.1 in Sethi et al. (2004), there exists a constant γ_r such that

$$\mathsf{E}[\tilde{T}_{n+1} - \hat{T}_{n+1}(\theta_n - \theta_{n-1})]^r \leq \gamma_r \quad \text{and} \quad \mathsf{E}[\hat{T}_{n+1}(\theta_n - \theta_{n-1}) - T_n]^r \leq \gamma_r.$$

This completes the proof.

Case 2. $\theta_n < \theta_{n-1}$. In this case, it follows from the definition of T_{n+1} that

$$T_{n+1} - T_n = \inf\left\{t \geq T_n + \frac{\theta_{n-1} - \theta_n}{d} : x^{(n+1)}(t) = \theta_n, \right.$$
$$\left. M(t) < d \quad \text{and} \quad M(t-) \geq d\right\} - T_n.$$

Using again Lemma 3.1 and Corollary 3.1 in Sethi et al. (2004) we have the lemma.

Finally, $\mathsf{E}[\tau_1^*]^r < \gamma_r$ directly follows from $\tau_1^* = T_1$. Q.E.D.

THEOREM 3.1 *Suppose that Assumptions* (A1)–(A5) *hold, then*

$$\sum_{n=1}^{\infty} a_n^{1-\delta} \varepsilon_{n+1}$$

converges a.s., where δ is a constant in $[0,\ 1/2)$.

Proof. Let

$$\tau_{n+1} = \inf\left\{t > T_n : \int_{T_n}^t [M(s) - d]\, ds \geq 0 \quad \text{and} \quad M(t) < d\right\} - T_n,$$

and

$$f_{n+1}(\theta_n) = \int_{T_n}^{T_n + \tau_{n+1}} [h(x^*(\theta_n, s)) + c(u^*(\theta_n, s))]\, ds, \qquad (8.22)$$

where $u^*(\theta_n, \cdot)$ is given by (8.2) with $\theta = \theta_n$, and

$$\frac{\mathrm{d}x^*(\theta_n, s)}{\mathrm{d}s} = u^*(\theta_n, s) - d, \quad x^*(\theta_n, T_n) = \theta_n.$$

Using (8.18) and (8.22),

$$\sum_{n=1}^{\infty} a_n^{1-\delta} \varepsilon_{n+1} = \sum_{n=1}^{\infty} a_n^{1-\delta} (f_{n+1}(\theta_n) - [\mathsf{E}\tau_1^*] \cdot f(\theta_n))$$

$$+ \sum_{n=1}^{\infty} a_n^{1-\delta}(f_{n+1} - f_{n+1}(\theta_n)). \qquad (8.23)$$

Let $\mathcal{F}_n = \sigma(M(s) : s \leq T_n)$, then

$$\mathsf{E}\left[f_{n+1}(\theta_n)|\mathcal{F}_n\right] = \mathsf{E}\left[\int_{T_n}^{T_n+\tau_{n+1}} [h(x^*(\theta_n, s)) + c(u^*(\theta_n, s))]\,\mathrm{d}s \,\middle|\, \mathcal{F}_n\right]$$

$$= \left(\mathsf{E}\int_0^{\tau_1^*} [h(x(\theta, s)) + c(u(\theta, s))]\,\mathrm{d}s\right)\bigg|_{\theta=\theta_n}$$

$$= [\mathsf{E}\tau_1^*] \cdot \mathsf{E}\left[f(\theta_n)|\mathcal{F}_n\right].$$

This consequently implies $\{f_{n+1}(\theta_n) - [\mathsf{E}\tau_1^*] \cdot f(\theta_n), \mathcal{F}_n, n \geq 0\}$ is a martingale difference sequence. Note that for $t \in [T_n,\ T_{n+1}]$,

$$|\theta_n| \leq b \text{ and } |x^{(n+1)}(t)| \leq |x(T_n)| + (m \vee d) \cdot (t - T_n).$$

It follows from Assumptions (A2) and (A3) that there exists a positive constant C_1 such that

$$|f_{n+1}(\theta_n)| \leq C_1 \cdot \tau_{n+1} + C_h \left(|\theta_n| + \tau_{n+1} \cdot (m \vee d)\right)^{K_h} \cdot \tau_{n+1}, \quad (8.24)$$

$$[\mathsf{E}\tau_1^*] \cdot |f(\theta_n)| \leq \mathsf{E}\left[C_1 \cdot \tau_1^* + C_h \left(|\theta_n| + \tau_1^* \cdot (m \vee d)\right)^{K_h} \cdot \tau_1^*\right]. \quad (8.25)$$

Hence, from Lemma 3.2, there exist positive constants C_2 and C_3 such that

$$\sum_{n=1}^{\infty} a_n^{2(1-\delta)} \mathsf{E}[(f_{n+1}(\theta_n) - [\mathsf{E}\tau_1^*] \cdot f(\theta_n))^2 | \mathcal{F}_n]$$

$$\leq \sum_{n=1}^{\infty} a_n^{2(1-\delta)} \left(C_2 + C_3 \cdot \mathsf{E}[(\tau_{n+1})^2 + (\tau_{n+1})^{2K_h+2}]\right) < \infty,$$

where Assumption (A4) is applied in the last inequality. By the local convergence theorem of martingales (see Corollary 2.8.5 in Stout (1974)),

$$\sum_{n=1}^{\infty} a_n^{1-\delta}(f_{n+1}(\theta_n) - [\mathsf{E}\tau_1^*] \cdot f(\theta_n)) \quad \text{converges a.s.} \qquad (8.26)$$

Now we consider the second term in the right-hand of (8.23). First we have

$$\sum_{n=1}^{\infty} a_n^{1-\delta}(f_{n+1} - f_{n+1}(\theta_n))$$

$$= \sum_{n=1}^{\infty} a_n^{1-\delta}(f_{n+1} - f_{n+1}(\theta_n)) \cdot I_{\{|\theta_n - \theta_{n-1}| \geq (n-1)^{-\delta_1}\}}$$

$$+ \sum_{n=1}^{\infty} a_n^{1-\delta}(f_{n+1} - f_{n+1}(\theta_n)) \cdot I_{\{|\theta_n - \theta_{n-1}| < (n-1)^{-\delta_1}\}}, \quad (8.27)$$

where δ_1 is a positive constant which will be specified later on. From the definition of θ_n, we have

$$|\theta_{n+1} - \theta_n| \leq a_n \cdot |f_{n+1}|.$$

Similar to (8.24),

$$\begin{aligned}
|f_{n+1}| &\leq C_1 \cdot (T_{n+1} - T_n) + C_h \left[|\theta_n|\right. \\
&\quad \left. + (T_{n+1} - T_n) \cdot (m \vee d)\right]^{K_h} \cdot (T_{n+1} - T_n).
\end{aligned}$$

Using this inequality, we have

$$\begin{aligned}
|\theta_{n+1} - \theta_n| &\leq a_n \cdot \{C_1 \cdot (T_{n+1} - T_n) + C_h \left[|\theta_n|\right. \\
&\quad \left. + (T_{n+1} - T_n) \cdot (m \vee d)\right]^{K_h} \cdot (T_{n+1} - T_n)\}. (8.28)
\end{aligned}$$

Using Assumption (A4), Lemma 3.2 and (8.28)

$$\begin{aligned}
&\mathsf{Pr}\{|\theta_{n+1} - \theta_n| \geq n^{-\delta_1}\} \\
&\leq n^{2\delta_1} \mathsf{E}[|\theta_{n+1} - \theta_n|^2] \\
&\leq n^{2\delta_1} a_n^2 \mathsf{E}\{C_1 \cdot (T_{n+1} - T_n) \\
&\quad + C_h \left[|\theta_n| + (T_{n+1} - T_n) \cdot (m \vee d)\right]^{K_h} \cdot (T_{n+1} - T_n)\}^2 \\
&\leq C_4 n^{-2(1-\delta_1)}, \quad (8.29)
\end{aligned}$$

for some $C_4 > 0$. Using Hölder's inequality, for any $r > 1$,

$$\begin{aligned}
&\mathsf{E}\left[\sum_{n=1}^{\infty} a_n^{1-\delta}|f_{n+1} - f_{n+1}(\theta_n)|I_{\{|\theta_n - \theta_{n-1}| \geq (n-1)^{-\delta_1}\}}\right] \\
&= \sum_{n=1}^{\infty} a_n^{1-\delta} \mathsf{E}\left[|f_{n+1} - f_{n+1}(\theta_n)|I_{\{|\theta_n - \theta_{n-1}| \geq (n-1)^{-\delta_1}\}}\right] \\
&\leq \sum_{n=1}^{\infty} a_n^{1-\delta} \left(\mathsf{E}|f_{n+1} - f_{n+1}(\theta_n)|^r\right)^{\frac{1}{r}} \times \\
&\qquad \times \left(\mathsf{Pr}\{|\theta_n - \theta_{n-1}| \geq (n-1)^{-\delta_1}\}\right)^{1-\frac{1}{r}} \\
&\leq \sum_{n=1}^{\infty} a_n^{1-\delta} C_4 \cdot (n-1)^{-2(1-\delta_1)(1-\frac{1}{r})} \times \\
&\qquad \times \left(\mathsf{E}|f_{n+1} - f_{n+1}(\theta_n)|^r\right)^{\frac{1}{r}}. \quad (8.30)
\end{aligned}$$

Using again Lemma 3.2 and (8.24), we know that there exists a positive constant C_5 such that

$$\mathsf{E}|f_{n+1} - f_{n+1}(\theta_n)|^r < C_5.$$

We choose $r = 2$ and $\delta_1 = 1/2$. In view of (8.30), we have

$$\mathsf{E}\left[\sum_{n=1}^{\infty} a_n^{1-\delta}|f_{n+1} - f_{n+1}(\theta_n)|I_{\{|\theta_n-\theta_{n-1}|\geq(n-1)^{-\delta_1}\}}\right] < \infty.$$

This implies that

$$\sum_{n=1}^{\infty} a_n^{1-\delta}(f_{n+1} - f_{n+1}(\theta_n))I_{\{|\theta_n-\theta_{n-1}|\geq(n-1)^{-\delta_1}\}} \quad \text{converges a.s.} \quad (8.31)$$

Now we define

$$\hat{\tau}_{n+1} = \inf\{t > T_n : M(t) \geq d\} - T_n.$$

From the definition of T_n, we know that

$$M(T_n-) \geq d \quad \text{and} \quad M(T_n+) < d.$$

By the Markov property of $M(t)$, we have that

$$\mathsf{Pr}\left(\hat{\tau}_{n+1} < \frac{(n-1)^{-\delta_1}}{d}\right) \leq C_6(n-1)^{-\delta_1},$$

for some constant $C_6 > 0$. Note that

$$\mathsf{E}\sum_{n=1}^{\infty} a_n^{1-\delta}|f_{n+1} - f_{n+1}(\theta_n)| \cdot I_{\{|\theta_n-\theta_{n-1}|<(n-1)^{-\delta_1}\}} \cdot$$
$$\cdot I_{\{\hat{\tau}_{n+1}<(1/d)(n-1)^{-\delta_1}\}}$$
$$\leq \sum_{n=1}^{\infty} a_n^{1-\delta}\mathsf{E}|f_{n+1} - f_{n+1}(\theta_n)| \cdot I_{\{\hat{\tau}_{n+1}<(1/d)(n-1)^{-\delta_1}\}}.$$

Similar to (8.30), using again Hölder's inequality but with $r > (1-2\delta)^{-1}$, we have

$$\mathsf{E}\sum_{n=1}^{\infty} a_n^{1-\delta}|f_{n+1} - f_{n+1}(\theta_n)| \cdot I_{\{|\theta_n-\theta_{n-1}|<(n-1)^{-\delta_1}\}} \cdot$$
$$\cdot I_{\{\hat{\tau}_{n+1}<(1/d)(n-1)^{-\delta_1}\}} < \infty,$$

which gives

$$\sum_{n=1}^{\infty} a_n^{1-\delta}(f_{n+1} - f_{n+1}(\theta_n)) \cdot I_{\{|\theta_n - \theta_{n-1}| < (n-1)^{-\delta_1}\}} \cdot I_{\{\hat{\tau}_{n+1} < (1/d)(n-1)^{-\delta_1}\}}$$

converges a.s. $\hspace{10cm}$ (8.32)

Finally we consider

$$\sum_{n=1}^{\infty} a_n^{1-\delta}(f_{n+1} - f_{n+1}(\theta_n)) \cdot I_{\{|\theta_n - \theta_{n-1}| < (n-1)^{-\delta_1}\}} \cdot I_{\{\hat{\tau}_{n+1} \geq (1/d)(n-1)^{-\delta_1}\}}.$$

From the definitions of T_{n+1} and τ_{n+1}, conditional on

$$\{\theta_{n-1} \geq \theta_n\} \cap \{|\theta_n - \theta_{n-1}| < (n-1)^{-\delta_1}\} \cap \{\hat{\tau}_{n+1} \geq (1/d)(n-1)^{-\delta_1}\},$$

we have

$$T_{n+1} - T_n = \inf\Big\{t : M(t) < d \text{ and}$$
$$\int_{T_n + \frac{\theta_{n-1} - \theta_n}{d}}^t [M(s) - d]\, \mathrm{d}s \geq 0\Big\} - T_n, \quad (8.33)$$

$$\tau_{n+1} = \inf\Big\{t \geq T_n : M(t) < d \text{ and } \int_{T_n + \frac{\theta_{n-1} - \theta_n}{d}}^t [M(s) - d]\, \mathrm{d}s$$
$$\geq -\int_{T_n}^{T_n + \frac{\theta_{n-1} - \theta_n}{d}} [M(s) - d]\, \mathrm{d}s\Big\} - T_n, \quad (8.34)$$

$$|x^{(n+1)}(s) - x^*(\theta_n, s)| \leq (\theta_{n-1} - \theta_n)$$
$$+ \Big|\int_{T_n}^{T_n + \frac{\theta_{n-1} - \theta_n}{d}} [M(z) - d]\, \mathrm{d}z\Big|, \quad s \in [T_n, T_{n+1}), \quad (8.35)$$

$$u(\theta_n, s) = u^*(\theta_n, s), \quad s \in \Big[T_n + \frac{\theta_{n-1} - \theta_n}{d}, \ T_{n+1}\Big). \quad (8.36)$$

Similarly, conditional on

$$\{\theta_{n-1} < \theta_n\} \cap \{|\theta_n - \theta_{n-1}| < (n-1)^{-\delta_1}\} \cap \{\hat{\tau}_{n+1} \geq (1/d)(n-1)^{-\delta_1}\},$$

we have

$$T_{n+1} - T_n = \inf\{t : M(t) < d \text{ and}$$
$$\int_{T_n}^t [M(s) - d]\, \mathrm{d}s \geq \theta_n - \theta_{n-1}\} - T_n, \quad (8.37)$$

$$\tau_{n+1} = \inf\Big\{t : M(t) < d \text{ and } \int_{T_n}^t [M(s) - d]\, \mathrm{d}s \geq 0\Big\} - T_n, \quad (8.38)$$

$$|x^{(n+1)}(s) - x^*(\theta_n, s)| \leq (\theta_n - \theta_{n-1}), \quad s \in [T_n, T_n + \tau_{n+1}), \quad (8.39)$$

$$u(\theta_n, s) = u^*(\theta_n, s), \quad s \in [T_n, T_n + \tau_{n+1}). \quad (8.40)$$

Using the Markov property of $M(t)$, from (8.33), (8.34), and (8.37)-(8.38), we have

$$
\mathsf{E}\left\{[\tau_{n+1} - (T_{n+1} - T_n)] \cdot \right.
$$

$$
\left. \cdot I_{\{\{\theta_{n-1} \geq \theta_n\} \cap \{|\theta_n - \theta_{n-1}| < (n-1)^{-\delta_1}\} \cap \{\hat{\tau}_{n+1} \geq (1/d)(n-1)^{-\delta_1}\}\}}\right\}
$$

$$
\leq C_7(n-1)^{-\delta_1}, \tag{8.41}
$$

$$
\mathsf{E}\left\{[(T_{n+1} - T_n) - \tau_{n+1}] \cdot \right.
$$

$$
\left. \cdot I_{\{\{\theta_{n-1} < \theta_n\} \cap \{|\theta_n - \theta_{n-1}| < (n-1)^{-\delta_1}\} \cap \{\hat{\tau}_{n+1} \geq (1/d)(n-1)^{-\delta_1}\}\}}\right\}
$$

$$
\leq C_7(n-1)^{-\delta_1}, \tag{8.42}
$$

for some constant $C_7 > 0$. Going alone the line of the proof of (8.32), using (8.35)-(8.36) and (8.41), we can prove

$$
\sum_{n=1}^{\infty} a_n^{1-\delta}(f_{n+1} - f_{n+1}(\theta_n)) \cdot
$$

$$
\cdot I_{\{\{|\theta_n - \theta_{n-1}| < (n-1)^{-\delta_1}\} \cap \{\hat{\tau}_{n+1} \geq (1/d)(n-1)^{-\delta_1}\} \cap \{\theta_{n-1} \geq \theta_n\}\}}
$$

converges a.s. $\tag{8.43}$

In a same way, by (8.39)–(8.40) and (8.42),

$$
\sum_{n=1}^{\infty} a_n^{1-\delta}(f_{n+1} - f_{n+1}(\theta_n)) \cdot
$$

$$
\cdot I_{\{\{|\theta_n - \theta_{n-1}| < (n-1)^{-\delta_1}\} \cap \{\hat{\tau}_{n+1} \geq (1/d)(n-1)^{-\delta_1}\} \cap \{\theta_{n-1} < \theta_n\}\}}
$$

converges a.s. $\tag{8.44}$

Combining (8.31), (8.32), and (8.43)-(8.44) yield the theorem.   Q.E.D.

THEOREM 3.2 *If Assumptions (A1)–(A6) hold, then* $\lim_{n\to\infty} \theta_n = \theta^0$ *a.s. and* $|\theta_n - \theta^0| = o(a_n^\delta)$ *a.s. for* $\delta \in [0, \frac{1}{2})$.

Proof. $\lim_{n\to\infty} \theta_n = \theta^0$ follows from Lemma 3.1 and Theorem 2.2.1 in Chen (2002), and the second claim follows from Lemma 3.2 and Theorem 3.1.1 in Chen (2002).   QED.

THEOREM 3.3 *Suppose that*

$$
\mathsf{E}\left[\int_0^{\tau_1^*} (h(x(\theta, s)) + c(u(\theta, s)))\, ds\right]^2
$$

is continuous at θ^0 *and Assumptions (A1)–(A6) hold, then*

$$
\frac{\theta_n - \theta^0}{\sqrt{a_n}} \Rightarrow \mathcal{N}(0, \sigma^2), \quad n \to \infty
$$

where $\mathcal{N}(0, \sigma^2)$ *is a normal random variable with mean zero and variance* σ_1^2, *and*

$$\sigma^2 = \frac{\sigma_1^2}{2\gamma \cdot [\mathsf{E}\tau_1^*] + \beta},$$

$$\sigma_1^2 = \mathsf{E}\left[\int_0^{\tau_1^*} \left(h(x(\theta^0, s)) + c(u(\theta^0, s)) \right) ds\right]^2.$$

Proof. By Theorem 3.2, $|\theta_n - \theta^0| = o(a_n^{1/4})$ a.s. Thus after a finite number of truncations, (8.13)-(8.18) will become the usual Robbins-Monto algorithm (see Chen (2002)), so we can get that there exists an n_0 (may depend on the sample path) such that for $n \geq n_0$,

$$\theta_{n+1} = \theta_n - a_n([\mathsf{E}\tau_1^*] \cdot f(\theta_n) + \varepsilon_{n+1}). \tag{8.45}$$

Using Assumption (A5), we have,

$$\begin{aligned}
(\tfrac{a_n}{a_{n+1}})^{\frac{1}{2}} &= (\tfrac{a_n - a_{n+1}}{a_{n+1}} + 1)^{\frac{1}{2}} \\
&= 1 + \tfrac{1}{2}a_n(a_{n+1}^{-1} - a_n^{-1}) + O((\tfrac{a_n - a_{n+1}}{a_{n+1}})^2) \\
&= 1 + \tfrac{1}{2}\beta a_n + o(a_n).
\end{aligned} \tag{8.46}$$

It follows from (8.45) and (8.46) that for $n \geq n_0$

$$\begin{aligned}
\frac{\theta_n - \theta^0}{\sqrt{a_{n+1}}} &= \varphi_{n,n_0} \frac{\theta_{n_0} - \theta^0}{\sqrt{a_{n_0}}} + \sum_{i=n_o}^{n} \varphi_{n,i+1}\sqrt{a_i}\varepsilon_{i+1} \\
&\quad + \sum_{i=n_o}^{n} \varphi_{n,i+1}a_i\varepsilon_{i+1} \cdot (\tfrac{\beta\sqrt{a_i}}{2} + o(\sqrt{a_i})) \\
&\quad + \sum_{i=n_o}^{n} \varphi_{n,i+1}\sqrt{a_i}(1 + \tfrac{1}{2}\beta a_i + o(a_i)) \cdot O(|\theta_i - \theta^0|^2).
\end{aligned} \tag{8.47}$$

where

$$\varphi_{n,i} = \begin{cases} (1 + a_n A_n) \cdots (1 + a_i A_i), & \text{for } i \leq n \\ 1, & \text{for } i = n+1 \\ 0, & \text{for } i \geq n+2 \end{cases} \tag{8.48}$$

and

$$\begin{aligned}
A_n &= [\mathsf{E}\tau_1^*]\gamma + \tfrac{\beta}{2} + [\mathsf{E}\tau_1^*]\gamma \cdot \tfrac{\beta a_n}{2} + o(1) + [\mathsf{E}\tau_1^*]\gamma \cdot o(a_n) \\
&\to A := [\mathsf{E}\tau_1^*]\gamma + \tfrac{\beta}{2}, \ n \to \infty.
\end{aligned} \tag{8.49}$$

By some tedious algebraic calculations, (see Sections 3.1 and 3.3 of Chen (2002)), there are constants $\lambda_0 > 0, \lambda > 0$, the $\varphi_{n,i}, n \geq 0, i \geq 0$ defined by (8.48) satisfy the following properties:

$$|\varphi_{n,k}| \leq \lambda_0 \exp\{-\lambda \sum_{j=k}^{n} a_j\}, \tag{8.50}$$

$$\sup_n \sum_{i=1}^{n} a_i |\varphi_{n,i}|^r < \infty, \quad r > 0, \tag{8.51}$$

$$\lim_{n \to \infty} \sum_{i=1}^{n} a_i \varphi_{n,i+1}^2 = \int_0^\infty e^{2At} dt = \frac{1}{2[\mathsf{E}\tau_1^*]\gamma + \beta}. \tag{8.52}$$

By (8.50),

$$\varphi_{n,n_0} \frac{\theta_{n_0} - \theta^0}{\sqrt{a_{n_0}}} \quad \text{converges to 0, a.s.} \tag{8.53}$$

For any fixed n_1, by (8.50) again

$$\sum_{i=n_0}^{n_1} \varphi_{n,i+1} a_i \varepsilon_{i+1} \left(\frac{\beta \sqrt{a_i}}{2} + o(\sqrt{a_i}) \right) \to 0, a.s.$$

While for any given $\varepsilon > 0$, we may take n_1 sufficiently large such that for $n > n_1$, $|\varepsilon_{n+1}(\frac{\beta \sqrt{a_n}}{2} + o(\sqrt{a_n}))| < \varepsilon$. Therefore

$$\sum_{i=n_1+1}^{n} a_i |\varphi_{n,i+1}| \cdot |\varepsilon_{i+1} \cdot \left(\frac{\beta \sqrt{a_i}}{2} + o(\sqrt{a_i}) \right)|$$

$$\leq \varepsilon \sum_{i=n_1+1}^{n} a_i |\varphi_{n,i+1}|$$

$$\leq \varepsilon \sup_n \sum_{i=n_1+1}^{n} a_i |\varphi_{n,i+1}| \to 0 \ a.s.$$

Then

$$\sum_{i=n_o}^{n} \varphi_{n,i+1} a_i \varepsilon_{i+1} \cdot \left(\frac{\beta \sqrt{a_i}}{2} + o(\sqrt{a_i}) \right) \quad \text{converges to 0 a.s.} \tag{8.54}$$

For the given n_1 and ε above, we may assume that for $i \geq n_1$

$$\left| (1 + \frac{1}{2}\beta a_i + o(a_i)) \frac{O(|\theta_i - \theta^0|^2)}{\sqrt{a_i}} \right| < \varepsilon$$

because $|\theta_n - \theta^0| = o(a_n^{1/4})$. Analogous to (8.54), we get that

$$\sum_{i=n_o}^{n} \varphi_{n,i+1} \sqrt{a_i} (1 + \frac{1}{2}\beta a_i + o(a_i)) \cdot O(|\theta_i - \theta^0|^2)$$

$$\text{converges to 0, a.s.} \tag{8.55}$$

Hence to complete the proof of theorem, it suffices to prove that

$$\sum_{i=n_0}^{n} \varphi_{n,i+1}\sqrt{a_i}\varepsilon_{i+1} \Rightarrow \mathcal{N}(0,\sigma^2), \quad n \to \infty. \tag{8.56}$$

From (8.50) we can prove that

$$\sum_{i=0}^{n_0-1} \varphi_{n,i+1}\sqrt{a_i}\varepsilon_{i+1} \to 0, \quad a.s. \text{ as } n \to \infty.$$

So to prove (8.56), it suffices to prove that

$$\sum_{i=0}^{n} \varphi_{n,i+1}\sqrt{a_i}\varepsilon_{i+1} \Rightarrow \mathcal{N}(0,\sigma^2), \quad n \to \infty. \tag{8.57}$$

Now we decompose the right side of (8.57) into two parts:

$$\sum_{i=0}^{n} \varphi_{n,i+1}\sqrt{a_i}\varepsilon_{i+1} = \sum_{i=0}^{n} \varphi_{n,i+1}\sqrt{a_i}(f_{i+1}(\theta_i) - [\mathsf{E}\tau_1^*]f(\theta_i))$$

$$+ \sum_{i=0}^{n} \varphi_{n,i+1}\sqrt{a_i}(f_{i+1} - f_{i+1}(\theta_i)).$$

By Lemma 3.3.1 and Theorem 3.3.1 in Chen (2002), and (8.52), we have

$$\sum_{i=0}^{n} \varphi_{n,i+1}\sqrt{a_i}(f_{i+1}(\theta_i) - [\mathsf{E}\tau_1^*]f(\theta_i)) \Rightarrow \mathcal{N}(0,\sigma^2), \quad n \to \infty. \tag{8.58}$$

Therefore, to prove (8.57) we only need to prove

$$\sum_{i=0}^{n} \varphi_{n,i+1}\sqrt{a_i}(f_{i+1} - f_{i+1}(\theta_i)) \text{ converges to zero}$$

in probability as $n \to \infty$. $\tag{8.59}$

We decompose it into two parts as follows:

$$\sum_{i=0}^{n} \varphi_{n,i+1}\sqrt{a_i}(f_{i+1} - f_{i+1}(\theta_i))$$

$$= \sum_{i=0}^{n} \varphi_{n,i+1}\sqrt{a_i}(f_{i+1} - f_{i+1}(\theta_i))I_{\{|\theta_i-\theta_{i-1}|\geq(i-1)^{-\delta_1}\}}$$

$$+ \sum_{i=0}^{n} \varphi_{n,i+1}\sqrt{a_i}(f_{i+1} - f_{i+1}(\theta_i))I_{\{|\theta_i-\theta_{i-1}|<(i-1)^{-\delta_1}\}}.$$

156

Based on this decomposition, going along the same line of (8.27), we can prove (8.59). Therefore, we get the theorem.   QED.

REMARK 3.1 In the Robbins-Monto algorithm defined by (8.13)-(8.18), the integral interval $[T_n, T_{n+1}]$ depends on the threshold value θ_{n-1}. This dependence makes f_{n+1} to be complicated. To avoid this complexity, we can modify T_{n+1} as

$$
\begin{aligned}
T_{n+1} &= \inf\{t > \xi_n : x^{(n+1)}(t) = \theta_n, M(t-) \geq d, M(t) < d\}, \\
\xi_n &= \inf\{t > T_n : x^{(n+1)}(t) = \theta_n, M(t-) \geq d, M(t) < d\}, \\
f_{n+1} &= \int_{\xi_n}^{T_{n+1}} \left[h(x^{(n+1)}(s)) + c(u(\theta_n, s)) \right] \, \mathrm{d}s.
\end{aligned}
$$

Clearly, the interval $[\xi_n, T_{n+1}]$ does not depend on θ_{n-1}. Using this algorithm, going along the same lines, The results on the convergence, the convergence rate and the central limit theorem can be proved.

4.   Concluding Remarks

In this paper we use an infinitesimal perturbation analysis to approximate the optimal threshold value. Specifically, the classic Robbins-Monro algorithm is adopted as an identification algorithm to estimate the optimal threshold value. It is of much interest to examine the complex models such as multiproduct flexible manufacturing system, and flowshop manufacturing system.

Acknowledgments

This research is supported by a Distinguished Young Investigator Grant from the National Natural Sciences Foundation of China, and a grant from the Hundred Talents Program of the Chinese Academy of Sciences.

References

T. Bielecki and P.R. Kumar. (1988). "Optimality of zero-inventory policy for unreliable manufacturing system," *Operations Research,* **Vol.36**, 532-541.

H.F. Chen. (2002). *Stochastic Approximation and Its Applications.* Dordrecht/ Boston/ London: Kluwer Academic Publishers.

P. Glasserman. (1991). *Gradient Estimation via Perturbation Analysis.* Boston, MA, Kluwer.

Y.C. Ho and X.R. Cao. (1991). *Perturbation Analysis of Discrete Event Dynamic Systems.* Boston, MA, Kluwer.

J.Q. Hu and D. Xiang. (1993). "the queueing equivalence to a manufacturing system with failures," *IEEE Transactions on Automatic Control*, **Vol.38**, pp.499-522.

G. Liberopoulos and J.Q. Hu. (1995). "On the ordering of hedging points in a class of manufacturing flow control models," *IEEE Transactions on Automatic Control*, **Vol.40**, 282-286.

H. Robbins and S. Monro. (1951). "A stochastic approximation method," *Ann. Math. Statist.*, **Vol.22**, 400-407.

S.M. Ross. (1996). *Stochastic Processes* John Wiley and Sons,Inc. Second Edition.

S.P. Sethi, H. Zhang and Q. Zhang (2004). *Average-Cost Control of Stochastic Manufacturing Systems*, Springer-Verlag, New York.

A. Sharifnia. (1988). "Production control of a manufacturing system with multiple machine states," *IEEE Transactions on Automatic Control*, **AC-33**, 620-625.

W.F. Stout. (1974). *Almost Sure Convergence*. New York: Academic.

Q.Y. Tang and E.K. Boukas. (1999). "Adaptive control for manufactuing systems using infintesimal perturbation analysis," *IEEE Trans. Automat. Contr.*, **Vol.44**, 1719-1725.

H.M. Yan, G. Yin, and S.X.C. Lou. (1994). "Using stochastic optimization to determine threshold values for control of unreliable manufacturing systems," *Journal of Optimization Theory and Applications*, **Vol.83**, 511-539.

G. Yin and Q. Zhang. (1997). *Continuous–Time Markov Chains and Applications*, Springer.

Chapter 9

A STOCHASTIC CONTROL APPROACH TO OPTIMAL CLIMATE POLICIES

Alain Haurie

Hon. Prof. University of Geneva and Director of ORDECSYS, Place de l'Etrier 4, CH-1224 Chêne-Bougeries, Switzerland

alain.haurie@ordecsys.com

Abstract   The purpose of this paper is to show how a discrete event stochastic control paradigm similar to the one proposed by Suresh Sethi in the realm of finance, economic planning or manufacturing systems can be used to analyze the important issue of optimal timing in global climate change policies. One proposes a stochastic economic growth model to study the optimal schedule of greenhouse gases (GHG) emissions abatement in an economy that can invest in RD&D, in an existing or a future backstop technology and faces uncertainty in both climate and technical progress dynamics.

1. Introduction

In this paper one proposes a stochastic control formalism to address the problem of defining the proper timing of climate mitigation policies in the search for the optimal tradeoff between economic development and long term sustainability. The fundamental economics of climate change has been described by Nordhaus (1994), Nordhaus (1996), Manne & al. (1995), Toth (2003) in a formalism of optimal economic growth à la Ramsey (1928) which is much akin to an infinite horizon optimal control formalism as indicated by Haurie (2003). In these approaches the economic and climate dynamics are represented as deterministic systems coupled in the same integrated assessment model. The economic activity generates GHG[1] emissions; concentrations of GHG produce a radiative

*This work was partly supported by the NCCR-Climate program of the Swiss NSF

forcing that triggers a temperature change; a damage function represents the loss of output due to this temperature change. The optimal policy is thus obtained as a tradeoff between the cost of emissions abatement and the cost of the damages caused by temperature change. However a large uncertainty remains in both the climate sensitivity[2] and the damages caused by climate change.

Another avenue of research consists in developing a cost-effectiveness approach where, based on a precautionary approach, one will impose a global cap on the cumulative emissions of the economy over the whole planning horizon. Indeed this cap should be adapted to the knowledge one gains when time passes concerning the climate sensitivity. In this paper one adopts this modeling approach to study the optimal timing and mixing of GHG abatement policies. This can be summarized by the following questioning: (i) When should the economy make the maximum effort in abating GHG emissions? (ii) What is the optimal mix of abatement vs RD&D effort in the optimal policies? To contribute to answering these questions one proposes a piecewise deterministic control formalism that permits the explicit consideration of different sources of uncertainty in the design of optimal climate policies. The model is in the vein of those discussed by Haurie (2003) and takes its inspiration from the work of Prof. Suresh Sethi and his collaborators in the field of manufacturing, finance and economic planning (Sethi (1997), Sethi & Thompson (2000), Sethi & Zhang (1994)).

The paper is organized as follows. In section 2 one proposes an optimal economic growth model, under a global cumulative emissions constraint. The model allows for uncertainty in the climate sensitivity and in the endogenous technical progress triggered by RD&D investment. In section 3 one reformulates the search for a cost-effective policy as the solution of a control problem with an "isoperimetric" constraint and one gives the dynamic programming conditions satisfied by an optimal policy. In section 4 one proposes an interpretation of the optimality conditions in terms of timing of abatement decisions. In section 3.4.3 one concludes by an indication on a possible implementation of a numerical technique to solve this class of problems.

2. The economic growth model

In this section one proposes an optimal economic growth formalism to model a cost-effective climate policy under uncertainty. One assumes that the decision makers have uncertain knowledge of the true climate sensitivity and that technical progress which will bring new clean (backstop) technologies is also a stochastic process.

2.1   The multi-sector economic growth model

Consider an economy described by an economic growth model similar to the one proposed by Ramsey (1928). It produces an homogenous good that can be either consumed or invested in different types of capital. The capital of type 1, denoted K_1 is the current productive capital; it generates a high amount of emissions e_1 as a by-product, whereas the capital of type 2, denoted K_2 is clean and it generates a much lower amount of emissions e_2 as a by-product. However the second type of capital (also referred to as the "backstop technology") will be available only when a technological breakthrough happens as a result of RD&D investment. let K_3 denote the cumulative "knowledge" capital accumulated in the economy, through this type of investment. Accumulating this knowledge provides two benefits as it increases the probability of acceding to a backstop technology and it also increases the probability of knowing what is the true value of the climate sensitivity. This economic model is summarized below.

Welfare: Discounted sum of utility derived from consumption

$$\int_0^\infty e^{-\rho t} L(t) \log[c(t)]\, dt,$$

where

$$c(t) = \frac{C(t)}{L(t)}$$

denotes per capita consumption.

Population dynamics: Assume an exogenous population growth

$$\dot{L}(t) = g(t)L(t). \tag{9.1}$$

with $g(t) \to 0$ when $t \to \infty$.

Production functions: Assume the following form for the production[3] function

$$
\begin{aligned}
F(t, e, K_1, K_2, L_j) = \ &\max_{e_1, e_2} \Big\{ L^\alpha (A_1(t) e_1^{\beta^1} K_1^{\gamma^1} \\
&+ A_2(t) e_2^{\beta^2} K_2^{\gamma^2}) : e = e_1 + e_2 \Big\}. \tag{9.2}
\end{aligned}
$$

where $A^k(t), k = 1, 2$ are C^1 functions describing the autonomous technical progress, with $\dot{A}^k(t) \geq 0$ and $\lim_{t \to \infty} \dot{A}^k(t) = 0$. The output

$$Y = F(t, e, K_1, K_2, L),$$

thus depends[4] on emissions e, dirty capital K_1 available, clean capital K_2 available and labor L.

Capital dynamics: Capital accumulation is represented by the usual equations with constant depreciation rates

$$\dot{K}_i(t) = I_i(t) - \mu_i K_i(t),$$
$$K_i(0) = K_i^o \quad i = 1, 3. \tag{9.3}$$

For the clean productive capital a binary variable $\xi \in \{0, 1\}$ is introduced which indicates if the clean technology is available ($\xi = 1$) or not yet ($\xi = 0$). The accumulation of clean capital is described by

$$\dot{K}_2(t) = \xi(t) I_2(t) - \mu_2 K_2(t),$$
$$K_2(0) = K_2^0. \tag{9.4}$$

Breakthrough dynamics: The initial value $\xi(0) = 0$ indicates that there is no access to the clean capital at initial time. The switch to the value 1 occurs at a random time which is controlled through the global accumulation of RD&D capital. More precisely one introduces a jump rate function[5] $q_b(t, K_3(t))$ which will serve to determine the elementary probability of a switch

$$P[\xi(t + dt = 1 | \xi(t) = 0, K^3(t)]$$
$$= q_b(t, K^3(t)) \, dt + o(dt). \tag{9.5}$$

Allocation of output: The flexible good can be consumed or invested

$$C = Y - I^1 - I^2 - I^3. \tag{9.6}$$

2.2 The climate sensitivity issue

One considers that the scientific community is arriving at a good understanding of the climate change effect of GHGs. To avoid irreparable damages to the ecology of the planet one will have to impose a constraint on the long term accumulation of GHGs in the atmosphere. However there is still uncertainty about the true value of the climate sensitivity. (Currently values ranging from 1.5 °C to 4.5 °C are considered in climate models.) Therefore the real magnitude of the needed abatement effort may only be known some time in the future. Assume that three possible values[6] are considered at initial time $t_0 = 0$ with a priori probabilities π_ℓ, $\ell = 1, 2, 3$, as shown for example in Table 9.1, with

$$\pi^\ell \geq 0, \quad \sum_{\ell=1,2,3} \pi^\ell = 1.$$

To gain knowledge about the true value of climate sensitivity, the in-

value	probability
1.5 °C	π^1
3 °C	π^2
4.5 °C	π^3

Table 9.1. A priori probabilities of climate sensitivity values

ternational community may develop a research activity by investing in the research capital (laboratories, advanced research programs, etc.). The time θ at which the true climate sensitivity is known, is a random (Markov) time with intensity depending on the accumulated research capital stock

$$P[\theta \in (t, t + dt)|\theta \geq t, K_3(t)] = q_c(t, K_3(t))dt + o(dt) \text{ where}$$
$$\lim_{dt \to 0} \frac{o(dt)}{dt} = 0, \tag{9.7}$$

where the jump rate function $q_c(t, K_3(t))$ is known.

2.3   The long term emissions constraint

Introduce now a long term constraint on cumulative emissions that will limit the economic growth. GHGs are long lived and their effect on climate change is relatively slow compared to the economic dynamics. Therefore one represents the limit to growth in GHG accumulation by the following constraint on the total discounted sum of emissions

$$\int_0^\infty e^{-\rho t} e(t)\, dt \leq \bar{E}^\ell, \quad \ell = 1, 2, 3 \tag{9.8}$$

where $\bar{E}^\ell$ is a given bound corresponding to the climate sensitivity level $\ell = 1, 2, 3$, $e(t)$ represents the total emissions at time t and ρ is a pure time preference rate.

2.3.1   Justification of the discounted sum of emissions.

This extends to an infinite horizon setting the representation of the impact of climate changes on the world economies proposed by Labriet & Loulou (2003) who established a direct link between cumulative emissions and damages. In the present model, the time horizon being infinite one proposes to represent the damage as a function of the total discounted sum of emissions. To provide some justification for the use of such a constraint assume that the damage at time t is a linear function[7] of the total emissions up to time t

$$d(t) = \alpha \int_0^t e(s)\, ds. \tag{9.9}$$

It is also reasonable to assume that the emission rate $e(s)$ is bounded.

Now consider that the planner wants to limit the total discounted damage, represented as

$$D = \int_0^\infty e^{-\rho t} d(t)\, dt = \int_0^\infty e^{-\rho t} \left(\alpha \int_0^t e(s)\, ds \right) dt. \tag{9.10}$$

Integrating (9.10) by parts one obtains

$$D = \left[\alpha \int_0^t e(s)\, ds \times \left(-\frac{e^{-\rho t}}{\rho} \right) \right]_0^\infty + \frac{\alpha}{\rho} \int_0^\infty e^{-\rho t} e(t)\, dt. \tag{9.11}$$

The term between square brackets vanishes and it remains

$$D = \frac{\alpha}{\rho} \int_0^\infty e^{-\rho t} e(t)\, dt. \tag{9.12}$$

Therefore the constraint on the discounted sum of emissions is equivalent to a constraint on the discounted sum of damages.

2.3.2   An almost sure constraint.

It will be convenient to introduce a new state variable $\delta(t)$ with state equation

$$\delta(t) = \int_0^t e^{-\rho s} \bar{e}(s)\, ds. \tag{9.13}$$

or in differential equation form

$$\dot{\delta}(t) = e^{-\rho t} \bar{e}(t), \tag{9.14}$$
$$\delta(0) = 0. \tag{9.15}$$

Because we cannot have $e(t) < 0$ at any time the constraints will be expressed equivalently as

$$\delta(t) \leq E^\ell \quad \forall t \geq 0, \quad \ell = 1, 2, 3. \tag{9.16}$$

Consider a situation where at time θ one knows that the climate sensitivity is of type ℓ whereas $\delta(\theta) = E^\theta$. This means that the future emission path[8] should be such that

$$\int_\theta^\infty e^{-\rho s} e(s)\, ds \leq E^\ell - E^\theta, \tag{9.17}$$

or equivalently, if one makes the change of variable $\tau = s - \theta$

$$\int_0^\infty e^{-\rho\tau} e(\tau)\, d\tau \leq e^{\rho\theta}(E^\ell - E^\theta). \tag{9.18}$$

2.4 The timing issue

The stochastic economic growth model proposed above permits the study of the optimal timing of GHG emissions abatement and of RD&D investment when one imposes a constraint on the accumulated emissions. Indeed the decision variables are the emission rates $e(t)$ and the investment rates $I(t)$ in the different types of capital. The model captures the fundamental tradeoffs between economic development and climate control, immediate action (precautionary principle) and delayed action with better knowledge (wait and see). One recognizes here the terms of the current debate concerning international climate policy.

3. A stochastic isoperimetric control problem

In this section the optimal economic growth problem introduced above is reformulated as a stochastic control problem and the dynamic programming equations characterizing the optimal policy are derived.

3.1 State, controls and policies

Define the state variable

$$s = (\mathbf{K}, \delta, \xi, \zeta)$$

where $\mathbf{K} \in \mathbb{R}^{+3}$ represents the capital stocks, $\delta \in \mathbb{R}^+$ is the total discounted emissions already accounted for, $\xi \in \{0, 1\}$ indicates the eventual availability of the advanced (clean) technology and

$$\zeta \in \{0, 1, 2, 3\}$$

represents the state of knowledge concerning climate sensitivity values (0 indicates initial uncertainty, 1,2,3 indicate knowledge that one of the 3 possible sensitivity values is the true one).

At initial time $t^0 = 0$ the state

$$s^0 = (\mathbf{K}^0, \delta^0, \xi^0, \zeta^0)$$

is such that $K_2^0 = 0$ and $\xi = 0$ since the advanced technology is not yet available, while $\delta^0 = 0$ since one starts accounting for emission, and $\zeta^0 = 0$ as one does not know the exact sensitivity of climate. It will be convenient to introduce a special notation for the continuous state variable $x = (\mathbf{K}, \delta)$ and for the discrete variables $\eta = (\xi, \zeta)$. The control variables are denoted $u = (e, \mathbf{I})$ where

$$\mathbf{I}(t) = (I_i(t))_{i=1,2,3}.$$

One summarizes the capital and emissions accumulation dynamics under the general state equations

$$\dot{x}(t) = f^{\eta(t)}(t, x(t), u(t)) \tag{9.19}$$

Here the dependence on the discrete variable η takes care of the constraint that

$$I_2(t) = 0 \text{ if } \xi(t) = 0.$$

3.2   The performance criterion

Given the state variables and $x = (K_1, K_2, K_3, \delta)$, and the control variable $u = (e, I_1, I_2, I_3)$ the instantaneous utility of consumption is determined. Therefore one can introduce the reward function

$$\tilde{L}(t, x, u) = L \log[(F(t, e, K_1, K_2, L) - I_1 - I_2 - I_3)/L]. \tag{9.20}$$

The controls are subject to the constraints $e(t) \geq 0$, $I_i(t) \geq 0$, $i = 1, 2, 3$ and this is summarized in general notations by $u(t) \in U$.

Consider the sequence of random times τ^0, τ^1, τ^2 where $\tau^0 = 0$ is the initial time and τ^1, τ^2 are the jump times of the

$$\eta(\cdot) = (\xi(\cdot), \zeta(\cdot))$$

process. Denote s^0, s^1, s^2 the state observed at jump times τ^0, τ^1, τ^2. A policy γ is a mapping that associates with a jump time value τ and an observed state s a control $u(\cdot) : [\tau, \infty) \to U^{\eta(t)}(s)$ that will be used until the next jump occurs. This corresponds to the concept of piecewise deterministic control. Associated with a policy γ and an initial state s^0 there is an expected reward defined by

$$J(\gamma; s^0) = \mathrm{E}_\gamma \left[\int_0^\infty e^{-\rho t} \tilde{L}(t, x(t), u(t)) \, dt \right], \tag{9.21}$$

where the expectation is taken w.r.t. the probability measure induced by the policy γ.

3.3 Optimal admissible policies

A policy γ is admissible if it generates a control which satisfies almost surely the emission constraints (9.8). Due to this last type of constraints one may call this problem a *stochastic isoperimetric problem*.

An optimal policy maximizes the expected reward (9.21) among all the admissible policies. When $\zeta(t) = 1, 2, 3$ the climate sensitivity is known and the constraint is a standard isoperimetric constraint. When $\zeta(t) = 0$ there is uncertainty on climate sensitivity. Because the constraints must be satisfied almost surely we must therefore impose that $\delta(t) \leq \min_{\ell=1,2,3} E^\ell$ as long as $\zeta(t) = 0$.

An elegant way to take into account this constraint is to define an extended reward function

$$\mathcal{L}^{\eta(t)}(t, x(t), u(t)) = \begin{cases} \tilde{L}(t, x(t), u(t)) & \text{when } \delta(t) < E^{\zeta(t)} \\ -\infty & \text{when } \delta(t) \geq E^{\zeta(t)}, \end{cases} \quad (9.22)$$

when $\zeta(t) = 1, 2, 3$. One thus introduces, by the way of a nondifferentiable reward an infinite penalty of having over-emitted when the climate sensitivity is known.

3.4 The dynamic programming equations

One writes the dynamic programming equations by considering the jump times of the discrete jump process $\eta(\cdot)$. At each jump time, given the state reached at that jump, one defines the optimization problem which determines the *reward-to-go* value function. According to this model formulation there are 3 jump times, including the initial time.

3.4.1 After the last jump. Assume that the last jump occurs at time τ^2 and that the cumulated emissions up to that time are given by δ. From time τ^2 onwards there is no more uncertainty, one knows the true climate sensitivity and the new advanced technology is available. The emission schedule $e(\cdot) : [\tau^2, \infty) \to \mathbb{R}^+$ is therefore subject to the constraint

$$\int_{\tau^2}^{\infty} e^{-\rho t} e(t) \, dt \leq \bar{E}^\ell - \delta. \quad (9.23)$$

At time τ^2, given the state

$$s^2 = (\eta^2, x^2) = (1, \ell, K_1^2, K_2^2, K_3^2, \delta^2)$$

let us define the value function

$$V_{1,\ell}^2(x^2)$$

168

as the solution of the optimization problem

$$V_{1,\ell}^2(x^2) = \max_{u(\cdot)} e^{\rho \tau^2} \int_{\tau^2}^\infty e^{-\rho t} \mathcal{L}^{(1,\ell)}(t, x(t), u(t))\, dt \qquad (9.24)$$

subject to the state equations

$$\dot{x}(t) = f^{(1,\ell)}(t, x(t), u(t)) \quad u(t) \in U \quad t \geq \tau^2; \quad x(\tau^2) = x^2. \qquad (9.25)$$

Note that the isoperimetric constraint is taken care of by the use of the extended reward $\mathcal{L}^{(1,\ell)}(x, u)$. Hence the value function $V_{1,\ell}^2(x^2)$ is itself defined over $\mathbb{R} \cup -\infty$, with

$$V_{1,\ell}^2(x^2) = -\infty \quad \text{if } \delta^2 \geq E^\ell.$$

3.4.2   After the first jump.   The first jump occurs at time τ^1. At this jump time the discrete state can switch from $(0,0)$ to $(1,0)$ which means that the backstop technology becomes available before one knows exactly what the true climate sensitivity is; or it can switch from $(0,0)$ to $(1,\ell)$, where $\ell = 1,2,3$ which means that one learns about the true climate sensitivity before the backstop technology becomes available.

State $s^1 = (1,0,x^1)$.   In the first type of transition, let $s_{1,0}^1 = (K^1, \delta^1, 1, 0)$ be the system's state right after the jump time. The stochastic control problem to solve can be described as follows

$$V_{1,0}^1(x^1) = \max_{u(\cdot)} \mathbb{E}_{K_3(\cdot)} e^{\rho \tau^1} \left[\int_{\tau^1}^{\tau^2} e^{-\rho t} \mathcal{L}^{(1,0)}(t, x(t), u(t))\, dt \right.$$
$$\left. + e^{-\rho \tau^2} V_{1,\zeta(\tau^2)}^2(x(\tau^2)) \right]$$
s.t.
$$\dot{x}(t) = f^{(1,0)}(t, x(t), u(t)) \quad u(t) \in U \quad t \geq \tau^1; \quad x(\tau^1) = x^1.$$

Here the second jump time τ^2 is stochastic. The associated jump rate is $q_c(t, K^3(t))$ at any time $t \geq \tau^1$. One has denoted $s^2(\tau^2)$ the random state reached after the second jump time.

Using standard probability reasoning for this type of problem one obtains the equivalent infinite horizon deterministic control problem

$$V_{1,0}^1(x^1) = \max_{u(\cdot)} e^{\rho \tau^1} \int_{\tau^1}^\infty e^{-\rho t + \int_0^t q_c(s, K^3(s))\, ds} \big(\mathcal{L}^{(1,0)}(t, x(t), u(t))$$
$$+ q_c(t, K^3(t)) \big(\sum_{\ell=1}^3 \pi_\ell V_{1,\ell}^2(x(t)) \big)\, dt \qquad (9.26)$$
s.t.
$$\dot{x}(t) = f^{(1,0)}(t, x(t), u(t)) \quad u(t) \in U \quad t \geq \tau^1; \quad x(\tau^1) = x^1.$$

Since the value functions

$$V_{1,\ell}^2(x), \quad \ell = 1, 2, 3$$

take value in $\mathbb{R} \cup -\infty$ the function $V_{1,0}^1(x^1)$ is also defined on $\mathbb{R} \cup -\infty$ and this will induce the observance of the isoperimetric constraint.

State $s^1 = (0, \ell, x^1)$. In the second type of transition, let

$$s^1 = (0, \ell, \mathbf{K}^1, \delta^1, 0, \ell)$$

be the system's state right after the jump time when the true climate sensitivity ($\ell = 1, 2, 3$) has been revealed. The stochastic control problem to solve can be described as follows

$$
\begin{aligned}
V_{0,\ell}^1(x^1) \quad = \quad & e^{\rho \tau^1} \max_{u(\cdot)} \mathrm{E}_{K^3(\cdot)} \left[\int_{\tau^1}^{\tau^2} e^{-\rho t} \mathcal{L}^{(0,\ell)}(t, x(t), u(t))\, dt \right. \\
& \left. + e^{-\rho \tau^2} V_{1,\varsigma(\tau^2)}^2 (x^2(\tau^2)) \right]
\end{aligned}
$$

s.t.

$$\dot{x}(t) \quad = \quad f^{(0,\ell)}(t, x(t), u(t)) \quad u(t) \in U \quad t \geq \tau^1; \quad x(\tau^1) = x^1.$$

The second jump time τ^2 is still stochastic. The associated jump rate is $q_b(t, K^3(t))$ at any time $t \geq \tau^1$. One has denoted $s^2(\tau^2)$ the random state reached after the second jump time.

Using the same standard reasoning one obtains the equivalent infinite horizon deterministic control problem

$$
\begin{aligned}
V_{0,\ell}^1(x^1) \quad = \quad & \max_{u(\cdot)} e^{\rho \tau^1} \int_{\tau^1}^{\infty} e^{-\rho t + \int_0^t q_c(s, K^3(s))\, ds} \left(\mathcal{L}^{(0,\ell)}(t, x(t), u(t)) \right. \\
& \left. + \; q_b(t, K^3(t)) V_{1,\ell}^2(x(t)) \right) dt
\end{aligned}
\tag{9.27}
$$

s.t.

$$\dot{x}(t) \quad = \quad f^{(0,\ell)}(t, x(t), u(t)) \quad u(t) \in U \quad t \geq \tau^1; \quad x(\tau^1) = x^1.$$

Again, since the functions $V_{1,\ell}^2(x)$ take value in $\mathbb{R} \cup -\infty$ the function $V_{0,\ell}^1(x^1)$ is also defined on $\mathbb{R} \cup -\infty$ and this will induce the observance of the isoperimetric constraint.

3.4.3 At the initial time.
At initial time the discrete state is $(0,0)$, i.e. one does not know the true climate sensitivity and one does not have access to the backstop technology.

$$V_{0,0}^0(x^0) \quad = \quad \max_{u(\cdot)} \mathrm{E}_{K^3(\cdot)} \left[\int_{\tau^1}^{\tau^2} e^{-\rho t} L(t, x(t), u(t))\, dt \right.$$

$$+ e^{-\rho \tau^1} V^1_{\xi(\tau^1)}(x^1(\tau^1))\Big]$$

s.t.

$$\dot{x}(t) \;=\; f^{(0,0)}(t, x(t), u(t)) \quad u(t) \in U \quad t \geq 0; \quad x(0) = x^0.$$

Still the same classical probability reasoning yields the associated infinite horizon control problem,

$$
V^0_{0,0}(x^0) \;=\; \max_{u(\cdot)} \int_0^\infty e^{-\rho t + \int_0^t (q_c(s, K^3(s)) + q_b(s, K^3(s)))\, ds} \Big(L(t, x(t), u(t))
$$
$$
+ \; q_b(t, K^3(t)) V^1_{1,0}(x(t))
$$
$$
+ \; q_c(t, K^3(t)) \sum_{\ell=1}^{3} \pi_\ell V^1_{0,\ell}(x(t)) \Big)\, dt \qquad (9.28)
$$

s.t.

$$\dot{x}(t) \;=\; f^{(0,0)}(t, x(t), u(t)) \quad u(t) \in U \quad t \geq 0; \quad x(0) = x^0.$$

Since the functions $V^1_{0,\ell}(x(t))$ take value in $\mathbb{R} \cup -\infty$ the function $V^0_{0,0}(x^0)$ is also defined on $\mathbb{R} \cup -\infty$ and this will induce the observance of the isoperimetric constraint.

4. Interpretation

The interpretation of the control formulation given above concerns principally the initial time $t = 0$, when one has to implement the first control and which represents the time when one negotiates an international climate policy. It is interesting for that purpose to analyze the components of the integrand of the associated infinite horizon control problem (9.28).

There is first an endogenous, state dependent discount rate

$$
-\rho t + \int_0^t (q_c(s, K^3(s)) + q_b(s, K^3(s)))\, ds.
$$

There is also an extended reward function

$$
L(t, x(t), u(t)) + q_b(t, K^3(t)) V^1_{1,0}(x(t) + q_c(t, K^3(t)) \sum_{\ell=1}^{3} \pi_\ell V^1_{0,\ell}(x(t)
$$

which takes into account the possible futures when there will be a switch in the discrete state (better knowledge of the climate sensitivity or access to an improved technology). The decisions to abate (choice of $e(t)$) or to invest in research (choice of $I_3(t)$) are then dictated by the optimal tradeoff between the different contributions to the reward. The reward introduced in the auxiliary infinite horizon control problem (9.28) gives the correct information to the decision maker in order to produce the optimal timing of actions. This reward depends on the value functions obtains through the solving of other auxiliary infinite horizon control problems, as defined by (9.26), (9.27). These value functions depend on the value function (9.24) obtained by solving the optimal control problem when all uncertainty has disappeared. This shows how the assessment of future utility trickles down to the present and permits the evaluation of the actions to undertake now.

5.   Conclusion

The model proposed in this brief paper contains several elements that appear in the current debate about the implementation of a global climate policy. There is uncertainty about the sensitivity of climate; the solution may reside in the introduction of new carbon free technologies that are not yet available; abating too early may limit unduly the economic development whereas, not abating enough may cause immense damages if the climate sensitivity is high.

The solution of the associated dynamic programming problem shows how these different aspects will be taken into account in the design of an optimal trade-off. The model that has been sketched in this short paper could be parameterized as in the DICE, RICE or MERGE models, to represent the world economy. A numerical solution could be relatively easily obtained, by implementing a policy improvement algorithm. This would imply a repeated solution of many control problems with candidate value functions. The numerical solutions obtained could give some insights on the delicate problem of timing in GHG abatement abatement policies. This could be an interesting contribution of stochastic control theory to an important societal and economic problem.

Notes

1. Greenhouse gas.

2. Recall that the climate sensitivity parameter is the average surface atmospheric temperature change triggered by a doubling of GHG concentration compared with preindustrial level.

3. This choice is based on the interpretation of emissions as production factors. If one reduces emissions for a fixed level of capital one decreases the output level. The marginal productivity of emissions decreases when the emission level increases. The backstop technology is characterized by a much higher productivity of emissions.

4. One makes the usual assumptions concerning the parameters $\beta^i, \gamma^i, i = 1, 2$ which are linked to the marginal productivity of the production factors.

5. We always assume the required regularity, e.g. continuous differentiability w.r.t. time and state.

6. Indeed this number of 3 is chosen for convenience in the exposition.

7. Damages are usually represented as a nonlinear function of the SAT change which is due to the radiative forcing of GHG concentrations (itself a nonlinear function of these concentrations). In Labriet & Loulou (2003) it has been observed that the damage functions used in the literature can be accurately summarized by a linear dependence on the total cumulative emissions.

8. If one considers, from initial time 0 on a constant emission rate that satisfies the constraint one obtains $\bar{e}^\ell = \frac{E^\ell}{\rho}$. From time θ onward, with the above condition at θ, the sustainable constant emission rate would be $\bar{e}^\ell = \frac{e^{\rho\theta}(E^\ell - E^\theta)}{\rho}$.

References

Haurie A., Integrated assessment modeling for global climate change:an infinite horizon viewpoint, *Environmental Modeling and Assesment*, Vol.-6, No. 1, pp 7-34, 2003.

Labriet M. and Loulou R., Coupling climate damages and GHG abatement costs in a linear programming framework, *Environmental Modeling & Assessment*, 8, 261–274, 2003.

A. Manne, R. Mendelsohn, and R. Richels, Merge: A model for evaluating regional and global effects of GHG reduction policies, *Energy Policy*, 1:17–34, 1995.

A. Manne and R. Richels, An Alternative Approach to Establishing Trade-offs Among Greenhouse Gases. *Nature*, 410:675–677, 2001.

Nordhaus W.D., *Managing the Global Commons: The Economics of Climate Change*, MIT Press, Cambridge, Mass., 1994.

Nordhaus W.D. and Yang Z., A regional dynamic general-equilibrium model of alternative climate change strategies, *American Economic Review*, 86, 4, 741–765, 1996.

Ramsey F., A mathematic theory of saving, *Economic Journal*, 38, 543–549, 1928.

Sethi, S.P., *Optimal Consumption and Investment with Bankruptcy*, Kluwer Academic Publishers, Norwell, MA, 1997

Sethi, S.P. and Thompson, G.L., *Optimal Control Theory: Applications to Management Science and Economics*, Second Edition, Kluwer Academic Publishers, Boston, 2000

Sethi, S.P. and Zhang, Q., *Hierarchical Decision Making in Stochastic Manufacturing Systems*, in series Systems and Control: Foundations and Applications, Birkhuser Boston, Cambridge, MA, 1994

Toth F., Climate Policy in Light of Climate Science: the ICLIPS Project, *Climatic Change*, 56, 1–2, 7–36. 2003.

Chapter 10

CHARACTERIZATION OF JUST IN TIME SEQUENCING VIA APPORTIONMENT

Joanna Józefowska
Institute of Computing Science
Poznań University of Technology
Poznań, Poland
joanna.jozefowska@cs.put.poznan.pl

Łukasz Józefowski
Institute of Computing Science
Poznań University of Technology
Poznań, Poland
joseph@man.poznan.pl

Wiesław Kubiak
Faculty of Business Administration
Memorial University of Newfoundland
St. John's, Canada
wkubiak@mun.ca

Abstract   The just in time sequencing is used to balance workloads throughout just in time supply chains intended for low-volume high-mix family of products. It renders supply chains more stable and carrying less inventories of final products and components but at the same time it ensures less shortages. A number of algorithms have been proposed in the literature to optimize just in time sequencing. This paper characterizes these algorithms via characteristics developed by the apportionment theory.

Keywords: Just-in-Time sequencing, apportionment theory

1. Introduction

The problem of sequencing different models of the same product for production by a just in time system is refereed to as a *just in time sequencing* problem. The just in time sequencing was introduced and perfected by the Toyota Production System where it was originally used to distribute production volume and mix of models as evenly as possible over the production sequence, see Monedn (Monden, 1983), and also Groenevelt (Groenevelt, 1993), and Vollman, Berry and Wybark (Vollman et al., 1992). The just in time sequencing has proven to be a universal and robust tool used to balance workloads throughout just in time supply chains intended for low-volume high-mix family of products, Kubiak (Kubiak, 2005). It renders supply chains more stable and carrying less inventories of final products and components but at the same time it ensures less shortages.

In the just in time sequencing problem, there are n different models with positive integer demands $d_1, ..., d_n$ for models $1, \ldots, n$ respectively to be sequenced. The total demand is then $D = \sum_{i=1}^{n} d_i$. The problem is to find a sequence $s_1, \ldots, s_D$, where model i occurs exactly d_i times that minimizes a certain measure of deviation of the *actual* production level of each model from its *ideal* level of production. The ideal level for model i at t, $t = 1, \ldots, D$, equals tr_i, where $r_i = \frac{d_i}{D}$. The actual level for model i at t, denoted by x_{it}, is simply the number of copies of model i in the prefix $s_1 \ldots s_t$ of the sequence. The deviation between the actual and the ideal levels of i at t is defined as $|x_{it} - tr_i|$. Miltenburg (Miltenburg, 1989) suggests the total deviation defined as follows

$$f(x) = \sum_{i=1}^{n} \sum_{t=1}^{D} |x_{it} - tr_i| \qquad (10.1)$$

as the objective to minimize, whereas Steiner and Yeomans (Steiner and Yeomans, 1993) subsequently propose maximum deviation defined as follows

$$g(x) = \max_{it} |x_{it} - tr_i| \qquad (10.2)$$

as another objective to minimize. Formally, the just in time sequencing problem is to find nonnegative integers x_{it}, $t = 1, \ldots, D$ and $i = 1, \ldots, n$ that minimize either $f(x)$ or $g(x)$ subject to the constrains (10.3), (10.4), and (10.5) defined as follows

$$\sum_{i=1}^{n} x_{it} = t \quad t = 1, \ldots, D \tag{10.3}$$

$$0 \leq x_{it+1} - x_{it} \leq 1 \quad n = 1, \ldots, n; t = 1, \ldots, D-1 \tag{10.4}$$

$$\sum_{t=1}^{D} x_{it} = d_i \quad i = 1, \ldots, n. \tag{10.5}$$

Kubiak and Sethi (Kubiak and Sethi, 1991) and (Kubiak and Sethi, 1994) show, using the *level curve* concept, the reduction of the problem of minimizing $f(x)$ subject to (10.3), (10.4), and (10.5) to the assignment problem. Their algorithm is the first efficient, that is polynomial in D and n, algorithm for the problem. Subsequently, Steiner and Yeomans, present an efficient algorithm for the problem of minimizing $g(x)$ subject to (10.3), (10.4), and (10.5). Their algorithm is based on the idea of level curves introduced by Kubiak and Sethi (Kubiak and Sethi, 1991) and it essentially reduces the sequencing problem to the matching problem in convex graphs.

Independently, Tijdeman (Tijdeman, 1980) while studying a problem called the *chairman assignment* problem effectively gives an algorithm that finds a solution x such that $g(x) < 1$. This solution, however, does not necessarily minimize $g(x)$.

Inman and Bulfin (Inman and Bulfin, 1991) give an algorithm to minimize

$$f(y) = \sum_{i=1}^{n} \sum_{j=1}^{d_i} \left(y_{ij} - \tau_{ij} \right)^2, \tag{10.6}$$

where $\tau_{ij} = \frac{2j-1}{2r_i}$ and y_{ij} is the position of copy j of model i, $i = 1, \ldots, n$, $j = 1, \ldots, d_i$, in the sequence.

These four algorithms will be characterized in this paper using characteristics of apportionment methods.

Bautista, Companys and Corominas (Bautista et al., 1996) are the first to observe that the algorithm of Inman and Bulfin (Inman and Bulfin, 1991) is the Webster *divisor* method of apportionment. They also point out a very strong link that exists between the just in time sequencing and the apportionment problem. The latter addresses the fundamental question of how to divide the seats of a legislature fairly according to the populations of states, Balinski and Young (Balinski and Young, 1982).

Balinski and Shahidi (Balinski and Shahidi, 1998) (see also Balinski and Ramirez (Balinski and Ramirez, 1999)) propose an elegant approach to the just in time sequencing via *axiomatics* originally developed for the apportionment problem. The axiomatic method of apportionment theory relies on some socially desirable characteristics that one requires an apportionment to posses. These characteristics include, for instance, quota satisfaction, house and population monotonicity but also many others. They have been shown crucial for the solutions of the apportionment problem, (Balinski and Young, 1982), to posses. However, the famous *Impossibility Theorem* of Balinski and Young puts a clear limitation on which characteristics do not contradict one another by showing that having solutions that satisfy quota and that are at the same time population monotone is generally impossible.

This paper will characterize the four algorithms: the algorithm of Kubiak and Sethi, the algorithm of Steiner and Yeomans, the algorithm of Tijdeman, and the algorithm of Inman and Bulfin via the characteristics introduced in the apportionment theory. These characteristics will be described in detail in Section 2. The material there will closely follow the exposition given in the book by (Balinski and Young, 1982). Section 3 will present the transformation between the just in time sequencing problem and the apportionment problem. Section 4 will characterize the Steiner and Yeomans algorithm. Section 5 will characterize the Tijdeman algorithm. Section 6 will characterize the Kubiak and Sethi algorithm. Section 7 will summarize the characteristics of the just in time sequencing algorithms. Finally, Section 8 will present concluding remarks and open problems.

2.   The apportionment problem

The apportionment problem has its roots in the proportional election system designed for the House of Representatives of the United States where each state receives seats in the House proportionally to its population, Balinski and Young (Balinski and Young, 1982).

The instance of the problem is defined by the house size h and an integer vector of state populations:

$$p = (p_1, p_2, p_3, \ldots, p_s). \tag{10.7}$$

An apportionment of h seats among s states is an integer vector

$$a = (a_1, a_2, a_3, \ldots, a_s) \geq 0 \tag{10.8}$$

$$\sum_{i=1}^{s} a_i = h.$$

The solution for the apportionment problem is found by a method $M(p, h)$ that for any vector p and house size h returns a vector a, which is written as $a \in M(p, h)$. We refer to this method as an apportionment method or algorithm.

2.1   Characterization of apportionments

Balinski and Young (Balinski and Young, 1982) propose the following characteristics of an apportionment method:

- satisfying the quota (staying within the quota),

- house monotone,

- population monotone,

- divisor,

- parametric,

- uniform (rank-index).

Their details will be described in the subsequent sections.

2.2   Satisfying the quota

The apportionment of a_i seats to state i satisfies the *lower quota* for population vector p and house size h if and only if

$$a_i + 1 > \frac{p_i h}{\sum_{k=1}^{s} p_k} \tag{10.9}$$

and it satisfies the *upper quota* if and only if

$$a_i - 1 < \frac{p_i h}{\sum_{k=1}^{s} p_k}. \tag{10.10}$$

The a_i satisfies the *quota* if it satisfies simultaneously the lower and the upper quota. The apportionment vector a satisfies the quota if and only if it satisfies simultaneously the lower and the upper quota for all states.

2.3 House monotone method

The house monotonicity requirement stipulates that for an apportionment vector a and the house size h if this size increases by 1, then the actual apportionment of any state cannot decrease. Any apportionment method that gives an apportionment vector a for the house size h and the population vector p, and an apportionment vector $a' \geq a$ for the house of size $h' = h + 1$ and the same population vector p is said to be *house monotone*. All divisor methods defined later in Section 2.6 are house monotone. There are however important, at least historically, apportionment methods that are *not* house monotone. The Hamilton method, known also as the largest reminder method, described in detail in Balinski and Young (Balinski and Young, 1982) is an example of a method that is not house monotone.

2.4 House monotone method that stays within the quota

Historically, the first method that stays within the quota and that is simultaneously house monotone is the *Quota method* proposed by Balinski and Young (Balinski and Young, 1975). However, a more general method is proposed by Still (Still, 1979), his algorithm works as follows:

For the house size $h = 0$, assign each state 0 seats. For the house size $h > 0$ assign one additional seat to one of the states from the eligible set $E(h)$ defined as follows.

The eligible set $E(h)$ for any house size $h > 0$ consists of all states i that pass the following two tests:

1. The *upper quota test*. The number of seats that state i has in a house of size h before the additional seat is assigned is less then the *upper quota* for state i at house size h.

2. The *lower quota test*. Let h_i be the house size at which state i first becomes entitled to obtain the next seat, i.e. h_i is the smallest house size $h' \geq h$ at which the lower quota of state i is greater or equal $a_{i(h-1)} + 1$, or

$$h_i = \left\lceil \frac{a_{i(h-1)+1}}{p_i} \sum_{i=1}^{n} p_i \right\rceil.$$

For each house size g in the interval $h \leq g \leq h_i$ define $s_i(g, i) = a_{i(h-1)} + 1$ (the number of seats that state i has in a house of size h

before an additional seat is assigned $+1$); for $j \neq i$ $s_j(g,i) = max \{$ *number of seats that state j has at house size h before an additional seat is assigned, lower quota of state j* $\}$. If there is no house size g, $h \leq g \leq h_i$, for which $\sum_j s_j(g,i) > g$, then state i satisfies the *lower quota test.*

The eligible set $E(h)$ for house size h consist of all the states which may receive the available seat without causing a violation of quota, either for h or any larger house size. This can be written as follows:

E(h)={ *state i: state i passes the upper quota test and state i passes the lower quota test* }.

Still proved, that $E(h)$ is never empty and that all methods that provide solutions that are *house monotone* and *stay within the quota* belong to this general method. The states from the eligible set $E(h)$ can be chosen in a number of ways for example by using quota-divisor methods. The class of *quota-divisor* methods is based on the divisor methods defined in Section 2.6. The crucial detail that differentiates the *quota-divisor* methods from the *divisor* methods is that in the former the states selected by the quota-divisor methods must be eligible, that is they must come form $E(h)$. This algorithm is defined as follows

1 $M(p,0) = 0$

2 If $a \in M(p,h)$ and $k,i \in E(h)$ satisfies $\frac{p_k}{d(a_k)} = \max_i \frac{p_i}{d(a_i)}$, then $b \in M(p,h+1)$ with $b_k = a_k + 1$ for $i = k$ and $b_i = a_i$ for $i \neq k$,

where $E(h)$ is the eligible set defined above.

2.5   Population monotone method

Balinski and Young (Balinski and Young, 1982) introduce population monotone apportionment methods. These methods are designed to ensure that if state i's population increases and j's decreases, then state i gets no fewer seats and state j gets no more seats with the new populations than they do with the original populations and unchanged house size h. Formally, if for any two vectors of populations $p, p' > 0$ and vectors of apportionments $a \in M(p,h)$, $a' \in M(p',h')$

182

$$\frac{p'_{i'}}{p'_{j'}} \geq \frac{p_i}{p_j} \Rightarrow \left\{ \begin{array}{c} (a'_{i'} \geq a_i \vee a'_{j'} \leq a_j) \\ or \\ \frac{p'_{i'}}{p'_{j'}} = \frac{p_i}{p_j} \text{ and } a'_{i'}, a'_{j'} \text{ can be substituted for } a_i, a_j \text{ in a.} \end{array} \right\}.$$

(10.11)

Balinski and Young (Balinski and Young, 1982) show that any population monotone method is house monotone as well but not the other way around. Their famous Impossibility Theorem shows that the failure to stay within the quota is the price that any population monotone apportionment method must pay for its desirable qualities.

Theorem 1 *It is impossible for an apportionment method to be population monotone and stay within the quota at the same time for any reasonable instance of the problem ($s \geq 4$ and $h \geq s + 3$).*

2.6   Divisor method

An important way of finding proportional share of h seats is to find an ideal district size or divisor x to compute the quotients q_i^x of each state

$$q_i^x = \frac{p_i}{x} \tag{10.12}$$

and to round them according to some rule. The sum of all quotients must equal h. There are many ways to round the quotient. Any rounding procedure can be described by specifying a dividing point $d(a)$ in each interval of quotients $[a, a + 1]$ for each non negative integer a. Balinski and Young define in (Balinski and Young, 1982) a d-rounding of any positive real z, $[z]_d$, to be an integer a such that $d(a - 1) \leq z \leq d(a)$, which is unique unless $z = d(a)$, in which case the value is either a or $(a+1)$. It is also required that $d(a) < d(a+1)$. Any monotone increasing $d(a)$ defined for all integers $a \geq 0$ and satisfying $a \leq d(a) \leq a + 1$ is called a divisor criterion. The divisor method based on d is thus defined as follows:

$$M(p, h) = \{a \; : \; a_i = \left[\frac{p_i}{x}\right]_d \text{ and } \sum_{i=1}^{s} a_i = h \text{ for some } x\}. \tag{10.13}$$

Using (10.13) and having $d(a-1) \leq \frac{p_i}{x} \leq d(a)$ the method can be defined alternatively as follows.

$$M(p,h) = \{a \; : \; \min_{a_i > 0} \frac{p_i}{d(a_i - 1)} \geq \max_{a_i \geq 0} \frac{p_j}{d(a_j)}, \sum_{i=1}^{s} a_i = h\} \qquad (10.14)$$

where $\frac{p_i}{0}$ is defined such that $p_i > p_j$ implies $\frac{p_i}{0} > \frac{p_j}{0}$.
Table 10.1 shows the $d(a)$ function of the best known divisor methods.

Methods name	Adams	Dean	Hill	Webster	Jefferson
d(a)	a	$\frac{a(a+1)}{a+1/2}$	$\sqrt{a(a+1)}$	$a + 1/2$	$a + 1$

Table 10.1. The best known divisor methods

To summarize, a divisor method is defined as follows:

1 $M(p, 0) = 0$,

2 If $a \in M(p, h)$ and k satisfies $\frac{p_k}{d(a_k)} = \max_i \frac{p_i}{d(a_i)}$, then $b \in M(p, h + 1)$ with $b_k = a_k + 1$ for $i = k$ and $b_i = a_i$ for $i \neq k$.

Any divisor method is population monotone, and thus house monotone. Therefore, by the Impossibility Theorem no divisor method stays within the quota.

2.7   Parametric method

Parametric method ϕ^δ is a divisor method with $d(a) = a + \delta$, where $0 \leq \delta \leq 1$. The parametric methods are cyclic. That is for two instances of the just in time sequencing problem $D_1 = d_1, d_2, ..., d_n$ and $D_2 = kD_1 = kd_1, kd_2, ..., kd_n$, the sequence for problem D_2 is obtained by k repetitions of the sequence for problem D_1.

2.8   Uniform (rank-index) method

An apportionment method is said to be uniform if it ensures that an apportionment $a = (a_1, a_2, ..., a_s)$ of h seats of the house among states with populations $p = (p_1, p_2, ..., p_s)$ will stay the same when it is restricted to any subset S of these states and the house size $\sum_{i \in S} a_i = h'$. In other words, according to Balinski and Young (Balinski and Young, 1982), if for every t, $2 \leq t \leq s$, $(a_1, ..., a_s) \in M((p_1, ..., p_s), h)$

184

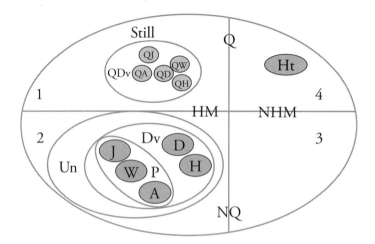

Figure 10.1. Classification of apportionment methods.

implies $(a_1, ..., a_t) \in M((p_1, ..., p_t), \sum_{i=1}^{t} a_i)$ and if also $(b_1, ..., b_t) \in M((p_1, ..., p_t), \sum_{i=1}^{t} a_i)$, then $(b_1, ..., b_t, a_{t+1}, ..., a_s) \in M((p_1, ..., p_s), h)$. Each uniform method can be obtained by using a *rank-index* function. A rank-index function $r(p, a)$ is any real-valued function of rational p and integer $a \geq 0$ that is decreasing in a, i.e. $r(p, a - 1) > r(p, a)$. Let F be the set of all solutions defined as follows.

1 For $h = 0$ let $f(p, 0) = 0$.

2 If $f(p, h) = a$, then $f(p, h + 1)$ is found by giving $a_i + 1$ seats to some state i such that $r(p_i, a_i) \geq r(p_j, a_j)$ and a_j seats to each $j \neq i$.

The *rank-index* method based on $r(p, a)$ is defined as follows:

$$M(p, h) = \{a \; : \; a = f(p, h) \; for \; some \; f \in F\} \qquad (10.15)$$

Balinski and Young (Balinski and Young, 1982) prove that every divisor method is *uniform*. Because every *uniform* method is *rank-index* then it is also *house monotone*.

3.   Classification of the apportionment methods

The chart in Figure 1.1 summarizes our discussion in Section 2. The abbreviations used in the chart are explained below:

1,2,3,4 = quarters 1, 2, 3 and 4 respectively.

HM = **H**ouse **M**onotone methods, quarters 1 and 2.

NHM = **N**ot **H**ouse **M**onotone methods, quarters 3 and 4.

Q = stay within the **Q**uota methods, quarters 1 and 4.

NQ = **N**ot within the **Q**uota methods, quarters 2 and 3.

Un = **U**niform methods.

Dv = **D**ivisor methods.

P = **P**arametric methods.

Still = methods obtained with the Still algorithm, quarter 1.

QDv = **Q**uota-**D**ivisor methods.

J,W,A,H,D = **J**efferson, **W**ebster, **A**dams, **H**ill and **D**ean divisor methods.

QJ, QW, QA, QH, QD = **Q**uota-**J**efferson, **Q**uota-**W**ebster, **Q**uota-**A**dams, **Q**uota-**H**ill, **Q**uota-**D**ean quota methods obtained using Stills algorithm.

Ht = **H**amilton method (Largest Remainder method).

4.   Transformation

In the transformation between the just in time sequencing and apportionment problems, state i corresponds to model i and the demand d_i for model i corresponds to population p_i of state i. The cumulative number of units x_{it} of i completed by t corresponds to the number a_i of seats apportioned to state i in a house of size t. The following is the summary of the correspondences between the two problems:

$$\text{number of states } s \longleftrightarrow \text{ number of models } n$$
$$\text{state } i \longleftrightarrow \text{ model } i$$
$$\text{population } p_i \text{ of state } i \longleftrightarrow \text{ demand } d_i \text{ for model } i$$
$$\text{size of house } h \longleftrightarrow \text{ position in sequence } t$$
$$\text{for a house of size } h, \, a_i \longleftrightarrow x_{it}$$
$$\text{total population } P = \sum_{i=1}^{s} p_i \longleftrightarrow \text{ total demand } D = \sum_{i=1}^{n} d_i.$$

5.   The classification of sequencing algorithms

In this section, we show the location of the Inman-Bulfin algorithm (Inman and Bulfin, 1991), the Steiner-Yeomans (Steiner and Yeomans, 1993) algorithm, the Tijdeman (Tijdeman, 1980) algorithm and the Kubiak-Sethi (Kubiak and Sethi, 1994) algorithm in the chart in Figure 10.1.

5.1 Inman-Bulfin algorithm

Bautista, Companys and Corominas (Bautista et al., 1996) observe that the Inman and Bulfin algorithm (Inman and Bulfin, 1991) to minimize function (10.6) is equivalent to the Webster *divisor* method. Consequently, sequences produced by the Inman-Bulfin algorithm have all the characteristics of the Webster apportionment method solutions, see Figure 1.1.

5.2 Steiner-Yeomans algorithm

The algorithm is based on the following theorem of Steiner and Yeomans (Steiner and Yeomans, 1993).

Theorem 2 *A just in time sequence with*

$$\min \max_{it} |x_{it} - tr_i| \leq T \qquad (10.16)$$

exists if and only if there exists a sequence that allocates the j-th copy of i in the interval $[E(i,j), L(i,j)]$ *where*

$$E(i,j) = \left\lceil \frac{1}{r_i}(j - T) \right\rceil, \qquad (10.17)$$

$$L(i,j) = \left\lfloor \frac{1}{r_i}(j - 1 + T) + 1 \right\rfloor. \qquad (10.18)$$

The algorithm tests the values of T from the following list in ascending order

$$T = \frac{D - d_{max}}{D}, \frac{D - d_{max+1}}{D}, \dots, \frac{D - 1}{D}. \qquad (10.19)$$

For each T, the algorithm calculates the $E(i,j)$ and $L(i,j)$ for each pair (i,j), $i = 1, \dots, n$ and $j = 1, \dots, d_i$. Finally, it assigns positions $t = 1, \dots, D$ starting with $t = 1$ and ending with $t = D$ to the yet unassigned but still *available* at t pairs (i,j) following the ascending order of their $L(i,j)$. A pair (i,j) is available at t if and only if $E(i,j) \leq t \leq L(i,j)$. If some pairs can not be assigned by the algorithm, then the value of T is rejected as infeasible. Otherwise, T is feasible. Brauner and Crama (Brauner and Crama 2001) show the following theorem.

Theorem 3 *At least one of the values on the list (10.19) is feasible.*

Therefore, we have

$$\min_{it} \max |x_{it} - tr_i| \le 1 - \frac{1}{D}. \tag{10.20}$$

We observe that if T' is feasible then all T, $T' \le T \le 1 - \frac{1}{D}$ are feasible as well. The smallest feasible T is denoted by T^* and it is referred to as optimum.

5.3   Steiner-Yeomans algorithm is a quota-divisor method

We are now ready to show that the Steiner-Yeomans algorithm is a quota-divisor method of apportionment. We have the following theorem.

Theorem 4 *The Steiner-Yeomans algorithm with T, $T^* \le T < 1$ and a tie $L(i,j) = L(k,l)$ between i and k broken by choosing the one with*

$$\min\{\frac{1}{r_i}(j - 1 + T), \frac{1}{r_k}(l - 1 + T)\}$$

is a quota-divisor method with

$$d(a) = a + T.$$

Proof: First, we observe that i can only receive its j-th seat in a house of size h that falls between $E(i,j)$ and $L(i,j)$. More precisely h satisfies the following inequality

$$\frac{j - T}{r_i} \le E(i,j) \le h \le L(i,j) \le \frac{j - 1 + T}{r_i} + 1. \tag{10.21}$$

Replacing j by a_i in (10.21), we obtain

$$a_i - 1 < a_i - T \le hr_i \le a_i - 1 + T + r_i < a_i + 1, \tag{10.22}$$

for $T < 1$. Therefore, a_i stays within the quota for h.
Second, for position $t + 1$ the algorithm chooses i with

$$\min_i \left\{ \frac{j - 1 + T}{r_i} \right\} \tag{10.23}$$

or equivalently

$$\max_i \left\{ \frac{d_i}{j - 1 + T} \right\}. \tag{10.24}$$

However, $x_{it} = j - 1$, thus

$$\max_i \left\{ \frac{d_i}{x_{it} + T} \right\} \tag{10.25}$$

defines a divisor method with $d(x) = x + T$, $T < 1$. This divisor method always stays within the quota which proves the theorem.□

It is worth observing that the Steiner-Yeomans algorithm is in fact a quota-parametric method with $\delta = T$.

5.4   Tijdeman algorithm

Tijdeman (Tijdeman, 1980) introduced a problem called the *chairman assignment* problem. The chairman assignment problem is defined as follows. Suppose k states form a union $S = (S_1, S_2, \ldots, S_k)$ and every year a union chairman has to be selected in such a way that at any time the accumulated number of chairmen from each state is proportional to its weight, S_i has a weight λ_i with $\sum_{i=1}^{k} \lambda_i = 1$. We denote the state designating the chairman in the jth year by ω_j. Hence $\omega = \{\omega_j\}_{j=1}^{\infty}$ is a sequence in the alphabet S. Let $A_\omega(i, t)$ denote the number of chairmen representing S_i in the first t years and define

$$D(\omega) = \sup_{it} |\lambda_i t - A_\omega(i, t)|. \tag{10.26}$$

The problem is to choose ω in such a way that $D(\omega)$ is minimal. Tijdeman (Tijdeman, 1980) proves the following theorem.

Theorem 5 *Let λ_{it} be a double sequence of non-negative numbers such that $\sum_{1 \leq i \leq k} \lambda_{it} = 1$ for $t = 1, \ldots$. For an infinite sequence S in $\{1, \ldots, n\}$ let x_{it} be the number of i's in the t-prefix of S. Then there exists a sequence S in $\{1, \ldots, n\}$ such that*

$$\max_{it} \left| \sum_{1 \leq j \leq t} \lambda_{it} - x_{it} \right| \leq 1 - \frac{1}{2(n - 1)}.$$

Let us define $\lambda_{it} = r_i = \frac{d_i}{D}$ for $t = 1, \ldots$. Then, this theorem ensures the existence of an infinite sequence S such that

$$\max_{it} |t r_i - x_{it}| \leq 1 - \frac{1}{2(n - 1)}.$$

To ensure that the required number of copies of each product is in D-prefix of S consider the D-prefix of S and suppose that there is i with $x_{iD} > d_i$. Then, there is j with $x_{jD} < d_j$. It can be easily checked that replacing the last i in the D-prefix by j does not increase the absolute maximum deviation for the D-prefix. Therefore, we can readily obtain a D-prefix where each i occurs exactly d_i times and with maximum deviation not exceeding $1 - \frac{1}{2(n-1)}$. We consequently have the following upper bound stronger that the one in (1.20).

$$\min \max_{it} |x_{it} - tr_i| \leq 1 - \frac{1}{2n-2} \tag{10.27}$$

The sequence satisfying this bound is built as follows. Let $J_t, t = 1, \ldots, D$ be a set of models satisfying the following condition at t:

$$\sigma_i = tr_i - x_{it-1} \geq \frac{1}{2n-2}, \tag{10.28}$$

where x_{it-1} is the cumulative number of units of model i scheduled between 1 and $t - 1$. Apportion t to model i from the set J_t with the minimal value of

$$\frac{1 - \frac{1}{2n-2} - \sigma_i}{r_i}. \tag{10.29}$$

5.5 Tijdeman algorithm is a quasi quota-divisor method

The inequality defining the set J_t of eligible states (10.28) in the Tijdeman algorithm defines the *new upper quota* test. In (10.28), i has x_{it-1} seats in a house of size $h = t - 1$. Thus we can rewrite (10.28) as follows:

$$(t+1)r_i - x_{it} \geq \frac{1}{2n-2} \tag{10.30}$$

or equivalently

$$tr_i \geq x_{it} - 1 + 1 - r_i + \frac{1}{2n-2}. \tag{10.31}$$

Replacing x_{it} by a_i and t by h in the inequality (10.31) we get the *new upper quota* as follows

$$a_i - 1 + 1 - r_i + \frac{1}{2n-2} < h \frac{p_i}{\sum_{k=1}^{s} p_k}. \tag{10.32}$$

This however implies

$$a_i - 1 < q_i = h \frac{p_i}{\sum_{k=1}^{s} p_k}$$

since $1 - r_i + \frac{1}{2n-2} > 0$. Therefore, (10.32) implies (10.10) and the upper quota is satisfied. However, the following lemma shows that the eligible sets in the Tijdeman algorithms are narrower than those of defined by the original upper quota test from (10.10).

Lemma 1 *The set of eligible i's satisfying the new upper quota test in (10.32) is a proper subset of the set of eligible i's satisfying the upper quota test in (10.10).*

Proof: Consider an instance with $n = 3$ models: P_1, P_2, and P_3 with demands 2, 3, and 7 units, respectively. The Tijdeman algorithm makes only models P_2 and P_3 eligible for one more unit at $t = 1$. Namely, for $(t = 1, x_{1t-1} = 0, x_{2t-1} = 0, x_{3t-1} = 0, r_1 = \frac{2}{12}, r_2 = \frac{3}{12}, r_3 = \frac{7}{12})$ we have

P_1: $tr_i - x_{1t-1} = \frac{2}{12} - 0 < \frac{1}{2n-2} = \frac{1}{4}$

P_2: $tr_i - x_{2t-1} = \frac{3}{12} - 0 = \frac{1}{2n-2} = \frac{1}{4}$

P_3: $tr_i - x_{3t-1} = \frac{7}{12} - 0 > \frac{1}{2n-2} = \frac{1}{4}$

and the inequality (10.28) is violated for P_1. On the other hand all three models are eligible for one more unit at $t = 1$ since with this unit they will all satisfy the original upper quota test from the inequality (10.10). This proves the lemma. □

At each step t of the Tijdeman algorithm the model that satisfies the following condition is selected from the eligible set J_t:

$$\min_i \left\{ \frac{1 - \frac{1}{2n-2} - \sigma_i}{r_i} \right\} \tag{10.33}$$

or

$$\min_i \left\{ \frac{1 - \frac{1}{2n-2} - tr_i + x_{it-1}}{r_i} \right\}. \tag{10.34}$$

By replacing $(1 - \frac{1}{2n-2})$ by Δ in (10.34) we obtain an equivalent selection criterion

$$\max_i \left\{ \frac{d_i}{\Delta + x_{it-1}} \right\}. \qquad (10.35)$$

The (10.35) may suggest that the Tijdeman algorithm is a *quota-divisor* method with $d(a) = a + \Delta$ where $\Delta = 1 - \frac{1}{2n-2}$. However, we show in Lemma 2 that the algorithm narrows down the set of eligible models more than quota-divisor methods would normally do. Therefore, to distinguish the quota-divisor methods from the quota-divisor methods with narrower eligible sets we call the latter *quasi* quota-divisor methods.

Lemma 2 *The eligible set of i's satisfying the new upper quota test in (10.32) is a proper subset of the eligible set of i's in the quota-divisor method with $d(a) = a + \Delta$ where $\Delta = 1 - \frac{1}{2n-2}$.*

Proof: We consider the same three model instance as in Lemma 1. We compare the selection made for the first two positions by the Tijdeman algorithm with the one made by the quota-divisor method defined with $d(a) = a + \Delta$ where $\Delta = 1 - \frac{1}{2n-2}$.
Let us begin with the Tijdeman algorithm. For $t = 1$, we have $x_{1t-1} = 0, x_{2t-1} = 0, x_{3t-1} = 0$. We check which models make the eligible set defined by the inequality (10.28):

P_1: $tr_i - x_{1t-1} = \frac{2}{12} - 0 < \frac{1}{2n-2} = \frac{1}{4}$

P_2: $tr_i - x_{2t-1} = \frac{3}{12} - 0 = \frac{1}{2n-2} = \frac{1}{4}$

P_3: $tr_i - x_{3t-1} = \frac{7}{12} - 0 > \frac{1}{2n-2} = \frac{1}{4}$

Thus, P_2 and P_3 are eligible, P_1 is not. Moreover, P_3 minimizes (10.33) and receives position $t = 1$.
Then, for $t = 2$, we have $x_{1t-1} = 0, x_{2t-1} = 0, x_{3t-1} = 1$. We again check which models make the eligible set defined by the inequality (10.28):

P_1: $tr_i - x_{1t-1} = \frac{1}{3} - 0 = \frac{1}{3} > \frac{1}{2n-2} = \frac{1}{4}$

P_2: $tr_i - x_{2t-1} = \frac{1}{2} - 0 = \frac{1}{2} > \frac{1}{2n-2} = \frac{1}{4}$

P_3: $tr_i - x_{3t-1} = \frac{7}{6} - 1 = \frac{1}{6} < \frac{1}{2n-2} = \frac{1}{4}$

Now P_1 and P_2 are eligible, P_3 is not. Moreover, P_2 minimizes (10.33) and thus receives position $t = 2$.

Now, let us consider the quota-divisor method with $d(a) = a + 1 - \frac{1}{2n-2}$.
For $n = 3$ we obtain $d(a) = a + 1 - \frac{1}{4}$.
For $t = 1$, $x_{1t-1} = 0, x_{2t-1} = 0, x_{3t-1} = 0$. We check if the models belong to the eligible set $E(t)$ defined in Section 2.4.

The upper quota test:
Each of the three models passes the upper quota test since $a_1 = a_2 = a_3 = 1$ satisfies the following inequalities

$$a_1 - 1 = 1 - 1 < q_1 = 1 \cdot \frac{2}{12}$$

$$a_2 - 1 = 1 - 1 < q_2 = 1 \cdot \frac{3}{12}$$

$$a_3 - 1 = 1 - 1 < q_3 = 1 \cdot \frac{7}{12}$$

The lower quota test:
Model P_i, $i = 1, 2, 3$, is entitled to receive its first position by $t_i = \lceil \frac{1}{r_i} \rceil$. Thus, P_1 is entitled to receive its first position by $t_1 = 6$, P_2 by $t_2 = 4$, and P_3 by $t_3 = 2$ (in other words $h_1 = 6, h_2 = 4, h_3 = 2$). Consequently, at $t = 1$ we have the following lower quota tests in the Still algorithm:

For model P_1, $t - 1 = 0, t_1 = 6$ we have to step through $g = 1, 2, 3, 4, 5$

$$g = 1 : s(g, P_1) = 1, s(g, P_2) = 0, s(g, P_3) = 0 \Rightarrow \sum_i s(g, P_i) = 1 = g$$

$$g = 2 : s(g, P_1) = 1, s(g, P_2) = 0, s(g, P_3) = 1 \Rightarrow \sum_i s(g, P_i) = 2 = g$$

$$g = 3 : s(g, P_1) = 1, s(g, P_2) = 0, s(g, P_3) = 1 \Rightarrow \sum_i s(g, P_i) = 2 < g$$

$$g = 4 : s(g, P_1) = 1, s(g, P_2) = 1, s(g, P_3) = 2 \Rightarrow \sum_i s(g, P_i) = 4 = g$$

$$g = 5 : s(g, P_1) = 1, s(g, P_2) = 1, s(g, P_3) = 2 \Rightarrow \sum_i s(g, P_i) = 4 < g.$$

For P_2, $t - 1 = 0, t_2 = 4$ so we have to step through $g = 1, 2, 3$

$$g = 1 : s(g, P_1) = 0, s(g, P_2) = 1, s(g, P_3) = 0 \Rightarrow \sum_i s(g, P_i) = 1 = g$$

$$g = 2 : s(g, P_1) = 0, s(g, P_2) = 1, s(g, P_3) = 1 \Rightarrow \sum_i s(g, P_i) = 2 = g$$

$$g = 3 : s(g, P_1) = 0, s(g, P_2) = 1, s(g, P_3) = 1 \Rightarrow \sum_i s(g, P_i) = 2 < g.$$

For P_3, $t - 1 = 0$, $t_3 = 2$ so we have to step through $g = 1$

$$g = 1 : s(g, P_1) = 0, s(g, P_2) = 0, s(g, P_3) = 1 \Rightarrow \sum_i s(g, P_i) = 1 = g.$$

Clearly, for $t = 1$ all three models satisfy the lower quota test and so they do the upper quota test. Therefore, the quota-divisor method with $d(a) = a + 1 - \frac{1}{4}$ selects model P_3 since we have

$$\frac{7}{0 + 1 - \frac{1}{4}} > \frac{3}{0 + 1 - \frac{1}{4}} > \frac{2}{0 + 1 - \frac{1}{4}}.$$

For $t = 2$, $x_{1t-1} = 0$, $x_{2t-1} = 0$, $x_{3t-1} = 1$.

The upper quota test:

The three models pass the upper quota test since $a_1 = a_2 = 1$ and $a_3 = 2$ satisfies the following inequalities

$$a_1 - 1 = 1 - 1 < q_1 = 2 \cdot \frac{2}{12}$$

$$a_2 - 1 = 1 - 1 < q_2 = 2 \cdot \frac{3}{12}$$

$$a_3 - 1 = 2 - 1 < q_3 = 2 \cdot \frac{7}{12}.$$

The lower quota test:

Model P_1 is entitled to receive its first position by $t_1 = 6$, P_2 is entitled to receive its first position by $t_2 = 4$ (in other words $h_1 = 6, h_2 = 4, h_3 = 4$). Finally, model P_3 is entitled to receive its second position by $t_3 = \lceil \frac{1+1}{r_3} \rceil = \lceil \frac{24}{7} \rceil = 4$. Consequently, at $t = 2$ we have the following lower quota tests in the Still's algorithm:

For P_1, $t - 1 = 1$, $t_1 = 6$ so we have to step through $g = 2, 3, 4, 5$

$$g = 2 : s(g, P_1) = 1, s(g, P_2) = 0, s(g, P_3) = 1 \Rightarrow \sum_i s(g, P_i) = 2 = g$$

$$g = 3 : s(g, P_1) = 1, s(g, P_2) = 0, s(g, P_3) = 1 \Rightarrow \sum_i s(g, P_i) = 2 < g$$

$$g = 4 : s(g, P_1) = 1, s(g, P_2) = 1, s(g, P_3) = 2 \Rightarrow \sum_i s(g, P_i) = 4 = g$$

$$g = 5 : s(g, P_1) = 1, s(g, P_2) = 1, s(g, P_3) = 2 \Rightarrow \sum_i s(g, P_i) = 4 < g.$$

194

For P_2, $t - 1 = 1, t_2 = 4$ so we have to step through $g = 2, 3$ as well as through $g = 4, 5, 6$ to ensure that model P_1 receives one unit by $t_1 = 6$:

$$g = 2 : s(g, P_1) = 0, s(g, P_2) = 1, s(g, P_3) = 1 \Rightarrow \sum_i s(g, P_i) = 2 = g$$

$$g = 3 : s(g, P_1) = 0, s(g, P_2) = 1, s(g, P_3) = 1 \Rightarrow \sum_i s(g, P_i) = 2 < g$$

$$g = 4 : s(g, P_1) = 0, s(g, P_2) = 1, s(g, P_3) = 2 \Rightarrow \sum_i s(g, P_i) = 3 < g$$

$$g = 5 : s(g, P_1) = 0, s(g, P_2) = 1, s(g, P_3) = 2 \Rightarrow \sum_i s(g, P_i) = 3 < g$$

$$g = 6 : s(g, P_1) = 1, s(g, P_2) = 1, s(g, P_3) = 3 \Rightarrow \sum_i s(g, P_i) = 5 < g.$$

For P_3, $t - 1 = 1, t_3 = 4$ and we have to step through $g = 3, 4, 5, 6$ to assure that model P_1 receives one unit by $t_1 = 6$ and model P_2 by $t_2 = 4$:

$$g = 3 : s(g, P_1) = 0, s(g, P_2) = 0, s(g, P_3) = 2 \Rightarrow \sum_i s(g, P_i) = 2 < g$$

$$g = 4 : s(g, P_1) = 0, s(g, P_2) = 1, s(g, P_3) = 2 \Rightarrow \sum_i s(g, P_i) = 3 < g$$

$$g = 5 : s(g, P_1) = 0, s(g, P_2) = 1, s(g, P_3) = 2 \Rightarrow \sum_i s(g, P_i) = 3 < g$$

$$g = 6 : s(g, P_1) = 1, s(g, P_2) = 1, s(g, P_3) = 3 \Rightarrow \sum_i s(g, P_i) = 5 < g$$

At $t = 2$ all three models satisfy both the lower quota and the upper quota tests. Thus, the quota-divisor method with $d(a) = a + 1 - \frac{1}{4}$ will select either P_3 or P_2 since:

$$\frac{7}{1 + 1 - \frac{1}{4}} = \frac{3}{0 + 1 - \frac{1}{4}} > \frac{2}{0 + 1 - \frac{1}{4}}$$

However, we proved earlier that P_3 is not eligible for position $t = 2$ in the Tijdeman's algorithm. This proves the lemma. $\square$

5.6   Kubiak-Sethi algorithm

This algorithm is designed to minimize function (10.1) subject to constraints (10.3)-(10.4). The key idea of this algorithm consists in reduction to the assignment problem. For each unit of model i the ideal position is calculated according to the following formula

$$Z_j^{i*} = \left\lceil \frac{2j-1}{2r_i} \right\rceil \tag{10.36}$$

$n = 1, \ldots, n$ and $j = 1, \ldots, d_i$. Let C_{jt}^i be the cost of assigning the j-th unit of the model i to position t. If $t = Z_j^{i*}$, then the j-th unit of model i is produced in its ideal position in the sequence and thus the cost of such an assignment is $C_{jt}^i = 0$. If $t < Z_j^{i*}$, then the j-th unit of model i is produced too early. Consequently, the penalty ψ_{jl}^i (the excessive inventory cost) is incurred in each l between $l = t$ and $l = Z_j^{i*} - 1$. If $t > Z_j^{i*}$, then the j-th unit is produced too late. Consequently, the penalty ψ_{jl}^i (the shortage cost) is incurred in each l between $l = Z_j^{i*}$ and $l = t - 1$. Therefore, the cost of sequencing the j-th unit of model i in position t is calculated according to the following formula:

$$C_{jt}^i = \begin{cases} \sum_{l=t}^{Z_j^{i*}-1} \psi_{jl}^i & if \quad t < Z_j^{i*} \\ 0 & if \quad t = Z_j^{i*} \\ \sum_{l=Z_j^{i*}}^{t-1} \psi_{jl}^i & if \quad t > Z_j^{i*} \end{cases} \tag{10.37}$$

where

$$\psi_{jl}^i = ||j - lr_i| - |j - 1 - lr_i|| \tag{10.38}$$
$$(i,j) \in I = \{(i,j) : i = 1, \ldots, n; j = 1, \ldots, d_i\}, l = 1, \ldots, D.$$

Kubiak and Sethi (Kubiak and Sethi, 1994; Kubiak and Sethi, 1991) show that a solution minimizing function (10.1) subject to constraints (10.3)-(10.4) can be constructed from any optimal solution of the following assignment problem:

$$\min \sum_{t=1}^{D} \sum_{(i,j) \in I} C_{jt}^i x_{jt}^i \tag{10.39}$$

subject to

$$\sum_{(i,j) \in I} x_{jt}^i = 1, t = 1, \ldots, D$$

$$\sum_{t=1}^{D} x_{jt}^i = 1, (i,j) \in I$$

$$x^i_{jt} = 0 \text{ or } 1, k = 1, \ldots, D; (i, j) \in I$$

where

$$x^i_{jt} = \begin{cases} 1 & \text{if } (i, j) \text{ is assigned to period t,} \\ 0 & \text{otherwise.} \end{cases}$$

5.7   Characterization of Kubiak-Sethi algorithm

The characterization is based on the following three lemmas.

Lemma 3 *The Kubiak-Sethi algorithm does not stay within the quota.*

Proof: Corominas and Moreno (Corominas and Moreno, 2003) observe that no solution minimizing function (10.1) subject to constraints (10.3)-(10.4) stays within the quota for the instance of $n = 6$ models with their demands being $d_1 = d_2 = 23$ and $d_3 = d_4 = d_5 = d_6 = 1$. This proves the lemma since the Kubiak-Sethi algorithm minimizes (10.1). □

Lemma 4 *The Kubiak-Sethi algorithm is house monotone.*

Proof: The lemma follows from the constraint (10.4).□

Lemma 3 and Lemma 4 situate Kubiak-Sethi algorithm in the second quarter of the chart in Figure 10.1.

Theorem 6 *The Kubiak-Sethi algorithm is not uniform.*

Proof: Let us consider an instance consisting of $n = 5$ models with the following demand vector: $(7, 6, 4, 2, 1)$. All the optimal sequences obtained by the Kubiak-Sethi algorithm are listed below (sequences contain only indexes of models):

1 (1,2,3,1,2,4,1,3,2,1,5,2,3,1,2,4,1,3,2,1)

2 (1,2,3,1,4,2,1,3,2,1,5,2,3,1,2,4,1,3,2,1)

3 (1,2,3,1,2,4,1,3,2,1,5,2,3,1,4,2,1,3,2,1)

4 (1,2,3,1,4,2,1,3,2,1,5,2,3,1,4,2,1,3,2,1)

5 (1,2,3,4,1,2,1,3,2,1,5,2,3,1,2,4,1,3,2,1)

6 (1,2,3,4,1,2,1,3,2,1,5,2,3,1,4,2,1,3,2,1)

7 (1,2,3,4,1,2,1,3,2,1,5,2,3,1,2,1,4,3,2,1)

8 (1,2,3,1,2,4,1,3,2,1,5,2,3,1,2,1,4,3,2,1)

9 (1,2,3,1,4,2,1,3,2,1,5,2,3,1,2,1,4,3,2,1).

Moreover, there are 9 more that are the mirror reflection of the nine just presented sequences.

Let us now consider four model subproblem of the original problem $(7, 6, 4, 2, 1)$ made up of models 2, 3, 4 and 5 with demand vector $(6, 4, 2, 1)$. This subproblem has only a single optimal solution

$$(2, 3, 4, 2, 3, 2, 5, 2, 3, 2, 4, 3, 2).$$

It follows from the definition of uniformity given in Section 2.8 that if the Kubiak-Sethi algorithm is *uniform*, then it should also produce the sequence

$$\alpha = (2, 3, 2, 4, 3, 2, 5, 2, 3, 4, 2, 3, 2)$$

obtained from $(1, 2, 3, 1, 2, 4, 1, 3, 2, 1, 5, 2, 3, 1, 4, 2, 1, 3, 2, 1)$ by deleting model 1 but it does not since α is not optimal. Therefore, the Kubiak-Sethi algorithm is not uniform which ends the proof.□

Balinski and Shahidi (Balinski and Shahidi, 1998) point out an important practical feature of the uniform algorithms which is that the cancelation of some models from just in time production does not impact the order in which other models are sequenced for production by the uniform algorithm.

Finally, Balinski and Young (Balinski and Young, 1982) show that all population monotone methods are uniform, thus, by Theorem 6, the Kubiak-Sethi algorithm is not population monotone.

5.8   Characterization of just in time sequencing algorithms

The chart in Figure 1.3 summarizes our characterization of the Steiner-Yeomans, the Inman-Bulfin, and the Tijdeman and Kubiak-Sethi algorithms. The abbreviations used in the chart are explained below.

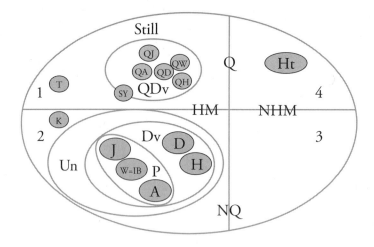

Figure 10.2. Characterization of just in time sequencing algorithms.

SY = Steiner-Yeomans algorithm

T = Tijdeman algorithm

IB = Inman-Bulfin algorithm

K = Kubiak-Sethi algorithm

6. Conclusions

We have shown that the Steiner-Yeomans algorithm is essentially a quota divisor method of apportionment. It remains open whether the algorithm is in fact equivalent to the Still algorithm. That is whether it is capable of producing any solution that the Still algorithm can produce for any instance of the problem. We conjecture that it is not.

We have also shown that the Tijdeman algorithm is a quasi quota divisor method which constraints the eligible sets more than the quota divisor method would normally do. Finally, we have shown that the Kubiak-Sethi algorithm is not uniform.

It also remains open if there exists any uniform method that stays within the quota. Balinski and Young (Balinski and Young, 1982) show that there exists no symmetric and uniform method that satisfies the quota. Since only symmetric methods seem desirable for the apportionment problem this result virtually closes the question by giving a negative answer in the context of the apportionment problem. However, the question remains open for the just in time sequencing. On the other hand any method that is uniform and weakly population monotone, we refer the reader to (Balinski and Young, 1982) for the definition of the latter, is a divisor method. Thus it is population monotone and, consequently, according to the Impossibility Theorem it can not stay within the quota.

Acknowledgment

The research of Wieslaw Kubiak has been supported by the Natural Science and Engineering Research Council of Canada Grant OGP0105675. The research of Joanna Józefowska and Łukasz Józefowski has been supported by the Polish Ministry of Education and Science Grant 3T11F 025 28.

References

Balinski, M., Ramirez, V. (1999). *Parametric methods of apportionment, rounding and production.* Mathematical Social Sciences 37, 107-122.

Balinski, M., Shahidi, N. (1998). *A simple approach to the product rate variation problem via axiomatics.* Operation Research Letters 22, 129–135.

Balinski, M., Young, H. (1975). *The Quota Method of Apportionment.* American Mathematical Monthly 82, 450–455.

Balinski, M., Young, H. (1982). *Fair Representation. Meeting the Ideal of One Man, One Vote,* Yale University Press.

Bautista, J., Companys, R., Corominas, A. (1996). *A Note on the Relation between the Product Rate Variation (PRV) Problem and the Apportionment Problem.* Journal of the Operational Research Society 47, 1410–1414.

Brauner, N., Crama, Y. (2004). *The maximum deviation just-in-time scheduling problem.* Discrete Applied Mathematics 134, 25-50.

Corominas, A., Moreno, N. (2003). *About the relations between optimal solutions for different types of min-sum balanced JIT optimisation problems.* Information Systems and Operational Research 41, 333–339.

Groenevelt, H., (1993). *The Just-in-Time Systems,* In: Graves, S.C., Rinnooy Kan, A.H.G., Zipkin, P.H., editors, Handbooks in Operations Research and Management Science Vol. 4, North Holland.

Inman, R., Bulfin, R. (1991). *Sequencing JIT mixed-model assembly lines.* Management Science 37, 901–904.

Kubiak, W. (2004). *Fair sequences.* In: Leung, J. Y-T., editor, *Handbook of Scheduling,* Chapman & Hall/CRC Computer and Information Science Series.

Kubiak, W. (2005). *Balancing Mixed-Model Supply Chains,* Chapter 6 in Volume 8 of the GERAD 25th Anniversary Series, Springer.

Kubiak, W., Sethi, S. (1991). *A note on: Level schedules for mixed-model assembly lines in just-in-time production systems.* Management Science 37, 121–122.

Kubiak, W., Sethi, S. (1994). *Optimal level schedules for flexible assembly lines in JIT production systems.* International Journal of Flexible Manufacturing Systems 6, 137–154.

Miltenburg, J. (1989). *Level schedules for mixed-model assembly lines in just-in-time production systems.* Management Science 35, 192–207.

Monden, Y., (1983). *Toyota Production Systems.* Industrial Engineering and Management Press, Norcross, GA.

Steiner, G., Yeomans, S. (1993). *Level Schedules for Mixed-model, Just-in-Time Processes.* Management Science 39, 728–735.

Still, J.W. (1979). *A class of new methods for Congressional Apportionment.* SIAM Journal on Applied Mathematics 37, 401–418.

Tijdeman, R. (1980). *The chairman assignment problem.* Discrete Mathematics 32, 323–330.

Vollman, T.E., Berry, W.L., Wybark, D.C. (1992). *Manufacturing Planning and Control Systems,* 3rd edition, IRWIN.

Chapter 11

LINEAR STOCHASTIC EQUATIONS IN A HILBERT SPACE WITH A FRACTIONAL BROWNIAN MOTION

B. Pasik-Duncan
Department of Mathematics
University of Kansas
Lawrence, KS 66049
bozenna@math.ku.edu

T. E. Duncan
Department of Mathematics
University of Kansas
Lawrence, KS 66049[*]
duncan@math.ku.edu

B. Maslowski
Institute of Mathematics
Czech Academy of Sciences
Pargue, Czech Republic
moslow@math.cas.cz

Keywords: Linear stochastic equations in Hilbert space, fractional Brownian motion, sample path properties of solutions, Ornstein-Uhlenbeck processes, stochastic linear partial differential equations

Abstract   A solution is obtained for a linear stochastic equation in a Hilbert space with a fractional Brownian motion. The Hurst parameter for the frac-

---

[*]Research supported in part by NSF grants DMS 0204669, DMS 050506, ANI 0125410, and GACR 201/04/0750

tional Brownian motion is not restricted. Sample path properties of the solution are obtained that depend on the Hurst parameter. An example of a stochastic partial differential equation is given.

1. Introduction

By comparison with the development of stochastic analysis of finite dimensional stochastic equations with fractional Gaussian noise, there are relatively few results available for the corresponding stochastic equations in an infinite dimensional space or for stochastic partial differential equations. For $H \in \left(\frac{1}{2}, 1\right)$, linear and semilinear equations with an additive fractional Gaussian noise, the formal derivative of a fractional Brownian motion are considered in [Dun00], [DPDM02], and [GA99]. Random dynamical systems described by such stochastic equations and their random fixed points are studied in [MS04]. A pathwise (or nonprobabilistic) approach is used in [MN02] to study a parabolic equation with a fractional Gaussian noise where the stochastic term is a nonlinear function of the solution. Strong solutions of bilinear evolution equations with a fractional Brownian motion are considered in [DMPD05] and [DJPD] and the same typ! e of equation is studied in [TTV03] where a fractional Feynman-Kac formula is obtained. A stochastic wave equation with a fractional Gaussian noise is considered in [Cai05] and a stochastic heat equation with a multiparameter fractional Gaussian noise is studied in [Hu01] and [HØZ04].

In this paper, stochastic linear evolution equations with an additive fractional Brownian motion are studied. In Section 2, some basic notions are recalled and the stochastic integral with respect to a cylindrical fractional Brownian motion in an infinite dimensional Hilbert space and deterministic integrands is introduced analogous to [AMN01] or [PT00] where the finite dimensional case is studied. Another approach to the stochastic integral here can use the results in [DJPD]. In Section 3, an infinite dimensional Ornstein-Uhlenbeck process is given as the solution of a linear stochastic equation in a Hilbert space. Some conditions are given so that the stochastic convolution integral, that defines the Ornstein-Uhlenbeck process as the mild solution of a linear equation, is well-defined and is suitably regular. For a stable linear equation, the existence of a limiting measure is verified. The case with the Hurst parameter H in $\left(0, \frac{1}{2}\right)$ i! s emphasized because this paper can be considered as complementary to [DPDM02] where $H \in \left(\frac{1}{2}, 1\right)$ is studied. An example of a linear stochastic partial differential equation is given.

2. Preliminaries

A cylindrical fractional Brownian motion in a separable Hilbert space is introduced, a Wiener-type stochastic integral with respect to this process is defined, and some basic properties of this integral are noted. Initially, some facts from the theory of fractional integration (e.g. [SKM93]) are described. Let $(V, \| \cdot \|, \langle \cdot, \cdot \rangle)$ be a separable Hilbert space and $\alpha \in (0,1)$. If $\varphi \in L^1([0,T], V)$, then the left-sided and right-sided fractional (Riemann-Liouville) integrals of φ are defined (for almost all $t \in [0,T]$) by

$$\left(I_{0+}^\alpha \varphi\right)(t) = \frac{1}{\Gamma(\alpha)} \int_0^t (t-s)^{\alpha-1} \varphi(s)\, ds$$

and

$$\left(I_{T-}^\alpha \varphi\right)(t) = \frac{1}{\Gamma(\alpha)} \int_t^T (s-t)^{\alpha-1} \varphi(s)\, ds$$

respectively where $\Gamma(\cdot)$ is the gamma function. The inverse operators of these fractional integrals are called fractional derivatives and can be given by their Weyl representations

$$\left(D_{0+}^\alpha \psi\right)(t) = \frac{1}{\Gamma(1-\alpha)} \left(\frac{\psi(t)}{t^\alpha} + \alpha \int_0^t \frac{\psi(t) - \psi(s)}{(t-s)^{\alpha+1}}\, ds \right)$$

and

$$\left(D_{T-}^\alpha \psi\right)(t) = \frac{1}{\Gamma(1-\alpha)} \left(\frac{\psi(t)}{(T-t)^\alpha} + \alpha \int_t^T \frac{\psi(s) - \psi(t)}{(s-t)^{\alpha+1}}\, ds \right)$$

where $\psi \in I_{0+}^\alpha \left(L^1([0,T], V)\right)$ and $\psi \in I_{T-}^\alpha \left(L^1([0,T], V)\right)$ respectively. Let $K_H(t,s)$ for $0 \le s \le t \le T$ be the kernel function

$$K_H(t,s) = c_H (t-s)^{H-\frac{1}{2}}$$
$$+ c_H \left(\frac{1}{2} - H \right) \int_s^t (u-s)^{H-\frac{3}{2}} \left(1 - \left(\frac{s}{u} \right)^{\frac{1}{2}-H} \right) du \quad (11.1)$$

where

$$c_H = \left[\frac{2H\Gamma\left(H + \frac{1}{2}\right)\Gamma\left(\frac{3}{2} - H\right)}{\Gamma(2 - 2H)} \right]^{\frac{1}{2}} \quad (11.2)$$

and $H \in (0,1)$. If $H \in \left(\frac{1}{2}, 1\right)$, then K_H has a simpler form as

$$K_H(t,s) = c_H \left(H - \frac{1}{2} \right) s^{\frac{1}{2} - H} \int_s^t (u-s)^{H-\frac{3}{2}} u^{H-\frac{1}{2}}\, du$$

A definition of a stochastic integral of a deterministic V-valued function with respect to a scalar fractional Brownian motion $(\beta(t), t \geq 0)$ is described. The approach follows from [AMN01], [DÜ99]. Another approach is given in [DJPD].

Let $\mathcal{K}_H^*$ be the linear operator given by

$$\mathcal{K}_H^* \varphi(t) = \varphi(t) K_H(T, t) + \int_t^T (\varphi(s) - \varphi(t)) \frac{\partial K_H}{\partial s}(s, t) \, ds \qquad (11.3)$$

for $\varphi \in \mathcal{E}$ where $\mathcal{E}$ is the linear space of V-valued step functions on $[0, T]$ and $\mathcal{K}_H^* : \mathcal{E} \to L^2([0, T], V)$. For $\varphi \in \mathcal{E}$,

$$\varphi(t) = \sum_{i=1}^{n-1} x_i \mathbb{1}_{[t_i, t_{i+1})}(t)$$

where $x_i \in V$, $i \in \{1, \ldots, n-1\}$ and $0 = t_1 < \cdots < t_n = T$.

Define

$$\int_0^T \varphi \, d\beta := \sum_{i=1}^n x_i \left(\beta_i(t_{i+1}) - \beta(t_i) \right). \qquad (11.4)$$

It follows directly that

$$\mathbb{E} \left\| \int_0^T \varphi \, d\beta \right\|^2 = |\mathcal{K}_H^* \varphi|_{L^2([0,T],V)}^2 \qquad (11.5)$$

Let $(\mathcal{H}, |\cdot|_{\mathcal{H}}, \langle \cdot, \cdot \rangle_{\mathcal{H}})$ be the Hilbert space obtained by the completion of the pre-Hilbert space $\mathcal{E}$ with the inner product

$$\langle \varphi, \psi \rangle_{\mathcal{H}} := \langle \mathcal{K}_H^* \varphi, \mathcal{K}_H^* \psi \rangle_{L^2([0,T],V)} \qquad (11.6)$$

for $\varphi, \psi \in \mathcal{E}$. The stochastic integral (11.4) is extended to $\varphi \in \mathcal{H}$ by the isometry (11.5). Thus $\mathcal{H}$ is the space of integrable functions and it is useful to obtain some more specific information. If $H \in \left(\frac{1}{2}, 1 \right)$, then it is easily verified that $\mathcal{H} \supset \tilde{\mathcal{H}}$ where $\tilde{\mathcal{H}}$ is the Banach space of Borel measurable functions with the norm $|\cdot|_{\tilde{\mathcal{H}}}$ given by

$$|\varphi|_{\tilde{\mathcal{H}}}^2 := \int_0^T \int_0^T \|\varphi(u)\| \|\varphi(v)\| \phi(u - v) \, du \, dv \qquad (11.7)$$

where $\phi(u) = H(2H - 1)|u|^{2H-2}$ and it is elementary to verify that $\tilde{\mathcal{H}} \supset L^p([0, t], V)$ for $p > \frac{1}{H}$, and, in particular, for $p = 2$ (e.g. [DPDM02]). If $\varphi \in \tilde{\mathcal{H}}$ and $H > \frac{1}{2}$, then

$$\mathbb{E} \left\| \int_0^T \varphi \, d\beta \right\|^2 = \int_0^T \int_0^T \langle \varphi(u), \varphi(v) \rangle \phi(u - v) \, du \, dv. \qquad (11.8)$$

If $H \in \left(0, \frac{1}{2}\right)$, then the space of integrable functions is smaller than for $H \in \left(\frac{1}{2}, 1\right)$. It is known that $\mathcal{H} \supset H^1([0,T], V)$ (e.g. [Hu05] Lemma 5.20) and $\mathcal{H} \supset C^\beta([0,T], V)$ for each $\beta > \frac{1}{2} - H$ (a more specific result is given in the next section). If $H \in \left(0, \frac{1}{2}\right)$, then the linear operator $\mathcal{K}_H^*$ can be described by a fractional derivative

$$\mathcal{K}_H^* \varphi(t) = c_H t^{\frac{1}{2} - H} D_{T-}^{\frac{1}{2} - H} \left(u_{H - \frac{1}{2}} \varphi \right) \tag{11.9}$$

and its domain is $\mathcal{H} = I_{T-}^{\frac{1}{2} - H} \left(L^2([0,T], V) \right)$ ([AMN01] Proposition 6).

DEFINITION 11.1 *Let $(\Omega, \mathcal{F}, \mathbb{P})$ be a complete probability space. A cylindrical process $\langle B, \cdot \rangle \colon \Omega \times \mathbb{R}_+ \times V \to \mathbb{R}$ on $(\Omega, \mathcal{F}, \mathbb{P})$ is called a standard cylindrical fractional Brownian motion with Hurst parameter $H \in (0,1)$ if*

1 *For each $x \in V \setminus \{0\}$, $\frac{1}{\|x\|} \langle B(\cdot), x \rangle$ is a standard scalar fractional Brownian motion with Hurst parameter H.*

2 *For $\alpha, \beta \in \mathbb{R}$ and $x, y \in V$,*

$$\langle B(t), \alpha x + \beta y \rangle = \alpha \langle B(t), x \rangle + \beta \langle B(t), y \rangle \quad a.s. \ \mathbb{P}.$$

Note that $\langle B(t), x \rangle$ has the interpretation of the evaluation of the functional $B(t)$ at x though the process $B(\cdot)$ does not take values in V.

For $H = \frac{1}{2}$, this is the usual definition of a standard cylindrical Wiener process in V. By condition 2 in Definition 11.1, the sequence of scalar processes $(\beta_n(t), t \geq 0, n \in \mathbb{N})$ are independent.

Associated with $(B(t), t \geq 0)$ is a standard cylindrical Wiener process $(W(t), t \geq 0)$ in V such that formally $B(t) = K_H \left(\dot{W}(t) \right)$. For $x \in V \setminus \{0\}$, let $\beta_x(t) = \langle B(t), x \rangle$. It is elementary to verify from (11.4) that there is a scalar Wiener process $(w_x(t), t \geq 0)$ such that

$$\beta_x(t) = \int_0^t K_H(t,s) \, dw_x(s) \tag{11.10}$$

for $t \in \mathbb{R}_+$. Furthermore, $w_x(t) = \beta_x \left((\mathcal{K}_H^*)^{-1} \mathbb{1}_{[0,t)} \right)$ where $\mathcal{K}_H^*$ is given by (11.3). Thus there is the formal series

$$W(t) = \sum_{n=1}^\infty w_n(t) e_n. \tag{11.11}$$

Now the stochastic integral $\int_0^T G \, d\beta$ is defined for an operator-valued function $G \colon [0,T] \to \mathcal{L}(V)$ is a V-valued random variable.

DEFINITION 11.2 *Let $G : [0,T] \to \mathcal{L}(V)$, $(e_n, n \in \mathbb{N})$ be a complete orthonormal basis in V, $g_{e_n}(t) := G(t)e_n$, $g_{e_n} \in \mathcal{H}$ for $n \in \mathbb{N}$, and B be a standard cylindrical fractional Brownian motion. Define*

$$\int_0^T G\,dB := \sum_{n=1}^\infty \int_0 Ge_n\,d\beta_n \qquad (11.12)$$

provided the infinite series converges in $L^2(\Omega)$.

The next proposition describes some $\mathcal{L}(V)$-valued functions G that satisfy Definition 11.2.

PROPOSITION 11.3 *Let $G\colon [0,T] \to \mathcal{L}(V)$ and $G(\cdot)x \in \mathcal{H}$ for each $x \in V$. Let $\Gamma_T\colon V \to L^2([0,T],V)$ be given as*

$$(\Gamma_T x)\,(t) = (\mathcal{K}_H^* Gx)\,(t) \qquad (11.13)$$

for $t \in [0,T]$ and $x \in V$. If $\Gamma_T \in \mathcal{L}_2(V, L^2([0,T],V))$—that is, Γ_T is a Hilbert-Schmidt operator—then the stochastic integral (11.12) is a well-defined centered Gaussian V-valued random variable with covariance operator $\tilde{Q}_T$ given by

$$\tilde{Q}_T x = \int_0^T \sum_{n=1}^\infty \langle (\Gamma_T e_n)\,(s), x\rangle\,(\Gamma_T e_n)\,(s)\,ds.$$

This integral does not depend on the choice of the complete orthonormal basis $(e_n, n \in \mathbb{N})$.

Proof The terms of the sum on the right-hand side of (11.12) are well-defined V-valued Gaussian random variables by the construction of the integral $\int_0^t g_{e_n}\,d\beta_n$ and these terms for $n \in \mathbb{N}$ are independent random variables. Furthermore,

$$\mathbb{E}\left\|\sum_{k=m}^\infty \int_0^T G(s)e_k\,d\beta_k(s)\right\|^2 = \sum_{k=m}^\infty \mathbb{E}\left\|\int_0^T G(s)e_k\,d\beta_k(s)\right\|^2$$

$$= \sum_{k=m}^\infty \int_0^T \|(\mathcal{K}_H^* G(s)e_k)^*\,(s)\|^2\,ds$$

$$= \sum_{k=m}^\infty \int_0^T \|(\Gamma_T e_k)\,(s)\|^2\,ds.$$

This last sum converges to zero as $m \to \infty$ because Γ_T is Hilbert-Schmidt.

To verify that (11.12) is a Gaussian random variable with covariance $\tilde{Q}_T$ note that for $\varphi \in \mathcal{H}$ and $x \in V$ it follows that

$$\int_0^t \varphi(s)\, d\beta_y(s) = \int_0^t \left(\mathcal{K}_H^* \varphi\right)(s)\, dw_x(s) \tag{11.14}$$

where $(w_x(t), t \geq 0)$ is the Wiener process given by (11.10). Thus the terms on the right-hand side of (11.12) are V-valued, zero mean independent Gaussian random variables with the covariance operator

$$\tilde{Q}_T^{(n)} x = \int_0^T \langle \left(\mathcal{K}_H^* Ge_n\right)(s), x \rangle \left(\mathcal{K}_H^* Ge_n\right)(s)\, ds \tag{11.15}$$

where $n \in \mathbb{N}$ and $x \in V$. It follows from (11.12) that

$$\begin{aligned}
\tilde{Q}_T(x) &= \sum_{n=1}^{\infty} \int_0^T \langle \mathcal{K}_H^* Ge_n, x \rangle \left(\mathcal{K}_H^* Ge_n\right)(s)\, ds \\
&= \int_0^T \sum_{n=1}^{\infty} \langle \left(\Gamma_T e_n\right)(s), x \rangle \left(\Gamma_T e_n\right)(s)\, ds.
\end{aligned} \tag{11.16}$$

The latter infinite series converges because Γ_T is Hilbert-Schmidt. The integral does not depend on the orthonormal basis that is chosen from (11.14) and the analogous result for stochastic integrals with respect to a cylindrical Wiener process. $\qquad \square$

REMARK 11.4 *Since* $\Gamma_T \in \mathcal{L}_2\left(V, L^2([0,T], V)\right)$, *it follows that the map* $x \mapsto (\Gamma_T x)(t)$ *is Hilbert-Schmidt on* V *for almost all* $t \in [0,T]$. *Let* Γ_T^* *be the adjoint of* Γ_T. *Then* Γ_T^* *is also Hilbert-Schmidt, and* Q_T *can be expressed as*

$$Q_T x = \int_0^T \left(\Gamma_T \left(\Gamma_T^* x\right)\right)(t)\, dt \tag{11.17}$$

for $x \in V$. *In fact, for* $x, y \in V$, *it follows by (11.16) that*

$$\begin{aligned}
\left\langle \tilde{Q}_T x, y \right\rangle &= \int_0^T \sum_{n=1}^{\infty} \langle \left(\Gamma_T e_n\right)(s), x \rangle \langle \left(\Gamma_T e_n\right)(s), y \rangle\, ds \\
&= \int_0^T \sum_{n=1}^{\infty} \langle e_n, \left(\Gamma_T^* x\right)(s) \rangle \langle e_n, \left(\Gamma_T^* y\right)(s) \rangle\, ds \\
&= \int_0^T \langle \left(\Gamma_T^* x\right)(s), \left(\Gamma_T^* y\right)(s) \rangle\, ds \\
&= \int_0^T \langle \left(\Gamma_T \left(\Gamma_T^* x\right)\right)(s), y \rangle\, ds
\end{aligned} \tag{11.18}$$

If $H \in \left(\frac{1}{2}, 1\right)$ and G satisfies

$$\int_0^T \int_0^T |G(u)|_{\mathcal{L}_2(V)} |G(v)|_{\mathcal{L}_2(V)} \phi(u - v) \, du \, dv < \infty$$

then

$$Q_T = \int_0^T \int_0^T G(u) G^*(v) \phi(u - v) \, du \, dv$$

where ϕ is given by (11.7) ([DPDM02], Proposition 2.2).

Now a result is given for the action of a closed (unbounded) linear operator on the stochastic integral.

PROPOSITION 11.5 *If $\tilde{A} \colon \mathrm{Dom}\left(\tilde{A}\right) \to V$ is a closed linear operator, $G \colon [0, T] \to \mathcal{L}(V)$ satisfies $G([0, T]) \subset \mathrm{Dom}\left(\tilde{A}\right)$ and both G and $\tilde{A}G$ satisfy the conditions for G in Proposition 11.3, then*

$$\int_0^T G \, dB \subset \mathrm{Dom}\left(\tilde{A}\right) \quad a.s. \ \mathbb{P}$$

and

$$\tilde{A} \int_0^T G \, dB = \int_0^T \tilde{A} G \, dB \quad a.s. \ \mathbb{P}. \tag{11.19}$$

Proof By the assumptions, it follows that $G e_n \in \mathcal{H}$ and $\tilde{A} G e_n \in \mathcal{H}$ for each $n \in \mathbb{N}$. So, by an approximation of the integrands by step functions, it easily follows that

$$\tilde{A} \int_0^T G e_n \, d\beta_n = \int_0^T \tilde{A} G e_n \, d\beta_n.$$

Since the integrals are Gaussian random variables, it follows that

$$\lim_{m \to \infty} \sum_{n=1}^m \int_0^T G e_n \, d\beta_n = \int_0^T G \, dB \tag{11.20}$$

in $L^2(\Omega)$ and almost surely. Similarly,

$$\lim_{m \to \infty} \tilde{A} \sum_{n=1}^m \int_0^T G e_n \, d\beta_n = \int_0^T \tilde{A} G \, d\beta.$$

Since $\tilde{A}$ is a closed linear operator, it follows that

$$\int_0^T G \, dB \in \mathrm{Dom}\left(\tilde{A}\right) \quad a.s. \ \mathbb{P}$$

and (11.16) is satisfied. $\qquad \square$

3. Fractional Ornstein-Uhlenbeck Processes

In this section, some properties of a fractional Ornstein-Uhlenbeck (O-U) process are investigated. This process is a mild solution of the linear stochastic equation

$$dZ(t) = AZ(t)\, dt + \Phi\, dB(t)$$
$$Z(0) = x \qquad\qquad (11.21)$$

where $Z(t), x \in V$, $(B(t), t \geq 0)$ is a cylindrical fractional Brownian motion (FBM) with the Hurst parameter $H \in (0,1)$, $\Phi \in \mathcal{L}(V)$, $A \colon \mathrm{Dom}(A) \to V$, $\mathrm{Dom}(A) \subset V$ and A is the infinitesimal generator of a strongly continuous semigroup $(S(t), t \geq 0)$ on V. A mild solution of (11.21) is given by

$$Z(t) = S(t)x + \int_0^t S(t-s)\Phi\, dB(s) \qquad (11.22)$$

where the stochastic integral on the right-hand side is given by Definition 11.2. Thus it is necessary to consider the existence and some other properties of the following stochastic convolution integral

$$\tilde{Z}(t) := \int_0^t S(t-s)\Phi\, dB(s) \qquad (11.23)$$

In a significant portion of this section, it is assumed that $(S(t), t \geq 0)$ is an analytic semigroup. In this case, there is a $\hat{\beta} \in \mathbb{R}$ such that the operator $\hat{\beta}I - A$ is uniformly positive on V. For each $\delta \geq 0$, $(V_\delta, \|\cdot\|_\delta)$ is a Hilbert space where $V_\delta = \mathrm{Dom}\left(\left(\hat{\beta}I - A\right)^\delta\right)$ with the graph norm topology so that

$$\|x\|_\delta = \left\|\left(\hat{\beta}I - A\right)^\delta x\right\|.$$

The shift $\hat{\beta}$ is fixed. The space V_δ does not depend on $\hat{\beta}$ because the norms are equivalent for different values of $\hat{\beta}$ satisfying the above condition.

The case $H \in \left(0, \frac{1}{2}\right)$ is primarily considered because the case $H \in \left(\frac{1}{2}, 1\right)$ has been treated in [DPDM02]. Only the main result in [DPDM02] is described.

PROPOSITION 11.6 *If $H \in \left(\frac{1}{2}, 1\right)$, $S(t)\Phi \in \mathcal{L}_2(V)$ for each $t > 0$ and*

$$\int_0^T \int_0^T u^{-\alpha} v^{-\alpha} |S(u)\Phi|_{\mathcal{L}_2(V)} |S(v)\Phi|_{\mathcal{L}_2(V)} \phi(u-v)\, du\, dv < \infty \qquad (11.24)$$

for some T_0 and $\alpha > 0$ where $\phi(u) = H(2H - 1)|u|^{2H-2}$, then there is a Hölder continuous V-valued version of the process $(\tilde{Z}(t), t \geq 0)$ with Hölder exponent $\beta < \alpha$. If $(S(t), t \geq 0)$ is analytic then there is a version of $(\tilde{Z}(t), t \geq 0)$ in $C^\beta([0,T], V_\delta)$ for each $T > 0$ and $\beta + \delta < \alpha$.

Proof This verification basically follows from the proofs of Propositions 3.2 and 3.3 of [DPDM02] with $\beta = 0$. It only remains to note that the same proofs yield the Hölder continuity for the sample paths of $\tilde{Z}$. Specifically, in Proposition 3.2 of [DPDM02], it is verified that

$$\tilde{Z}(t) = R_\alpha(Y)(t)$$

for $t \in [0, T]$ where

$$Y(t) := \int_s^t (t-s)^{-\alpha} S(t-s)\Phi\, dB(s)$$

$Y \in L^p([0,T], V)$ for each $p \geq 1$ and

$$(R_\alpha \varphi)(t) := \int_0^t (t-s)^{\alpha-1} S(t-s)\varphi(s)\, ds$$

It is well-known (e.g. Proposition A.1.1 [DPZ96]) that $R_\alpha : L^p[([0,T], V) \to C^\beta([0,T], V)$ for $\alpha > \frac{1}{p}$ and $\beta \in \left(0, \alpha - \frac{1}{p}\right)$ and $R_\alpha : L^p([0,T], V) \to C^{\alpha-\delta-\frac{1}{p}}([0,T], V_\delta)$ for $\delta > 0$ and $\delta < \alpha - \frac{1}{p}$. Since p is arbitrarily large, the proof is complete. $\qquad\square$

In the remainder of this section, the case $H \in \left(0, \frac{1}{2}\right)$ is considered.

LEMMA 11.7 *Let $H \in \left(0, \frac{1}{2}\right)$. If $(S(t), t \geq 0)$ is an analytic semigroup, then for each $x \in V$, $S(T - \cdot)\Phi x \in \mathcal{H}$ and*

$$|\mathcal{K}_H^* (S(T - \cdot)\Phi x)|^2_{L^2([0,T],V)}$$
$$\leq c \int_0^T \left(\frac{\|S(T-s)\Phi x\|^2}{(T-s)^{1-2H}} + \frac{\|S(T-s)\Phi x\|^2}{s^{1-2H}} \right.$$
$$\left. + \frac{\left|S\left(\frac{1}{2}(T-s)\right)\Phi x\right|^2}{(T-s)^{2\beta}} \right) ds \quad (11.25)$$

for each $\beta \in \left(\frac{1}{2} - H, \frac{1}{2}\right)$ and a constant $c = c_\beta$.

Proof By (11.9) there is the equality

$$\mathcal{K}_H^* (S(T - \cdot)\Phi x)(s) = c_H s^{\frac{1}{2}-H} D_{T-}^{\frac{1}{2}-H} \left(u_{H-\frac{1}{2}} S(T - \cdot)\Phi x \right)(s) \quad (11.26)$$

$$|\mathcal{K}_H^*\left(S(T-\cdot)\Phi x\right)|^2_{L^2([0,T],V)}$$

$$\leq c \int_0^T \left(\frac{\|S(T-s)\Phi x\|^2}{(T-s)^{1-2H}}\right.$$

$$\left. + s^{1-2H} \int_s^T \frac{\left\|r^{H-\frac{1}{2}}S(T-r)\Phi x - s^{H-\frac{1}{2}}S(T-s)\Phi x\right\|}{(r-s)^{\frac{3}{2}-H}} \, dr\right)^2 \, ds$$

$$\leq c \left(\int_0^T \left(\frac{\|S(T-s)\Phi x\|^2}{(T-s)^{1-2H}}\right.\right.$$

$$\left. + s^{1-2H}\|S(T-s)\Phi x\|^2 \int_s^T \frac{\left|r^{H-\frac{1}{2}} - s^{H-\frac{1}{2}}\right|}{(r-s)^{\frac{3}{2}-H}} \, dr\right)^2$$

$$\left. + s^{1-2H}\left(\int_s^T \frac{\|S(t-r)\Phi x - S(t-s)\Phi x\|^2 r^{H-\frac{1}{2}}}{(r-s)^{\frac{3}{2}-H}} \, dr\right)^2 \, ds\right)$$

$$\leq c \left(\int_0^T \left(\frac{\|S(T-s)\Phi x\|^2}{(T-s)^{1-2H}} + \frac{\|S(T-s)\Phi x\|^2}{s^{1-2H}}\right.\right.$$

$$\left.\left. + \left(\int_s^T \frac{\|S(T-r)\Phi x - S(T-s)\Phi x\|^2}{(r-s)^{\frac{3}{2}-H}} \, dr\right)^2\right) \, ds\right)$$

where c represents a generic constant that may differ at each use. To obtain the inequality above, the inequality

$$\left(\int_s^T \frac{\left|r^{H-\frac{1}{2}} - s^{H-\frac{1}{2}}\right|}{(r-s)^{\frac{3}{2}-H}} \, dr\right)^2 \leq c s^{-2+4H}$$

is used as well as $r^{H-\frac{1}{2}} \leq s^{H-\frac{1}{2}}$ because $r \geq s$. The first two integrands on the right-hand side of (11.26) correspond to those in (11.25) so it only remains to estimate the last term. By the analyticity of $(S(t), t \geq 0)$, it follows that

$$\int_0^T \left(\int_s^T \frac{\|S(t-r)\Phi x - S(T-s)\Phi x\|^2}{(r-s)^{\frac{3}{2}-H}} \, dr \right)^2 ds$$

$$\leq \int_0^T \left(\int_s^T \frac{|S(r-s)-I|_{\mathcal{L}(V_\beta,V)} \left| S\left(\frac{T-r}{2}\right)\right|_{\mathcal{L}(V,V_\beta)} \left\| S\left(\frac{T-r}{2}\right)\Phi x \right\|}{(r-s)^{\frac{3}{2}-H}} \, dr \right)^2 ds$$

$$\leq c \int_0^T \left(\int_s^T \frac{(r-s)^\beta \left\| S\left(\frac{T-r}{2}\right)\Phi x \right\|}{(T-r)^\beta (r-s)^{\frac{3}{2}-H}} \, dr \right)^2 ds$$

$$= c \int_0^T \left(\int_0^{T-s} \frac{\left\| S\left(\frac{T-s-r}{2}\right)\Phi x \right\|}{(T-s-r)^\beta} \frac{1}{r^{\frac{3}{2}-H-\beta}} \, dr \right)^2 ds$$

$$\leq c \left(\int_0^T \frac{dr}{r^{\frac{3}{2}-H-\beta}} \right)^2 \int_0^T \frac{\|S(t-r)\Phi x\|^2}{(T-r)^{2\beta}} \, dr \quad (11.27)$$

where the last inequality follows from the Young inequality. This final inequality gives the estimate for the last term on the right-hand side of (11.25) and completes the proof. □

The next result ensures that the stochastic convolution is a well-defined process.

THEOREM 11.8 *Let $S(t), t \geq 0$) be an analytic semigroup and $H \in \left(0, \frac{1}{2}\right)$. If for each $t \in (0, T)$ the operator $S(t)\Phi$ is Hilbert-Schmidt on V and*

$$\int_0^T \left(\frac{|S(s)\Phi|^2_{\mathcal{L}_2(V)}}{(T-s)^{1-2H}} + \frac{|S(s)\Phi|^2_{\mathcal{L}_2(V)}}{s^{2\beta}} \right) ds < \infty \quad (11.28)$$

for some $\beta \in \left(\frac{1}{2} - H, \frac{1}{2}\right)$, then the stochastic integral process $(\tilde{Z}(t), t \geq 0)$ given by (11.23) is well-defined and admits a measurable version.

Proof To verify that the stochastic integral with the integrand given by the operator-valued function $t \mapsto S(T-t)\Phi$ is well-defined using Proposition 11.3, it is necessary to show that

$$\sum_{n=1}^\infty \int_0^T \|\mathcal{K}_H^* S(T-s)\Phi e_n\|^2 \, ds < \infty \quad (11.29)$$

where $(e_n, n \in \mathbb{N})$ is a complete orthonormal basis in V. By Lemma 11.7, it follows that

$$\sum_{n=1}^{\infty} \int_0^T \left\| \mathcal{K}_H^* \left(S(T - \cdot)\Phi e_n \right)(s) \right\|^2 ds$$

$$\leq c \sum_{n=1}^{\infty} \int_0^T \left(\frac{\|S(T-s)\Phi x\|^2}{(T-s)^{1-2H}} + \frac{\|S(T-s)\Phi e_n\|^2}{s^{1-2H}} \right.$$

$$\left. + \frac{\left\| S\left(\frac{T-s}{2}\right)\Phi e_n \right\|^2}{(T-s)^{2\beta}} \right) ds$$

$$= c \int_0^T \left(\frac{|S(T-s)\Phi|^2_{\mathcal{L}_2(V)}}{(T-s)^{1-2H}} + \frac{|S(T-s)\Phi|^2_{\mathcal{L}_2(V)}}{s^{1-2H}} \right.$$

$$\left. + \frac{\left| S\left(\frac{T-s}{2}\right)\Phi \right|^2_{\mathcal{L}_2(V)}}{(T-s)^{2\beta}} \right) ds$$

$$\leq c \int_0^T \left(\frac{|S(s)\Phi|^2_{\mathcal{L}_2(V)}}{s^{2\beta}} + \frac{|S(s)\Phi|^2_{\mathcal{L}_2(V)}}{(T-s)^{1-2H}} \right) ds \quad (11.30)$$

for some generic constant c where $2\beta > 1 - 2H$. The right-hand side of the inequality (11.30) is finite by (11.28) so $(\tilde{Z}(t), t \in [0,T])$ is a well-defined process for $T > 0$. To verify the existence of a measurable version, the approach in Proposition 3.6 of [DPZ92] is used by showing the mean square continuity of $(\tilde{Z}(t), t \in [0,T])$ from the right. For $0 \leq T_2 \leq T_1 < T$ it follows that

$$\mathbb{E}\left[\|\tilde{Z}(T_1) - \tilde{Z}(T_2)\|^2 \right] \leq I_1 + I_2 \quad (11.31)$$

where

$$I_1 = 2\mathbb{E}\left[\left\| \int_{T_2}^{T_1} S(T_1 - t)\Phi \, dB(t) \right\|^2 \right]$$

and

$$I_2 = 2\mathbb{E}\left[\left\| (S(T_1 - T_2) - I) \int_0^{T_2} S(T_2 - r)\Phi \, dB(r) \right\|^2 \right].$$

Proceeding as in (11.30) where the interval $[0, T]$ is replaced by $[T_2, T_1]$ it follows that

$$I_1 \leq c \int_{T_2}^{T_1} \left(\frac{|S(T_1 - r)\Phi|^2_{\mathcal{L}_2(V)}}{(T_1 - r)^{2\beta}} + \frac{|S(T_1 - r)\Phi|_{\mathcal{L}_2(V)}}{(r - T_2)^{1-2H}} \right) dr \quad (11.32)$$

which tends to zero as $T_1 \downarrow T_2$ by (11.28).

Furthermore, it follows that

$$\left\| (S(T_1 - T_2) - I) \int_0^{T_2} S(T_2 - r)\Phi \, dB(r) \right\| \to 0 \quad \text{a.s. } \mathbb{P} \qquad (11.33)$$

as $T_1 \downarrow T_2$ by the strong continuity of the semigroup $(S(t), t \geq 0)$ and it easily follows that

$$\left\| (S(T_1 - T_2) - I) \int_0^{T_2} S(T_2 - r)\Phi \, dB(r) \right\|^2 \leq c \left\| \int_0^{T_2} S(T_2 - r)\Phi \, dB(r) \right\|^2$$

for a constant c that does not depend on T_1. Thus $I_2 \to 0$ as $T_1 \downarrow T_2$ by the Dominated Convergence Theorem. $\qquad \square$

COROLLARY 11.9 *Let $(S(t), t \geq 0)$ be an analytic semigroup and $H \in \left(0, \frac{1}{2}\right)$. If*

$$|S(t)\Phi|_{\mathcal{L}_2(V)} \leq ct^{-\gamma} \qquad (11.34)$$

for $t \in [0, T]$, $c \geq 0$ and $\gamma \in [0, H)$, then the integrability condition (11.28) is satisfied and thus $(\tilde{Z}(t), t \in [0, T])$ is a well-defined V-valued process with a measurable version.

Proof The condition (11.28) is

$$\int_0^T \left(\frac{1}{(T-s)^{1-2H} s^{2\gamma}} + \frac{1}{s^{2\beta + 2\gamma}} \right) ds < \infty \qquad (11.35)$$

which is satisfied if $2\beta + 2\gamma < 1$. On the other hand, it is necessary that $\beta > \frac{1}{2} - H$ so a suitable β can be chosen if $2\left(\frac{1}{2} - H\right) + 2\gamma < 1$, that is, $\gamma < H$. $\qquad \square$

The condition (11.34) is one that can be verified in many specific examples. The results in the remainder of this section are formulated using (11.34).

LEMMA 11.10 *Let $(S(t), t \geq 0)$ be an analytic semigroup and $H \in \left(0, \frac{1}{2}\right)$. If (11.34) is satisfied with $\gamma \in [0, H - \delta)$ for some $\delta \geq 0$, then the stochastic convolution integral takes values in V_δ (a.s. $\mathbb{P}$).*

Proof Since the linear operator $\tilde{A} = (\hat{\beta}I - A)^\delta$ is closed, by Proposition 11.5, it is only necessary to verify that the operator-valued function $t \mapsto \tilde{A}S(T - t)\Phi$ is integrable on $[0, T]$. It follows directly that

$$|\tilde{A}S(t)\Phi|_{\mathcal{L}_2(V)} \leq \left| \tilde{A}S\left(\frac{t}{2}\right) \right|_{\mathcal{L}(V)} \left| S\left(\frac{t}{2}\right)\Phi \right|_{\mathcal{L}_2(V)} \leq ct^{\gamma - \delta} \qquad (11.36)$$

for some $c > 0$ and all $t \in (0, T]$. Thus the proofs of Theorem 11.8 and Corollary 11.9 can be repeated replacing γ by $\gamma + \delta$. □

Now a sample path property of the stochastic convolution can be established.

THEOREM 11.11 *Let* $(S(t), t \geq 0)$ *be an analytic semigroup,* $H \in \left(0, \frac{1}{2}\right)$ *and (11.34) be satisfied. Let* $\alpha \geq 0$ *and* $\delta \geq 0$ *satisfy*

$$\alpha + \beta + \gamma < H. \tag{11.37}$$

Then there is a version of the process $(\tilde{Z}(t), t \in [0, T])$ *with* $C^{\alpha}([0, T], V_{\delta})$ *sample paths.*

Proof For $T_1 > T_2$ it follows that

$$\mathbb{E}\|\tilde{Z}(T_1) - \tilde{Z}(T_2)\|_{\delta}^2 \leq I_1 + I_2 \tag{11.38}$$

where

$$I_1 = 2\mathbb{E}\left\|\int_{T_2}^{T_1} S(T - r)\Phi \, dB(r)\right\|_{\delta}^2$$

and

$$I_2 = 2\mathbb{E}\left\|(S(T_1 - T_2) - I)\int_0^{T_2} S(T_2 - r)\Phi \, dB(r)\right\|_{\delta}^2.$$

Proceeding as in (11.30) or (11.32) with $S(T_1 - r)\Phi$ replaced by $\tilde{A}S(T_2 - r)\Phi$ and using (11.36) it follows that

$$
\begin{aligned}
I_1 &\leq c\int_{T_2}^{T_1} \left(\frac{|\tilde{A}S(T_1 - r)\Phi|_{\mathcal{L}_2(V)}^2}{(T_1 - r)^{2\beta}} + \frac{|\tilde{A}S(T_1 - r)\Phi|_{\mathcal{L}_2(V)}^2}{(r - T_2)^{1-2H}}\right) dr \\
&\leq c\int_{T_2}^{T_1} \left(\frac{1}{(T_1 - r)^{2\beta + 2\gamma + 2\delta}} + \frac{1}{(r - T_2)^{1-2H}(T_1 - r)^{2\gamma + 2\delta}}\right) dr \\
&\leq c\left(|T_1 - T_2|^{1-2\beta - 2\gamma - 2\delta} + |T_1 - T_2|^{1-2\delta - 2\gamma + H}\right) \tag{11.39}
\end{aligned}
$$

where $\beta > \frac{1}{2} - H$ so

$$I_1 \leq c|T_1 - T_2|^{2\alpha_1} \tag{11.40}$$

where $2\alpha_1 < 2H - 2\gamma - 2\delta$.

Now an upper bound is given for I_2. Since $(S(t), t \in [0, T])$ is analytic, it follows that for any $\alpha_2 > 0$,

$$I_2 \leq 2\,|S(T_1 - T_2) - I|_{\mathcal{L}(V_{\alpha_2 + \delta}, V_{\delta})}^{2\alpha_2}$$

$$\cdot \mathbb{E}\left\|\int_0^{T_2} S(T_2 - r)\Phi \, dB(r)\right\|_{\alpha_2 + \delta}^2 \tag{11.41}$$

Since $\tilde{Z}(T_2)$ is a Gaussian random variable the expectation on the right-hand side of (11.41) is finite if $\tilde{Z}(T_2)$ is a $V_{\alpha_2+\delta}$-valued random variable which is the case by Lemma 11.10 if $\alpha_2 + \delta + \gamma < H$. Combining this fact with (11.40) yields

$$\mathbb{E}\|\tilde{Z}(T_1) - \tilde{Z}(T_2)\|_\delta^2 \leq c|T_1 - T_2|^{2\alpha} \qquad (11.42)$$

for $\alpha < H - \gamma - \delta$ and $T_1, T_2 \in [0, T]$. Since the increment $\tilde{Z}(T_1) - \tilde{Z}(T_2)$ is a V_δ-valued Gaussian random variable

$$\mathbb{E}\|\tilde{Z}(T_1) - \tilde{Z}(T_2)\|_\delta^{2p} \leq c|T_1 - T_2|^{2\alpha p} \qquad (11.43)$$

for each $p \geq 1$. By the Kolmogorov criterion for sample path continuity (e.g. [DPZ92] Theorem 3.3) there is a $C^\lambda([0, T], V_\delta)$ version of the process $\left(\tilde{Z}(t), t \in [0, T]\right)$ for $\lambda < \frac{2\alpha p - 1}{2p}$ if $p > \frac{1}{2\alpha}$. Letting $p \to \infty$ shows that there is a $C^\alpha([0, T], V_\delta)$ version of $\left(\tilde{Z}(t), t \in [0, T]\right)$ for $\alpha < H - \gamma - \delta$. $\qquad \square$

An important special case arises when the operator Φ in (11.21) is a Hilbert-Schmidt operator on V. This condition implies that the fractional Brownian motion $(B(t), t \in [0, T])$ in (11.21) is a "genuine" V-valued process. In this case (11.34) is satisfied with $\gamma = 0$ which is described in the following corollary.

COROLLARY 11.12 *Let $(S(t), t \geq 0)$ be an analytic semigroup. If $\Phi \in \mathcal{L}_2(V)$ then the process $(\tilde{Z}(t), t \in [0, T])$ has a $C^\alpha([0, T], V_\delta)$ version for all $\alpha \geq 0, \delta \geq 0$ satisfying $\alpha + \delta < H$. In particular, there is $C^\alpha([0, T], V)$ version for $\alpha < H$.*

Note that Corollary 11.12 is satisfied for each $H \in (0, 1)$. The result for $H \in \left(\frac{1}{2}, 1\right)$ is a consequence of Proposition 2.1 in [DPDM02]. In the special case where $V = \mathbb{R}^n$, $V_\delta = V$ for all $\delta \geq 0$, $\Phi \in \mathcal{L}_2(V)$ trivially so $(\tilde{Z}(t), t \in [0, T])$ has a $C^\alpha([0, T], V)$ version for each $\alpha < H$. The trivial case where $\tilde{Z}(t) = B(t)$ demonstrates that for the present level of generality, the condition here is sharp.

The following result provides conditions for the existence of a limiting distribution.

PROPOSITION 11.13 *Let $(S(t), t \geq 0)$ be an analytic semigroup that is exponentially stable. Specifically,*

$$|S(t)|_{\mathcal{L}(V)} \leq Me^{-\omega t} \qquad (11.44)$$

for $t \geq 0$ and some $M > 0$ and $\omega > 0$ so (11.34) is satisfied with $\gamma \in [0, H)$ where $H \in (0, 1)$. Then for each $x \in V$ the family of Gaussian *probability measures* $(\mu_{Z(t)}, t \geq 0)$ *converges in the weak*-topology to a unique limiting measure* $\mu_\infty = N(0, Q_\infty)$ *where* $\mu_{Z(t)}$ *is the measure for* $Z(t)$,

$$Q_t = \int_0^t L(t,s)L^*(t,s)\,ds, \qquad (11.45)$$

$$Q_\infty = \lim_{t \to \infty} Q_t,$$

$L(t,s) \in \mathcal{L}(V)$, *and*

$$L(t,s)x = \mathcal{K}_H^*(S(t-\cdot)\Phi x)(s)$$

for $x \in V$ *and* $0 \leq s \leq t$.

Proof For $H \in \left(\frac{1}{2}, 1\right)$, the result is given in Proposition 3.4 of [DPDM02]. For $H \in \left(0, \frac{1}{2}\right)$, the methods in [DPZ92] for a Wiener process are suitably modified. Let

$$\hat{Z}(t) = \int_0^t S(r)\Phi\,dB(r)$$

and note that $\mu_{\hat{Z}(t)} = \mu_{\tilde{Z}(t)}$ for $t \geq 0$ because the covariance operators have the same form after a time reversal. Since $S(t)x \to 0$ in the norm topology as $t \to \infty$ by (11.44) it suffices to show that $(\hat{Z}(t), t \geq 0)$ is convergent in $L^2(\Omega, V)$. It is shown that the family of random variables $(\hat{Z}(t), t \geq 0)$ is Cauchy in $L^2(\Omega, V)$, that is,

$$\lim_{T_2 \to \infty, T_1 \geq T_2} \mathbb{E}\|\hat{Z}(T_1) - \hat{Z}(T_2)\|^2 = 0 \qquad (11.46)$$

To verify (11.46), it is sufficient to modify the proofs of Lemma 11.7 and Theorem 11.8 with the (local) time reversal of the semigroup $(S(t), t \geq 0)$ and use the fact that (11.34) and (11.44) imply that

$$|S(t)\Phi|_{\mathcal{L}_2(V)} \leq ce^{-\omega t}t^{-\gamma} \qquad (11.47)$$

for $t \geq 0$ and $c > 0$. Thus

$$\mathbb{E}\|\hat{Z}(T_1) - \hat{Z}(T_2)\|^2 = \int_{T_2}^{T_1} \sum_{n=1}^\infty \|\mathcal{K}_H^*(S(\cdot)\Phi e_n)(s)\|^2\,ds$$

and use inequalities that are analogous to (11.26), (11.27), and (11.30), together with (11.42) to obtain for $T_1 - T_2 \geq 1$, $T_2 \geq 1$,

$$\mathbb{E}\|\hat{Z}(T_1) - \hat{Z}(T_2)\|^2$$

$$\leq c \left(\frac{e^{-2\omega T_2}}{T_2^{2\gamma}} \int_0^{T_1-T_2} \frac{e^{-2\omega r}}{(T_1 - T_2 - r)^{1-2H}} \, dr \right.$$

$$+ e^{-2\omega T_2} \int_0^{T_1-T_2} \frac{e^{-2\omega r}}{r^{1-2H}} \, dr$$

$$\left. + \frac{e^{-2\tilde{\omega} T_2}}{T_2^{2\beta}} \left| S\left(\frac{T_2}{2}\right) \Phi \right|^2_{\mathcal{L}_2(V)} \int_0^\infty e^{-2\tilde{\omega} s} \, ds \right) \quad (11.48)$$

where the three terms on the right-hand side of (11.48) correspond to the three terms on the right-hand side of (11.25), so (11.46) follows. □

4.   An Example

Consider a $2m$th order stochastic parabolic equation

$$\frac{\partial u}{\partial t}(t, \xi) = [L_{2m}u](t, \xi) + \eta(t, \xi) \quad (11.49)$$

for $(t, \xi) \in [0, T] \times \mathcal{O}$ with the initial condition

$$u(0, \xi) = x(\xi) \quad (11.50)$$

for $\xi \in \mathcal{O}$ and the Dirichlet boundary condition

$$\frac{\partial^k u}{\partial v^k}(t, \xi) = 0 \quad (11.51)$$

for $(t, \xi) \in [0, T] \times \partial\mathcal{O}$, $k \in \{0, \ldots, m-1\}$, $\frac{\partial}{\partial v}$ denotes the conormal derivative, $\mathcal{O}$ is a bounded domain in $\mathbb{R}^d$ with a smooth boundary and L_{2m} is a $2m$th order uniformly elliptic operator

$$L_{2m} = \sum_{|\alpha| \leq 2m} a_\alpha(\xi) D^\alpha \quad (11.52)$$

and $a_\alpha \in C_b^\infty(\mathcal{O})$. For example, if $m = 1$ then this equation is called the stochastic heat equation. The process η denotes a space dependent noise process that is fractional in time with the Hurst parameter $H \in (0, 1)$ and, possibly, in space. The system (11.49)–(11.51) is modeled as

$$dZ(t) = AZ(t) \, dt + \Phi \, dB(t)$$

$$Z(0) = x \quad (11.53)$$

in the space $V = L^2(\mathcal{O})$ where $A = L_{2m}$,

$$\mathrm{Dom}(A) = \{\varphi \in H^{2m}(\mathcal{O}) \mid \frac{\partial^k}{\partial v^k}\varphi = 0 \text{ on } \partial D \text{ for } k \in \{0, \ldots, m-1\}\},$$

$\Phi \in \mathcal{L}(V)$ defines the space correlation of the noise process and $(B(t), t \geq 0)$ is a cylindrical standard fractional Brownian motion in V. For $\Phi = I$, the noise process is uncorrelated in space. It is well known that A generates an analytic semigroup $(S(t), t \geq 0)$. Furthermore

$$|S(t)\Phi|_{\mathcal{L}_2(V)} \leq |S(t)|_{\mathcal{L}_2(V)}|\Phi|_{\mathcal{L}(V)} \leq ct^{-\frac{d}{4m}} \qquad (11.54)$$

for $t \in [0, T]$. It is assumed that there is a $\delta_1 > 0$ such that

$$\mathrm{Im}(\Phi) \subset \mathrm{Dom}\left((\hat{\beta}I - A)^{\delta_1}\right) \qquad (11.55)$$

then for $r \geq 0$

$$|S(t)\Phi|_{\mathcal{L}_2(V)}$$
$$\leq |S(t)(\hat{\beta}I - A)^r|_{\mathcal{L}(V)}|(\hat{\beta}I - A)^{-r-\delta_1}|_{\mathcal{L}_2(V)}|(\hat{\beta}I - A)^{\delta_1}\Phi|_{\mathcal{L}(V)}$$
$$\leq ct^{-r} \quad (11.56)$$

for $t \in (0, T]$ if the operator $(\hat{\beta}I - A)^{-r-\delta_1}$ is a Hilbert-Schmidt operator on V, which occurs if

$$r + \delta_1 > \frac{d}{4m}. \qquad (11.57)$$

Thus, the condition (11.36) is satisfied with $\gamma = \frac{d}{4m}$ for arbitrary bounded operator Φ and if (11.55) is satisfied, then (11.36) is satisfied with $\gamma > \frac{d}{4m} - \delta_1$.

Summarizing the results of this section applied to the example (cf. Theorem 11.11 and Proposition 11.6) if

$$H > \frac{d}{2m} \qquad (11.58)$$

then for any $\Phi \in \mathcal{L}(V)$, the stochastic convolution process $(\tilde{Z}(t), t \in [0, T])$ is well-defined and has a version with $C^\alpha([0, T], V_\delta)$ paths for $\alpha \geq 0$, $\delta \geq 0$ satisfying

$$\alpha + \delta < H - \frac{d}{4m}. \qquad (11.59)$$

If moreover (11.55) is satisfied, then the previous conclusion is satisfied if

$$\alpha + \delta < H - \frac{d}{4m} + \delta_1. \qquad (11.60)$$

Note that these results are analogous to the case of a standard Wiener process $\left(H = \frac{1}{2}\right)$.

References

[AMN01]   E. Alòs, O. Mazet, and D. Nualart. Stochastic calculus with respect to Gaussian processes. *Ann. Probab.*, 29(2):766–801, 2001.

[Cai05]   P. Caithamer. The stochastic wave equation driven by fractional Brownian noise and temporally correlated smooth noise. *Stoch. Dyn.*, 5(1):45–64, 2005.

[DJPD]   T. E. Duncan, J. Jakubowski, and B. Pasik-Duncan. Stochastic integration for fraction Brownian motion in a Hilbert space. To appear in *Stochastic Dynamics*.

[DMPD05] T. E. Duncan, B. Maslowski, and B. Pasik-Duncan. Stochastic equations in Hilbert space with a multiplicative fractional Gaussian noise. *Stoc. Proc. Appl.*, 115:1357–1383, 2005.

[DPDM02] T. E. Duncan, B. Pasik-Duncan, and B. Maslowski. Fractional Brownian motion and stochastic equations in Hilbert spaces. *Stoch. Dyn.*, 2(2):225–250, 2002.

[DPZ92]   G. Da Prato and J. Zabczyk. *Stochastic equations in infinite dimensions*, volume 44 of *Encyclopedia of Mathematics and its Applications*. Cambridge University Press, Cambridge, 1992.

[DPZ96]   G. Da Prato and J. Zabczyk. *Ergodicity for infinite-dimensional systems*, volume 229 of *London Mathematical Society Lecture Note Series*. Cambridge University Press, Cambridge, 1996.

[DÜ99]   L. Decreusefond and A. S. Üstünel. Stochastic analysis of the fractional Brownian motion. *Potential Anal.*, 10(2):177–214, 1999.

[Dun00]   T. E. Duncan. Some stochastic semilinear equations in Hilbert space with fractional Brownian motion. In J. Menaldi, E. Rofman, and A Sulem, editors, *Optimal Control and Partial Differential Equations*, pages 241–247. IOS Press, Amsterdam, 2000.

[GA99]   W. Grecksch and V. V. Anh. A parabolic stochastic differential equation with fractional Brownian motion input. *Statist. Probab. Lett.*, 41(4):337–346, 1999.

[HØZ04]   Y. Hu, B. Øksendal, and T. Zhang. General fractional multiparameter white noise theory and stochastic partial differential equations. *Comm. Partial Differential Equations*, 29(1-2):1–23, 2004.

[Hu01]   Y. Hu. Heat equations with fractional white noise potentials. *Appl. Math. Optim.*, 43(3):221–243, 2001.

[Hu05]   Y. Hu. Integral transformations and anticipative calculus for fractional Brownian motions. *Mem. Amer. Math. Soc.*, 175(825):viii+127, 2005.

[MN02]   S. Moret and D. Nualart. Onsager-Machlup functional for the fractional Brownian motion. *Probab. Theory Related Fields*, 124(2):227–260, 2002.

[MS04]   B. Maslowski and B. Schmalfuss. Random dynamical systems and stationary solutions of differential equations driven by the fractional Brownian motion. *Stochastic Anal. Appl.*, 22(6):1577–1607, 2004.

[PT00]   V. Pipiras and M. S. Taqqu. Integration questions related to fractional Brownian motion. *Probab. Theory Related Fields*, 118(2):251–291, 2000.

[SKM93] S. G. Samko, A. A. Kilbas, and O. I. Marichev. *Fractional integrals and derivatives*. Gordon and Breach Science Publishers, Yverdon, 1993. Theory and applications.

[TTV03] S. Tindel, C. A. Tudor, and F. Viens. Stochastic evolution equations with fractional Brownian motion. *Probab. Theory Related Fields*, 127(2):186–204, 2003.

Chapter 12

HEDGING OPTIONS WITH TRANSACTION COSTS*

Wulin Suo
School of Business
Queen's University
Kingston, Ontario, Canada
wsuo@business.queensu.ca

Abstract

This paper studies the optimal investment problem for an investor with a HARA type utility function. We assume the investor already has an option in her portfolio, and she will setup her invesment/hedging strategy using bonds and the underlying stock to maximize her utility. When there are transaction costs, the investor's optimal investment/hedging strategy can be described by three regions: the buying region, the selling region, and the no transaction region. When her portfolio falls in the buying (selling) region, she will buy (sell) enough shares of the stock to make her portfolio lie in the no transaction region (NT). When her portfolio falls in the NT region, it is optimal for the investor to make no transaction. We introduce the concept of a viscosity solution to describe the indirect utility function. A numerical scheme is proposed to compute the indirect utility function. This in turn enables the asking price for an option to be computed.

1. Introduction

In this paper we study the problem of an investor who wants to hedge her liability arising from writing an option. Hedging is central to the option pricing theory. Arbitrage pricing arguments, such as those of Black and Scholes (1973), depend on the idea that an option can be perfectly hedged using the underlying asset, so making it possible to create a port-

---

*The author is grateful for the comments from Professors J.-C. Duan, John Hull, Yisong Tian and Alan White.

folio that replicates the option payoff exactly. The standard approach dealing with option hedging is through the Black-Scholes framework, i.e., assuming in a world of frictionless markets, the option holder can hedge her position by continuously re-balancing a portfolio consisting of the underlying asset and risk-free bonds (or money market account). The number of underlying asset the investor needs to hold in order to perfectly hedge her position is computed through the Black-Scholes option pricing formula, and is often referred to as the option's *delta*. However, this hedging strategy is infeasible in the real world since transaction costs are incurred whenever the portfolio is re-balanced. Since frequent trading of the underlying stock is assumed, transaction costs will make an important impact on the hedging of options.

The problem of hedging options in the presence of proportional costs have been studied by Leland (1985), Boyle and Vorst (1991), and Toft (1996), among others. The hedging strategy considered in these studies are based on the delta hedging strategy implied from the Black-Scholes option pricing framework. They study the hedging costs when the investor tries to perfectly hedge the option under consideration. These studies ignore the rationality of the investor in the sense that the costs incurred when hedging an option in such a way may be greater than the premium she receives for the option. Moreover, the investor may not want to perfectly hedge her liability in the option position because she may prefer to take on some of the risk in exchange for a higher return for her portfolio. In other words, the hedging strategy is not based on any optimality criteria. Since no-transaction cost is one of the crucial assumptions underlying the Black-Scholes option pricing paradigm, it should be reasonable to expect that the investor would realize that the Black-Scholes option pricing formula no longer holds, and therefore the delta hedging doctrine should be abandoned.

The optimality criterion for option hedging when the market has frictions (including transaction costs) can be defined in terms of expected utility. This approach seems more appropriate than exact replication since it reflects the tradeoffs that must be made between transaction costs and risk reduction. The approach is in a paradigm similar to that of Davis and Norman (1990), and Dumas and Luciano (1991). These papers describe optimal portfolio strategies to maximize expected utility over infinite time horizon, where transaction costs are incurred whenever trades are made on the stocks. They extend earlier works by Merton (1971) and Constantinides (1986). However, while these papers are concerned with optimal portfolio strategies, they are not focused on the problem of replicating (or hedging) options by means of the underlying asset.

In this paper we study the hedging strategy for an investor who already has an option in her portfolio. The investor's optimality criterion is characterized by a HARA type utility function (the methodology can also be applied to more general cases). The approach follows the work of Davis et al (1993). However, in addition to assuming a more general utility function, our attention is focused on the hedging behavior of the investor based on her current wealth level and risk preference. Since an option's price has to be determined from an equilibrium framework when the market has frictions, we will assume that the investor already has an option in her portfolio. We will briefly discuss the minimum premium the investor should charge in order to induce her to take the liability.

The rest of this paper is organized as follows. In Section 2, we will formulate the option hedging problem, and derive the investor's wealth process. The optimal hedging strategy in absence of transaction costs is also discussed. It turns out that when the investor has a positive net wealth, she would always "over" hedge her liability in the option position in the sense that her investment in the stock is always greater than the amount of delta derived from the Black-Scholes option pricing framework, as long as the growth rate of the stock is greater than the risk-free interest rate. In Section 3, we will consider the optimal hedging/investment problem when there are transaction costs. The Hamilton-Jacobi-Bellman equation for the indirect utility function of the invstment/hedging problem will be derived heuristically. In order to characterize the indirect utility function, we will introduce the concept of *viscosity solution* to the Hamilton-Jacobi-Bellman (HJB) equation, and demonstrate that the indirect utility function is indeed the unique solution in this sense. The optimal investment/hedging strategy is characterized in terms of the indirect utility function. In Section 4, we propose a numerical scheme to compute the indirect utility function based on the form of the optimal investment/hedging strategy we obtained in Section 3. Section 5 concludes.

2.   An optimal hedging problem without transaction costs

We assume that there are two assets available in the market: one is a stock whose price follows a geometric Brownian motion:

$$(12.1) \qquad \frac{dS}{S} = \alpha dt + \sigma dz.$$

The other asset is a risk-free bond or cash account where we assume that the interest rate is constant. In other words, one can borrow/lend at a constant rate of $r > 0$. We write the cash account as

(12.2)
$$B_t = B\exp(rt),$$

where B is the initial amount in the cash account. The investor is assumed to have a utility function $u(c)$, where $u(\cdot)$ is assumed to be twice continuously differentiable and concave. We assume that the investor's objective is to maximize her expected utility from the terminal wealth at time T.

We will also assume that the investor has a short position in a European call with a strike price K and maturity time T. She balances her portfolio in a way to maximize the expected utility at time T, at which time she will liquidate her positions in the stock and fulfil her obligations in the short position of the call if it ends in the money. In other words, at time T, her final wealth is

(12.3)
$$W_T = B_T + y_T^* - C_T,$$

where W_T is her total wealth, B_T is the value of her investment in the risk-free account, y_t^* is the cash value of her investment in the stock, i.e., the value of her stock holding minus the transaction cost (if there is any) when she liquidates them, and C_T is the liability from the short position in the option. For an European call option, we have

$$C_T = (S_T - K)^+.$$

We derive the investor's wealth equation. Consider the following investment problem: the investor has an initial wealth of W_0. She can invest the wealth in either the stock or the money market account. Let $h_0(t)$ and $h_1(t)$ be the amount she invests in the money market account and the number of shares she holds for the stock at time t, respectively. Her total wealth is thus

$$W(t) = h_0(t) + h_1(t)S(t).$$

Since there is no consumption before the maturity time T, we can assume that the portfolio is self-financing:

$$
\begin{aligned}
dW(t) &= dh_0(t) + h_1(t)dS(t) \\
&= rh_0(t)dt + h_1(t)\left[\alpha S(t)dt + \sigma S(t)dz_t\right] \\
&= \left[rh_0(t) + \alpha h_1(t)S(t)\right]dt + h_1(t)\sigma S(t)dz_t.
\end{aligned}
$$

Write $\pi(t)$ as the proportion of her wealth invested in the stock, i.e.,

$$\pi(t)W(t) = h_1(t)S(t),$$

then
$$(1 - \pi(t))W(t) = h_0(t),$$

and

$$
\begin{aligned}
dW(t) &= r(1 - \pi(t))W(t)dt + \alpha\pi(t)W(t)dt + \sigma\pi(t)W(t)dz_t \\
(12.4) \qquad &= (\alpha - r)\pi W(t)dt + rW(t)dt + \sigma\pi W(t)dz_t.
\end{aligned}
$$

The investor's problem is to maximize her expected utility at the maturity time:
$$\max_{\pi(\cdot)} Eu(W_T),$$

and the investor's indirect utility function is defined as follows: for an initial wealth W at time t,

$$U(t, W) = \max_{\pi(\cdot)} E_t u(W_T).$$

Let us introduce the following differential operator:

$$\mathcal{L}^{\pi} \equiv \frac{1}{2}\pi^2 W^2 \sigma^2 \frac{\partial^2}{\partial W^2} + [(\alpha - r)\pi W + rW]\frac{\partial}{\partial W},$$

then the Hamilton-Jacobi-Bellman (HJB hereafter) equation for the investment problem is

$$(12.5) \qquad \frac{\partial U}{\partial t} + \max_{\pi} \mathcal{L}^{\pi} U = 0$$

with the boundary condition

$$U(T, W) = u(W), \quad \forall\, W \geq 0.$$

Once the indirect utility function U is found, the optimal strategy π^* is given by

$$\pi W^2 \sigma^2 \frac{\partial^2 U}{\partial W^2} + (\alpha - r)W\frac{\partial U}{\partial W} = 0,$$

or

$$(12.6) \qquad \pi^* = -\frac{(\alpha - r)U_W}{W\sigma^2 U_{WW}}.$$

In order to solve the HJB equation, let us assume that the utility function is in the HARA class (hyperbolic absolute risk aversion):

$$(12.7) \qquad u(W) = \frac{W^\gamma}{\gamma}, \quad 0 < \gamma < 1.$$

This type of utility function is widely used in the optimal investment/consumption problems. Note that the relative risk aversion is given by

$$-W\frac{u''(W)}{u'(W)} = 1 - \gamma.$$

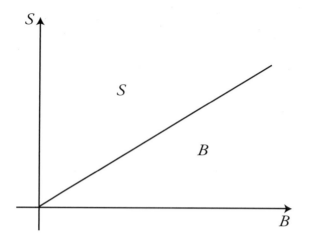

Figure 12.1. The case with no transaction costs

2.1 Optimal investment problem without options

In this case, the boundary condition of the HJB equation is thus

$$U(T, W) = \frac{W^\gamma}{\gamma}, \quad \forall\, W \geq 0.$$

Under these assumptions, the indirect utility function can be found as (see Appendix):

$$U(t, W) = g(t)\frac{W^\gamma}{\gamma},$$

where

(12.8)
$$g(t) = e^{\nu r(T-t)}$$

and

$$\nu = \frac{1}{2}\left(\frac{\alpha - r}{\sigma}\right)^2 \frac{1}{1 - \gamma} + r.$$

From (12.6), the optimal investment policy is given by

$$\pi^* = \frac{\alpha - r}{\sigma^2(1 - \gamma)}.$$

In this case, the investor's optimal investment strategy is to keep a fixed proportion π^* of her wealth in the stock. This means that, optimally, the investor acts in such a way that her portfolio holding is always on the line

$$S = \frac{\pi^*}{1 - \pi^*} B.$$

which is sometimes referred to as the *Merton line*.[1] It is worth mentioning that when $\alpha > r$, the investor always invests some portion of her wealth in the stock. Moreover, for this choice of the utility function, the proportion of the investor's wealth invested in the stock is independent of her wealth level.

This result is shown in Figure 12.1. When the portfolio falls in the region $\mathcal{B}$, the investor shall purchase enough shares of stock to keep her portfolio on the Merton line. Similarly, when her portfolio is in the region $\mathcal{S}$, she should sell some shares of the stock she is holding to move her portfolio on the Merton line. We will refer $\mathcal{B}$ and $\mathcal{S}$ as the *buying region* and *selling region* respectively.

2.2   When there is a short position in option

We now consider the case when there is a short position of European call already in her portfolio. We want to answer the following question: based on her wealth level and preference, what is her best investment strategy? Notice that the investor might not want to perfectly hedge her position in the option since she might prefer to have some risk in her portfolio in order to achieve a higher return. This is illustrated by the previous case when there is no option in her portfolio.

The terminal wealth in this case is

$$(12.9) \qquad \overline{W}_T = W_T - (S_T - K)^+.$$

The optimal investment/hedging strategy can be constructed in the following way. It is easy to show that for any time $t \leq T$ and wealth level $W \geq 0$, the indirect utility function is given by

$$
\begin{aligned}
\overline{U}(t, W) &= U(t, W - f(t, S)) \\
&= g(t) \frac{[W - f(t, S)]^\gamma}{\gamma},
\end{aligned}
$$

(12.10)

where $g(t)$ is given by (12.8) and $f(t, S)$ is the price of the call option under the Black-Scholes framework, i.e.,

$$f(t, S) = SN(d_1) - Ke^{-r(T-t)} N(d_2),$$

with

$$d_1 = \frac{\ln(S/K) + (r + \sigma^2/2)(T - t)}{\sigma\sqrt{T - t}}, \quad d_2 = d_1 - \sigma\sqrt{T - t}.$$

For a proof of this conclusion, see the Appendix.

Equation (12.10) suggests that the optimal strategy for the investor is to delta-hedge her short position in the European call, which needs an amount $f(0, S)$ initially, and then manage her remaining wealth optimally as before. While this is true, it is important to look at what her overall investment/hedging strategy is based on her wealth and risk preference.

In order to find the optimal investment/hedging strategy, recall that the stock price and the wealth processes are given by (12.1) and (12.4), respectively. The indirect utility is defined as

$$U(t, S, W) \equiv \max_{\pi} Eu(\overline{W}_T),$$

where $\overline{W}_T$ is given by (12.9). Unlike the previous case, we now have to include the current stock price as one of the variables for the indirect utility function. Similarly, let

$$
\begin{aligned}
\bar{\mathcal{L}}^{\pi} &\equiv \frac{1}{2}\pi^2 W^2 \sigma^2 \frac{\partial^2}{\partial W^2} + \frac{1}{2}\sigma^2 S^2 \frac{\partial^2}{\partial S^2} + \pi W S \sigma^2 \frac{\partial^2}{\partial S \partial W} \\
&+ [(\alpha - r)\pi W + rW]\frac{\partial}{\partial W} + \alpha S \frac{\partial}{\partial S},
\end{aligned}
$$

then the HJB equation for this problem becomes

$$U_t + \max_{\pi} \bar{\mathcal{L}}^{\pi} U = 0,$$

with the boundary condition given by

$$U(T, S, W) = \frac{(W - (S - K)^+)^{\gamma}}{\gamma}.$$

The optimal strategy is given by

$$\pi W^2 \sigma^2 U_{WW} + W S \sigma^2 U_{SW} + (\alpha - r)W U_W = 0,$$

or

(12.11) $$\pi^* = -\frac{\alpha - r}{\sigma^2 W} \cdot \frac{U_W}{U_{WW}} - \frac{S}{W} \cdot \frac{U_{SW}}{U_{WW}}.$$

Given the solution in (12.10), we have

$$
\begin{aligned}
U_W &= g(t)\gamma \left[W - f(t, S)\right]^{\gamma - 1}, \\
U_{WW} &= g(t)\gamma(\gamma - 1)\left[W - f\right]^{\gamma - 2}, \\
U_S &= -g(t)\gamma \left[W - f\right]^{\gamma - 1} \cdot f_S, \\
U_{SW} &= -\gamma(\gamma - 1)g(t)\left[W - f\right]^{\gamma - 2} \cdot f_S.
\end{aligned}
$$

The optimal investment/hedging strategy can be written as

$$(12.12) \qquad \pi^* = \frac{\alpha - r}{\sigma^2(1-\gamma)} \cdot \frac{W-f}{W} + \frac{S}{W} \cdot f_S.$$

It is interesting to notice that in this case, the proportion of the investor's wealth invested in the stock depends on her wealth level.

The wealth that is invested in the stock is given by

$$\pi^* W = \frac{\alpha - r}{\sigma^2(1-\gamma)} \cdot (W - f) + S \cdot f_S,$$

which implies that she will perfectly hedge her liability in the option only if her initial wealth equals the premium she receives from the short position in the option. Assuming $\alpha > r$ and that the investor has more wealth than just the option premium, the investor will always invest more than the amount that is necessary for the delta hedging strategy for the option derived from the Black-Scholes model.

3. The case when there are transaction costs

Now we consider the case when transaction costs are incurred when one buys/sells stocks. We still assume that there are two assets available: one is the risky asset, or a stock, whose price is described by a geometric Brownian motion

$$\frac{dS}{S} = \alpha dt + \sigma dz,$$

and the other asset is a money market account (or risk-free bond) with a risk-free interest of r.

As before, the investor's wealth is distributed among investments in both the stock and the money market account. Moreover, we assume that a proportional cost is incurred each time a transaction of buying/selling of the stock is made. More specifically, in order to purchase (sell) one share of the stock, he will pay (receive)

$$(1+\lambda)S \qquad ((1-\mu)S).$$

Her wealth can be described by

$$(12.13) \qquad dW_t = rW_t dt - (1+\lambda)S_t dL_t + (1-\mu)S_t dM_t,$$

where L_t and M_t are the cumulative amount of shares of the stock the investor has bought and sold up to time t, respectively. For technical reasons, we assume

- L and M are adapted to the information generated by the price process S;

- $L(\cdot)$ and $M(\cdot)$ are right continuous and have left limits;

- $L(0-) = M(0-) = 0$.

The total number of shares the investor holds at time t is given by

$$y(t) = L(t) - M(t).$$

An investment/hedging strategy is described by the pair (L, M).

3.1 The optimal investment problem without options

As before, we assume that the investor has a utility function given by (12.7) and she choose the investment/hedging strategy (L, M) to maximize her expected utility at time T:[2]

$$\max_{(L,M)} Eu(W_T).$$

In order to derive the optimal investment strategy, we define the indirect utility function as

$$U(t, S, y, W) \equiv \max_{(L,M)} E_t(u(W(T))),$$

where

- t is the current time;

- S is the spot price of the stock at time t;

- y is the net position of the investor's position in the stock;

- W is the investor's net wealth (or cash value) at time t.

When the utility function is assumed to be in the form (12.7), it is easy to show the following: for the special choice of the utility function u, for any $\rho > 0$, we have

$$U(t, \rho S, y, \rho W) = \rho^{\gamma} U(t, S, y, W).$$

The HJB equation

For any positive constants C_1 and C_2, consider the subset $\mathcal{S}_{C_1, C_2}$ of trading strategies in the following form

$$L(t) = \int_0^t l(s)ds, \quad M(t) = \int_0^t m(s)ds$$

such that
$$0 \le l(s) \le C_1, \quad 0 \le m(s) \le C_2.$$

The wealth process can be written as

$$
\begin{aligned}
dW &= [rW - (1+\lambda)l(t)S + (1-\mu)m(t)S]\,dt, \\
dy &= [l(t) - m(t)]\,dt, \\
\frac{dS}{S} &= \alpha dt + \sigma dz.
\end{aligned}
$$

Define

$$U^{C_1,C_2}(t,S,y,W) \equiv \max_{(L,M)\in\mathcal{S}_{C_1,C_2}} E_t(u(W(T))),$$

then the investment problem reduces to a case similar to the one considered in Section 2. As a result, the HJB equation for $U^{C_1,C_2}(t,S,y,W)$ take the following form:

$$(12.14) \qquad \frac{\partial U}{\partial t} + \max_{0 \le l \le C_1, \, 0 \le m \le C_2} \mathcal{L}^{l,m} U = 0,$$

where

$$
\begin{aligned}
\mathcal{L}^{l,m} &= \frac{1}{2}\sigma^2 S^2 \frac{\partial^2}{\partial S^2} + \alpha S \frac{\partial}{\partial S} \\
&\quad + [rW - (1+\lambda)lS + (1-\mu)Sm]\frac{\partial}{\partial W} + (l-m)\frac{\partial}{\partial y} \\
&= \frac{1}{2}\sigma^2 S^2 \frac{\partial^2}{\partial S^2} + \alpha S \frac{\partial}{\partial S} + rW \frac{\partial}{\partial W} \\
&\quad + \left[\frac{\partial}{\partial y} - (1+\lambda)S\frac{\partial}{\partial W}\right] l + \left[-\frac{\partial}{\partial y} + (1-\mu)S\frac{\partial}{\partial W}\right] m.
\end{aligned}
$$

From the HJB equation (12.14), we can see that the optimal investment strategy is given by

$$
l^* = \begin{cases} C_1 & \text{if } \dfrac{\partial U^{C_1,C_2}}{\partial y} - (1+\lambda)S\dfrac{\partial U^{C_1,C_2}}{\partial W} > 0, \\ 0 & \text{otherwise;} \end{cases}
$$

$$
m^* = \begin{cases} C_2 & \text{if } -\dfrac{\partial U^{C_1,C_2}}{\partial y} + (1-\mu)S\dfrac{\partial U^{C_1,C_2}}{\partial y} > 0, \\ 0 & \text{otherwise.} \end{cases}
$$

Letting $C_1, C_2 \to \infty$, we can expect that

$$U^{C_1,C_2} \quad \to \quad U,$$

$$\frac{\partial U^{C_1,C_2}}{\partial W} \rightarrow \frac{\partial U}{\partial W},$$

$$\frac{\partial U^{C_1,C_2}}{\partial y} \rightarrow \frac{\partial U}{\partial y},$$

$$\frac{\partial U^{C_1,C_2}}{\partial S} \rightarrow \frac{\partial U}{\partial S},$$

$$\frac{\partial^2 U^{C_1,C_2}}{\partial S^2} \rightarrow \frac{\partial^2 U}{\partial S^2}.$$

As a result, the HJB takes the form of the so called *variational inequality*:

$$\max\left\{ \frac{\partial U}{\partial y} - (1+\lambda)S\frac{\partial U}{\partial W}, -\frac{\partial U}{\partial y} + (1-\mu)S\frac{\partial U}{\partial W}, \right.$$

(12.15)

$$\left. \frac{\partial U}{\partial t} + \frac{1}{2}\sigma^2 S^2 \frac{\partial^2 U}{\partial S^2} + \alpha S\frac{\partial U}{\partial S} + rW\frac{\partial U}{\partial W} \right\} = 0.$$

From the derivation of this equation, we can see that the (S, y, W) space can be divided into three regions:

NT: This region is defined by

$$\frac{\partial U}{\partial y} - (1+\lambda)S\frac{\partial U}{\partial W} \leq 0,$$

$$-\frac{\partial U}{\partial y} + (1-\mu)S\frac{\partial U}{\partial W} \leq 0,$$

$$\frac{\partial U}{\partial t} + \frac{1}{2}\sigma^2 S^2 \frac{\partial^2 U}{\partial S^2} + \alpha S\frac{\partial U}{\partial S} + rW\frac{\partial U}{\partial W} = 0.$$

$\mathcal{S}$: This region is defined by

$$\frac{\partial U}{\partial y} - (1+\lambda)S\frac{\partial U}{\partial W} = 0,$$

$$-\frac{\partial U}{\partial y} + (1-\mu)S\frac{\partial U}{\partial W} < 0,$$

$$\frac{\partial U}{\partial t} + \frac{1}{2}\sigma^2 S^2 \frac{\partial^2 U}{\partial S^2} + \alpha S\frac{\partial U}{\partial S} + rW\frac{\partial U}{\partial W} \leq 0.$$

$\mathcal{B}$: This region is defined by

$$\frac{\partial U}{\partial y} - (1+\lambda)S\frac{\partial U}{\partial W} < 0,$$

$$-\frac{\partial U}{\partial y} + (1-\mu)S\frac{\partial U}{\partial W} = 0,$$

$$\frac{\partial U}{\partial t} + \frac{1}{2}\sigma^2 S^2 \frac{\partial^2 U}{\partial S^2} + \alpha S\frac{\partial U}{\partial S} + rW\frac{\partial U}{\partial W} \leq 0.$$

The region $\mathcal{S}$ is called the *selling region*. In order to see this, define a function in the following form: for $h > 0$,

$$g(h) = U(t, S, y + h, W - (1 + \lambda)Sh),$$

which represents the utility change when h number of shares of the stock are purchased at time t. If $(S, y, W) \in \mathcal{S}$, then

$$g'(h) = \frac{\partial U}{\partial y} - (1 + \lambda)S\frac{\partial U}{\partial W} = 0,$$

in other words, the investor will immediately sell enough shares of the stock so that her portfolio will lie on $\partial \mathcal{S}$.

Similarly, the region $\mathcal{B}$ is called the *buying region* because the investor will immediately buy enough share of the underlying stock so that her portfolio will lie on $\partial \mathcal{B}$.

Moreover, from the derivation of the HJB equation (12.15), we can see that if the investor's portfolio lies in the region NT, she would make no transaction, and for this reason, NT is called the *no-transaction region*.

Recall that for any constant $\rho > 0$, we have

$$U(t, S, \rho y, \rho W) = \rho^\gamma U(t, S, y, W),$$

and thus

$$\frac{\partial U}{\partial y}(t, S, \rho y, \rho W) = \rho^{\gamma - 1}\frac{\partial U}{\partial y}(t, S, y, W),$$

$$\frac{\partial U}{\partial W}(t, S, \rho y, \rho W) = \rho^{\gamma - 1}\frac{\partial U}{\partial W}(t, S, y, W).$$

As a consequence, we can conclude that if (S, y, W) is in one of the regions, say $\mathcal{S}$, then $(S, \rho y, \rho W)$ is also in that region. In other words, each of the regions takes the form of a wedge in the (y, W) variables for any fixed S. See Figure 12.2.

It is interesting to compare this result with the case when there are no transaction costs, where it is optimal for the investor to continuously rebalance her portfolio so it will always lie on the Merton line. When transaction costs exist, it will be costly to re-balance her portfolio constantly, and as a result, it is optimal for her to re-balance her portfolio only if it drifts too far from the Merton line.

3.2   When there is a short position in option

Now we consider the case when there is a short position of a European call in the investor's portfolio at the beginning. Similar to Section 2, we

236

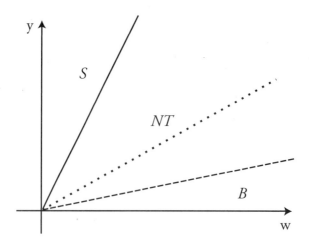

Figure 12.2. The case with transaction costs

use $\bar{U}$, $\bar{W}_T$ to denote the investor's indirect utility and terminal wealth in this case. In other words,

$$\bar{W}_T = W_T - (S_T - K)^+,$$

and

$$\bar{U}(t, S, y, W) = E_t[u(\bar{W}_T)|S_t = S, W_t = W, y_t = y].$$

The HJB equation for $\bar{U}$ is still the same as given in (12.15). However, the boundary condition is now given by

$$\bar{U}(T, S, y, W) = u(W - (S - K)^+).$$

When no option currently exist in the investor's portfolio at time 0, she can either synthetically create the option she wants, or by paying/recieving a certain price at time 0 to acquire/sell it. Assume that the investor is willing to hold a short position of a call option with maturity T and strike K for receiving a premium C upfront. The utility for the investor will be the following:

1 $U(0, S, 0, W)$ if she does not take the short position in call option;

2 $\bar{U}(0, S, 0, W + C)$ if she does take the short position in call option.

The price C for the investor to sell the call option should satisfy

$$\bar{U}(0, S, 0, W + C) \geq U(0, S, 0, W).$$

We define the *asking price* for this investor as

(12.16) $\hat{C} \equiv \min\{C \geq 0 : \bar{U}(0, S, 0, W + C) \geq U(0, S, 0, W)\}.$

In other words, the asking price is the minimum premium required for the investor to take the liability position in the call option.

We shall point out that $\hat{C}$ may not be the option price that the option is actually traded at since the latter requires some equilibrium in the market place.

3.3   The solution to the HJB equation

To show that the indirect utility function indeed satisfies the HJB equation (12.15), let us first assume that the indirect utility function $U(t, S, y, W)$ (for simplicity of notations, we write U instead of $\bar{U}$) is a smooth function with respect to the variables t, y, S and W. In other words, U is continuously differentiable with respect to the variables t, y and W, and twice continuously differentiable with respect to the variable S.

Now we show that U satisfies (12.15). For any admissible investment/hedging strategy (L, U), we have

$$\begin{cases} dW_t = rW_t dt - (1 + \lambda)S_t dL_t + (1 - \mu)S_t dU_t, \\ dS_t = \alpha S_t dt + \sigma S_t dz_t, \\ dy_t = dL_t - dU_t. \end{cases}$$

Applying Ito's formula (see Protter (1995), Theorem 33), we have, for any $s > t$,

$$
\begin{aligned}
U(s, S_s, y_s, W_s) \; = \; & U(t, S, y, W) \\
(12.17) \qquad & + \int_t^s \left(\frac{\partial}{\partial t} + \mathcal{L} \right) U(\theta, S_\theta, y_\theta, W_\theta) d\theta \\
(12.18) \qquad & + \int_t^s \left[\frac{\partial U}{\partial y} - (1 + \lambda)S_\theta \frac{\partial U}{\partial W} \right] dL_\theta \\
(12.19) \qquad & + \int_t^s \left[-\frac{\partial U}{\partial y} + (1 - \mu)S_\theta \frac{\partial U}{\partial W} \right] dU_\theta \\
(12.20) \qquad & + \int_t^s \frac{\partial U}{\partial S} \sigma S_\theta dz_\theta \\
& + \sum_{t \leq \theta < s} [U(\theta, S_\theta, y_\theta, W_\theta) - U(\theta, S_\theta, y_{\theta-}, W_{\theta-}) \\
(12.21) \qquad & - \nabla U(\theta, S_\theta, y_{\theta-}, W_{\theta-}) \cdot (\Delta y_\theta, \Delta W_\theta)],
\end{aligned}
$$

where

$$\mathcal{L} = \frac{1}{2}\sigma^2 S^2 \frac{\partial^2}{\partial S^2} + \alpha S \frac{\partial}{\partial S} + rW \frac{\partial}{\partial W},$$

$$\Delta y_\theta = y_\theta - y_{\theta-},$$

$$\Delta W_\theta = W_\theta - W_{\theta-},$$

$$\nabla U = \left(\frac{\partial U}{\partial y}, \frac{\partial U}{\partial W} \right).$$

It is obvious that the investor will never buy and sell at the same time when transaction costs are incurred. In other words, the processes L and M do not have jumps at the same time. For this reason, we can consider the following two cases separately:

- When L has a jump h at time θ, i.e., the investor purchase h shares of the stock at time θ, we have

$$\Delta W_\theta = W_{\theta-} - (1 + \lambda) S_\theta h,$$

$$\Delta y_\theta = y_{\theta-} + h.$$

In other words, when L has a jump, the processes (y, W) move along the direction of

$$(1, -(1 + \lambda)S),$$

which, as we have pointed earlier, implies that

$$U(\theta, S_\theta, y_\theta, W_\theta) - U(\theta, S_\theta, y_{\theta-}, W_{\theta-})$$
$$\leq \frac{\partial U}{\partial y}(\theta, S_\theta, y_{\theta-}, W_{\theta-})h - (1 + \lambda)S \frac{\partial U}{\partial W}(\theta, S_\theta, y_{\theta-}, W_{\theta-})h$$
$$= \nabla U(\theta, S_\theta, y_{\theta-}, W_{\theta-}) \cdot (\Delta y_\theta, \Delta W_\theta);$$

- Similarly, we can show that the above inequality holds when M has a jump.

As a result, the summation in (12.21) is non-positive.

Moreover, because the function U satisfies the equation (12.15), the integrands in (12.17), (12.18) and (12.19) are all non-positive. Noticing that the integral in (12.20) is a martingale, we have

$$EU(s, S_s, y_s, W_s) \leq U(t, S, y, W).$$

Taking $s = T$, it follows that

(12.22) $$U(t, S, y, w) \geq EU(T, S_T, y_T, W_T) = Eu(W_T).$$

On the other hand, given the function U that satisfies the equation (12.15), we can define the regions $\mathcal{S}$ and $\mathcal{B}$ as before. Introduce the following processes

$$(12.23) \quad dW_t = rW_t dt - (1+\lambda)S_t dL_t + (1-\mu)S_t dM_t,$$

$$(12.24) \quad dy_t = dL_t - dM_t,$$

$$(12.25) \quad L_t = \int_0^t 1_{\{(y_s, S_s, W_s) \in \partial\mathcal{B}\}} dL_s,$$

$$(12.26) \quad M_t = \int_0^t 1_{\{(y_s, S_s, W_s) \in \partial\mathcal{S}\}} dM_s.$$

With similar arguments as before, we can show that with this strategy, we have

$$U(t, S, y, W) = E_t u(W_T).$$

Therefore

$$U(t, S, y, W) = \max_{(L,M)} E_t u(W_T).$$

In other words, the function U is indeed the indirect utility function.

3.4 Viscosity solution to the HJB equation

We have assumed that the indirect utility function U is smooth in its variables. However, this may not be the case when there are transaction costs to the investment problem. For this reason, the concept of *viscosity solution* is needed.

We first define a functional (or operator) F on the smooth functions f:

$$
F(f) = \max\left\{ \frac{\partial f}{\partial y} - (1+\lambda)S\frac{\partial f}{\partial W}, -\frac{\partial f}{\partial y} + (1-\mu)S\frac{\partial f}{\partial W}, \right.
$$
$$
\left. \frac{\partial f}{\partial t} + \frac{1}{2}\sigma^2 S^2 \frac{\partial^2 f}{\partial S^2} + \alpha S\frac{\partial f}{\partial S} + rW\frac{\partial f}{\partial W} \right\}.
$$

Note that $F(f)$ is also a function of the variables (t, S, y, W).

A continuous function U is a *viscosity solution* to the HJB equation (12.15) if it satisfies the following conditions:

1 For any smooth function ϕ, if (t_0, S_0, y_0, W_0) is a local maximum point of the function $U - \phi$, then

$$F(\phi)(t_0, S_0, y_0, W_0) \geq 0.$$

240

2 For any smooth function ϕ, if (t_0, S_0, y_0, W_0) is a local minimum point of the function $U - \phi$, then

$$F(\phi)(t_0, S_0, y_0, W_0) \leq 0.$$

The concept of viscosity solution is primarily used in optimal control problems because their value functions are usually not smooth. For details on the concept of viscosity solutions, see Crandall, et all (1983). For its applications to optimal control problems, especially to those problems similar to the one considered in this paper, see Haussmann and Suo (1995).

It can be shown rigorously that the indirect utility function U is indeed the *unique* viscosity solution to the HJB equation (12.15). See Haussmann and Suo (1995) for the proof of this result under more general conditions.

4. Numerical method for solving the HJB equation

It is very difficult, if not impossible, to derive the analytical solution to the equation (12.15). For this reason, we propose a numerical method to compute the indirect utility function.

For any $t > 0$ and $s > t$, it can be shown that the following equation holds (usually called the dynamic programming principle, see Haussmann and Suo (1995) for a proof):

$$(12.27) \qquad U(t, S, y, W) = \max_{(L,M)} E_t[U(s, S_s, y_s, W_s)],$$

where the expectation is taken conditional on $(S_t = S, y_t = y, W_t = W)$.

Recall that the stock price follows a geometric Brownian motion. In order to compute the indirect utility function numerically, we first divide the time interval $[0, T]$ into N subintervals:

$$0 < \Delta_t < 2\Delta_t < \cdots < N\Delta_t = T,$$

where $\Delta_t = T/N$. We can approximate the random factor, which is represented by a Brownian motion, z by a binomial process z^*:

$$z^*(i+1) = \begin{cases} z^*(i) \cdot k_u, & \text{with probability } 1/2, \\ z^*(i) \cdot k_d, & \text{with probability } 1/2, \end{cases}$$

where k_u and k_d are determined such that the expected value and variance of the Brownian motion z on $[i\Delta_t, (i+1)\Delta_t]$ are matched. It follows that

$$k_u = \sqrt{\Delta_t}, \quad k_d = -\sqrt{\Delta_t}.$$

The stock price S is approximated by the binomial process

$$S^*(i+1) = \begin{cases} S^*(i) \cdot \exp(\alpha\Delta_t + \sigma\sqrt{\Delta_t}), & \text{with probability } 1/2, \\ S^*(i) \cdot \exp(\alpha\Delta_t - \sigma\sqrt{\Delta_t}), & \text{with probability } 1/2. \end{cases}$$

We will also assume that the minimum amount of stocks one can buy/sell in each transaction is Δ_y. The y-axis can then be divided into

$$-M\Delta_y < \cdots < -\Delta_y < 0 < \Delta_y < \cdots < M\Delta_y,$$

where M is chosen to be large enough so that the regions outside the interval $[-M\Delta_y, M\Delta_y]$ are economically irrelevant.

From (12.27), and the form of the optimal investment/hedging strategy described in (12.23)-(12.26), we can write,[3] for any t,

$$\begin{aligned} U^\Delta(t, S, y, W) &= \max_{(L,M)} E_t(U^\Delta(t + \Delta_t, S_{t+\Delta_t}, y_{t+\Delta_t}, W_{t+\Delta_t})) \\ &= \max\Big\{ U^\Delta(t, S, y + \Delta_y, W - (1+\lambda)S\Delta_y), \\ &\qquad U^\Delta(t, S, y - \Delta_y, W + (1-\mu)S\Delta_y), \\ &\qquad E(U^\Delta(t + \Delta_t, S_{t+\delta_t}, y, \exp(r\Delta_t)W) \Big\}. \end{aligned}$$

(12.28)

In other words, the optimal decision is described by three different actions:

1 sell minimum number of shares allowed;

2 buy minimum number of shares allowed; and

3 make no transactions.

As discussed earlier, when the portfolio falls in the sell region $\mathcal{S}$, it is optimal for the investor to sell enough (or the minimum number of) shares so that the portfolio will fall on the boundary of the no-transaction region. The opposite action should be taken when the portfolio falls in the buy region $\mathcal{B}$. When the portfolio falls in the region NT, it is optimal for the investor to make no transactions.

To be more specific, no transaction will be made if (for clarity of notations, we drop Δ from the approximate indirect utility function U^Δ)

$$\begin{aligned} \frac{\partial U}{\partial y} - (1+\lambda)S\frac{\partial U}{\partial W} &< 0, \\ -\frac{\partial U}{\partial y} + (1-\mu)S\frac{\partial U}{\partial W} &< 0. \end{aligned}$$

In this case,

$$
\begin{aligned}
U(t, S, y, W) &= E_t[U(t + \Delta_t, S_{t+\Delta_t}, y_{t+\Delta_t}, W_{t+\Delta_t})] \\
&= \frac{1}{2} U\left(t + \Delta_t, S \exp\left((\alpha - r)\Delta_t + \sigma\sqrt{\Delta_t}\right),\right. \\
&\quad \left. y_{t+\Delta_t}, \exp(r\Delta_t)W\right) \\
&\quad + \frac{1}{2} U\left(t + \Delta_t, S \exp\left((\alpha - r)\Delta_t - \sigma\sqrt{\Delta_t}\right),\right. \\
&\quad \left. y_{t+\Delta_t}, \exp(r\Delta_t)W\right).
\end{aligned}
$$

The action to buy a minimum number of shares will be taken if

$$
\frac{\partial U}{\partial y} - (1 + \lambda)S \frac{\partial U}{\partial W} = 0,
$$

in which case

$$
U(t, S, y, W) = U(t, S, y + \Delta_y, W - (1 + \lambda)S\Delta_y).
$$

The action to sell a minimum number of shares will be taken if

$$
-\frac{\partial U}{\partial y} + (1 - \mu)S \frac{\partial U}{\partial W} = 0,
$$

in which case

$$
U(t, S, y, W) = U(t, S, y - \Delta_y, W + (1 - \mu)S\Delta_y).
$$

Based on these results, we now propose a numerical scheme to compute the indirect utility function U:

- Divide the time interval $[0, T]$ and the y-axis as before. The W-axis can also be divided into K-intervals:

$$
0 < \Delta_W < \cdots < K\Delta_W.
$$

- Approximate the Brownian motion z by the binomial process z^*, and define, for (i, j, k, l),

$$
u(i, j, k, l) \equiv U(i\Delta_t, Sk_u^j k_d^{i-j}, k\Delta_y, l\Delta_W),
$$

where $0 \leq i \leq N$,$0 \leq j \leq i$, $-M \leq k \leq M$, and $0 \leq l \leq K$.

- At the maturity time T,

$$
u(N, j, k, l) = l\Delta_W - (Sk_u^j k_d^{N-j} - K)^+
$$

for all (j, k, l), where S is the stock price at $t = 0$.

- Recursively, for $i < N$, define

$$u^*(i,j,k,l) = \frac{1}{2}[u(i+1,j+1,k,l^+) + u(i+1,j,k,l^+)]$$

where l^+ is the integer such that corresponding mesh point for W is closest to the value $\exp(r\Delta_t)l\Delta_W$.[4]

- Check if the following relationships hold:

$$u^*(i,j,k,l) > u^*(i,j,k+1,l^+),$$
$$u^*(i,j,k,l) > u^*(i,j,k-1,l^-),$$

where l^+ (l^-) is the integer such that corresponding mesh point for W is closest to the value $l\Delta_W - (1+\lambda)S\Delta_y$ $(l\Delta_W + (1-\mu)S\Delta_y$, respectively). Such mesh points fall in the no-transaction region, so we can let

$$u(i,j,k,l) = u^*(i,j,k,l).$$

- Take such a point, say (j_0, k_0, l_0). Notice that U is a non-decreasing function in the variables y and W. Working recursively, we can check if

$$u^*(i,j,k,l) \leq u^*(i,j,k+1,l^+),$$

in which case, we can set

$$u(i,j,k,l) = u^*(i,j,k+1,l^+).$$

Similarly, if

$$u^*(i,j,k,l) \leq u^*(i,j,k-1,l^-),$$

we can set

$$u(i,j,k,l) = u^*(i,j,k-1,l^-).$$

In other words, we will make sure that for all (j,k,l),

$$u(i,j,k,l) \geq u(i,j,k+1,l^+),$$
$$u(i,j,k,l) \geq u(i,j,k-1,l^-).$$

This scheme can be easily implemented, and the indirect utility function can thus be computed numerically. With the indirect utility, we can study the optimal hedging strategy by studying the no-transaction region.

Similarly, the indirect utility for the investment problem without option can be computed numerically by changing the value of the function U at time T. The asking price of the option can thus be obtained through (12.16).

5.   Summary

In this paper, we study the optimal investment problem for an investor with a HARA type utility function. We show that when there are no transaction costs, and there is a short position in the investor's portfolio, she may prefer not to perfectly hedge the option, as in the Black-Scholes option pricing framework. In fact, when the growth rate of the stock is greater than the risk-free rate, the investor will always "over" hedge the option. When there is transaction cost in the investor's portfolio, the investor's optimal investment/hedging strategy can be described by three regions: the buying region, the selling region and the no transaction region. When her portfolio falls in the buying region, she will buy enough shares of the stock to make her portfolio lie in the no transaction region (NT). Similar results hold when her portfolio falls in the selling region. When her portfolio falls in the NT region, it is optimal for the investor to make no transaction, i.e., simply hold her portfolio. Viscosity solution is introduced to describe the indirect utility function. We also propose a numerical method to compute the indirect utility function, and the asking price for an option for be computed as a result.

The numerical scheme we proposed in this paper remains to be implemented in order to show numerically the properties of the hedging strategy.

6.   Appendix

Solving the HJB equation with HARA utility function:

We guess that the indirect utility function, or the solution to (12.5) is in the following form:

$$U(t, W) = g(t) \frac{W^\gamma}{\gamma}.$$

Then

$$U_t = g'(t) \frac{W^\gamma}{\gamma},$$
$$U_W = g(t) W^{\gamma-1},$$
$$U_W = g(t)(\gamma - 1) W^{\gamma-1},$$

and

$$\pi^* = -\frac{\alpha - r}{\sigma^2} \frac{U_W}{W U_{WW}} = \frac{\alpha - r}{\sigma^2(1 - \gamma)}.$$

Substitute these expressions into the HJB equation (12.5), we get

$$g'(t) + \nu\gamma g(t) = 0,$$

where

$$\nu \equiv \frac{1}{2}\left[\left(\frac{\alpha - r}{\sigma}\right)^2 + \frac{1}{1 - \gamma} + r\right].$$

The boundary condition reduces to

$$g(T) = 1.$$

The indirect utility can now be obtained by solving this equation.

Proof of (12.10):

In the case when there is no transaction cost, it is well-known that the call option can be replicated by a self-financing portfolio strategy that requires an initial cash of

$$c = SN(d_1) - e^{-r(T-t_0)}KN(d_2).$$

We denote this strategy by π^c. We will show that if we denote by $\bar{U}(t, S, W)$ as the indirect utility function for the investor at time t when her wealth is W and the current (or spot) stock price is S,, then

$$\bar{U}(0, S, W) = U(0, W - c).$$

In fact, if this is not the case, say,

$$\bar{U}(0, S, W) < U(0, W - c),$$

then let π_t be the optimal portfolio that requires an investment of $W - c$ at the beginning when there is no short position in call in her portfolio. By definition,

$$U(0, W - c) = EU(T, W_T^\pi),$$

Consider the new trading

$$\bar{\pi} = \pi + \pi^0,$$

which needs an initial investment of

$$(W - c) + c = W,$$

and has the terminal wealth of

$$W_T^\pi + (S_T - K)^+.$$

Thus we have

$$
\begin{aligned}
\bar{U}(0, S, W) &\geq EU(T, S_T, W_T^{\bar{\pi}}) \\
&= EU(T, S_T, W_T^{\pi} + (S_T - K)^+ - (S_T - K)^+) \\
&= EU(T, S_T, W_T^{\pi}) \\
&= E\left(\frac{(W_T^{\pi})^{\gamma}}{\gamma}\right) \\
&\geq U(0, W - c),
\end{aligned}
$$

which is a contradiction. In other words, we must have

$$
\bar{U}(0, S, W) \geq U(0, W - c).
$$

The other side of the inequality can be shown similarly.

In general, we can show that

$$
\bar{U}(t, S, W) = U(W - f(t, S)),
$$

where S is the spot price of the sock at time t, and $f(t, S)$ is the price of the call option when the when the spot stock price is S.

Notes

1. It is obvious that this result does not hold for general utility functions.

2. In the optimal control terminology, this type of problems are called *singular control* problems because the processes L and M may be singular functions, i.e., they may be continuous and yet non-differentiable.

3. It can be shown the function defined this way converges to the indirect utility function as $N \to \infty, \Delta_t, \Delta_y \to 0$, see Barles and Souganidis (1991) for the proof of a similar result.

4. For better approximation, we can use interpolations at $(i + 1)\Delta_t$ for this value

References

Black, F. and M. S. Scholes (1973), "The Pricing of Options and Corporate Liabilities." *Journal of Political Economy*, 81, pp637-59.

Boyle, P.P., and T. Vorst, (1990), "Option pricing in discrete time with transaction costs", *Journal of Finance*, 47, pp271-293.

Clewlow, L. and S. Hodges (1997), "OPtimal delta-hedging under transaction costs", *Journal of Economics Dynamics and Control*, 21, pp1353–1376

Constantinides, G.M. (1995), "Capital market equilibrium with transaction costs", *Journal of Political Economy*, 81, pp637-659.

Crandall, M.G., H. Ishii and P.L. Lions (1992), "User's guide to viscosity solutions to second order partial differential equations", *Bulletin of AMS*, 27, pp1-67.

Dumas, B and E. Luciaono (1991), "An exact solution to a dynammic portfolio choice problem under transaction costs", *Journal of Finance*, XLVI, pp577– 595

Davis, M.H.A. and A.R. Norman (1990), "Portfolio selection with transaction costs", *Mathematics of Operational Research*, 15, pp676–713

Davis, M.H.A, V.G. Panas and T. Zariphopoulou (1993), "European optionpricing with transaction costs", *SAIM J. Control and Optimization*, 31, pp490–493

Haussmann, U.G., and W. Suo (1995), "Singular stochastic controls II: The dynamic programming principle and applications", *SIAM J. Control and Optimization*, 33, pp937-959

Leland, H. (1995), "Option pricing with transaction costs", *Journal of Finance*, 40, pp1283-1031.

Merton, R. C. (1971), "Optimum Consumption and Portfolio Rules in a Continuous-Time Model." *Journal of Economic Theory*, 3, pp373-413.

Toft, K.B. (1996), "On the mean variance tradeoff in option replication", *Journal of Financial and Quantitative Analysis*, 31, pp233-263

Chapter 13

SUPPLY PORTFOLIO SELECTION AND EXECUTION WITH DEMAND INFORMATION UPDATES

Haifeng Wang

Center for Intelligent Networked Systems, Department of Automation
Tsinghua University, Beijing, 100084, China

wanghf99@mails.tsinghua.edu.cn

Houmin Yan

Department of Systems Engineering and Engineering Management
Chinese University of Hong Kong, Shatin, N.T., Hong Kong

yan@se.cuhk.edu.hk

Abstract   This paper considers a problem of multi-period supply portfolio selection and execution with demand information updates. A supply portfolio specifies a buyer's decision on selecting sourcing mix from among a group of suppliers. We develop a framework for optimal supply portfolio selection and execution. Further, we demonstrate that the optimal portfolio selection follows a base-stock policy and the option execution follows a modified base-stock policy. We also develop the structural properties of the optimal policy with respect to option contracts and inventories.

Keywords: Portfolio selection, option contract, base-stock policy, supply chain management

1.   Introduction

It has been well demonstrated that imperfect demand information influences buyer's decision about order quantities and a supplier's decision about production plans, especially when supply lead-time is significantly large. To facilitate the tradeoff between the supply lead-time and imperfect demand information, various forms of supply contracts exist. A supply contract provides flexibility to the buyer and early demand information to the supplier. However, the management of supply contracts is a challenging task to buyers, especially when the buyer has a number of supply contracts from which to choose. A supplier mix, i.e., purchase levels from different suppliers, is a supply portfolio to a buyer.

In this paper, we study option contracts. An option contract requires the early reservation of capacities, thus allowing the buyer to decide the exact purchase amount at a later time when an up to date demand information becomes available. We assume that the buyer has multiple suppliers from which to choose. At the same time, suppliers who are more flexible provide contracts with lower reservation prices and higher execution prices. In contrast, suppliers who are less flexible offer contracts with higher reservation prices and lower execution prices. Therefore, the buyer needs to reserve capacities from individual suppliers, known as the supply portfolio selection, and needs to decide the exact amounts to purchase from individual suppliers, which is known as supply portfolio execution.

Martinez-de-Albeniz and Simchi-Levi (2003) study the problem of supply portfolio selection and execution, where selection is made once at the beginning of the planning horizon. The same portfolio applies to the entire planning horizon, and the option execution happens at the end of each period when the demand becomes known. In this paper, we allow the buyer to select a unique portfolio for each period, and the option execution occurs before the demand becomes known.

There is a large body of literature on supply contracts. Eppen and Iyer (1997) study the "Backup Agreement" in fashion buying. Brown and Lee (2003) model the "Take-or-Pay" capacity reservation contract that is used in the semiconductor industry. Li and Kouvelis (1999) study a time-flexible contract that allows the firm to specify the purchase amount over a given period of time to meet deterministic demand. Tsay (1999) considers a quantity-flexible contract that couples the customer's commitment to purchase no less than a certain percentage. Cachon and Lariviere (2001) focus on the issue of information sharing in a supply chain using an option contract. For supply contract management, Cachon (2003) provides an excellent survey: we refer interested readers to this survey article and the references therein.

Another line of research is supply chain decisions with demand information updates. Iyer and Bergen (1997) study how a manufacturer-retailer channel affects the choice of production and marketing variables under a Quick-Response program. They analyze how the demand variance influences the total profit of the retailer. Donohue (2000) develops an information update model with two production modes: the more expensive production mode requires less production lead-time. Barnes-Schuster, Bassok and Anupindi (2002) develop a general model that involves two production modes together with one option contract and two-period demand. Sethi, Yan and Zhang (2004) study models with quantity flexible contracts which involves information updates and spot markets. Zhu and Thonemann (2004) study the benefits of sharing future demand information in a model with one retailer and multiple customers. A recent book by Sethi, Yan and Zhang (2005) provides an up to date review of models in inventory decisions with multiple delivery modes and demand information updates.

In this paper, we study a model of multi-period supply portfolio selection and execution with demand information updates. We characterize the portfolio selection and execution policies at the beginning and near the end of each period, respectively. We demonstrate that the portfolio selection follows a *base-stock policy* and the portfolio execution follows a *modified base-stock policy.*

In the next section, we introduce the notation and problem formulation. We develop the optimal policies of portfolio selection and option execution in Section 3. Concluding remarks are provided and future research directions are summarized in Section 4.

2. The Problem Formulation and Notations

In this section, we consider the problem of optimal supply portfolio selection and execution, where a buyer makes reservation and execution decisions with the initial and updated demand information. The sequence of this supplier portfolio selection and execution can be described as follows. At the beginning of each period, each supplier first presents the buyer with an option menu that indicates a unit reservation and execution price. Based on the demand information available at that time, the buyer makes a decision on how many units of the product to reserve from each supplier, which is known as the portfolio selection. Before the customer demand is realized, the buyer revisits the reservation plan with the updated demand information, and decides the exact amount to be purchased from each supplier, which is known as the option execution. Finally, the customer demand is realized, the unsatisfied customer demand is lost, and extra products are inventoried. The sequence of events and decisions are graphically illustrated in Figure 1.1. We first list the notation that is used in this paper.

T: length of the planning horizon

t: period index

n: number of available contracts for period t

$\mathbf{x_t}$: reserved capacities for period t, $\mathbf{x_t} = (x_1(t), \cdots, x_n(t))$

$\mathbf{q_t}$: amount of option exercised, $\mathbf{q_t} = (q_1(t), \cdots, q_n(t))$

I_t: demand information with a cumulative distribution function $F(\cdot)$, and density distribution function $f(\cdot)$

D_t: customer demand in period t, with unconditional and conditional distributions $H(z)$ and $H(z|I)$, respectively

$\beta(t)$: inventory level at the beginning of period t

$\mathbf{v_t}$: $\mathbf{v_t} = (v_1(t), v_2(t), \cdots, v_n(t))$, where $v_i(t)$ represents the unit reservation price for option i in period t

$\mathbf{w_t}$: $\mathbf{w_t} = (w_1(t), w_2(t), \cdots, w_n(t))$, where $w_i(t)$ represents the unit execution price for option i in period t

h_t: unit inventory holding cost in period t

r_t: unit revenue in period t, where $r_t > v_t + w_t$

s: unit salvage value at the end of the horizon

α: discount factor $(0 \le \alpha \le 1)$

$\pi_1^*(t, \beta(t))$: value function from the beginning of the 1st stage of period t with initial inventory level $\beta(t)$

$\pi_1(\mathbf{x_t}; t, \beta(t)|I_t)$: profit function from the beginning of the 1st stage of period t when the initial inventory level is $\beta(t)$ and the selected supply portfolio is $\mathbf{x_t}$

$\pi_2^*(t, \mathbf{x_t}|i)$: value function from the beginning of the 2nd stage of period t when the option capacity portfolio is $\mathbf{x_t}$ and the observed demand information is i

$\pi_2(\mathbf{q_t}; t, \mathbf{x_t}|i)$: profit function from the beginning of the 2nd stage of period t when option execution is $\mathbf{q_t}$ the option capacity portfolio is x_t and the demand information is i

To avoid trivial cases, we assume:

ASSUMPTION 2.1 *The contracts are listed in the order of the execution cost, i.e.* $w_1(t) < w_2(t) < \cdots < w_n(t)$.

With this assumption, it is clear that we have $v_1(t) > v_2(t) > \cdots > v_n(t)$. Otherwise, if there are two contracts such that $i < j$, and $v_i(t) \le$

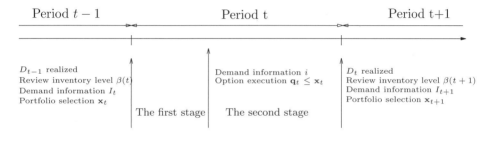

Figure 13.1. The sequence of events and decisions

$v_j(t)$, we conclude that contract i dominates contract j, because $w_i(t) < w_j(t)$.

With the above notation and preliminary analysis, we start to write the dynamic programming equations. Note that the inventory dynamics can be written as $\beta(t + 1) = (\beta(t) + q_1(t) + \cdots + q_n(t) - D_t)^+$ for $t = 1, \cdots, T$, where $x^+ = \min\{0, x\}$. The profit function for t is

$$
\begin{aligned}
\pi_1^*(t, \beta(t)) &= \max_{\mathbf{x_t} \geq 0} \{\pi_1(\mathbf{x_t}; t, \beta(t)|I_t)\} \\
&= \max_{\mathbf{x_t} \geq 0} \{-v_1(t)x_1(t) - \cdots - v_n(t)x_n(t) \\
&\quad + E_{I_t} \left[\max_{0 \leq \mathbf{q_t} \leq \mathbf{x_t}} \pi_2(\mathbf{q_t}; t, \mathbf{x_t}|i) \right] \},
\end{aligned}
\tag{13.1}
$$

where

$$
\begin{aligned}
\pi_2(\mathbf{q_t}; t, \mathbf{x_t}|i) &= E_{D_t} [r_t(D_t \wedge (\beta(t) + \cdots + q_n(t))) - w_1(t)q_1(t) - \cdots \\
&\quad - w_n(t)q_n(t) - h_t(\beta(t) + \cdots + q_n(t) - D_t)^+ \\
&\quad + \alpha \pi_1^* \left(t + 1, (\beta(t) + \cdots + q_n(t) - D_t)^+\right)].
\end{aligned}
\tag{13.2}
$$

The remaining inventory of last period is salvaged as:

$$
\begin{aligned}
&\pi_{T+1}^1((\beta(T) + q_1(T) + \cdots + q_n(T) - D_T)^+) \\
&= s(\beta(T) + q_1(T) + \cdots + q_n(T) - D_T)^+.
\end{aligned}
$$

3. The Optimal Portfolio Selection and Execution

To optimize the objective function of Equation (13.1), it is necessary to choose decision variables $\mathbf{x_t}$ and $\mathbf{q_t}$ for each t. Let us sketch the plan for the optimal portfolio selection and execution. We first assume that $\pi_1^*(t + 1, \beta(t + 1))$ is concave in $\beta(t + 1)$, and

$$\frac{d\pi_1^*(t+1, \beta(t+1))}{d\beta(t+1)} \bigg|_{\beta(t+1)=0} = v_1(t+1) + w_1(t+1)$$

and

$$\frac{d\pi_1^*(t+1, \beta(t+1))}{d\beta(t+1)} \bigg|_{\beta(t+1)=+\infty} < 0.$$

With these assumptions, we prove that $\pi_2(\mathbf{q_t}; t, \mathbf{x_t}|i)$ is concave in $q_k(t)$, for $q_j(t), j \neq k$. For any given $\mathbf{x_t}$ and the demand information i, we choose $\mathbf{q_t^*}(\mathbf{x_t}, i)$ to maximize $\pi_2(\mathbf{q_t}; t, \mathbf{x_t}|i)$. We then substitute $\mathbf{q_t^*}(\mathbf{x_t}, i)$ into $\pi_1(\mathbf{x_t}; t, \beta(t)|I_t)$ of Equation (13.1) and demonstrate that $\pi_1(\mathbf{x_t}; t, \beta(t)|I_t))$ is concave in $\mathbf{x_t}$ for any initial inventory $\beta(t)$. The next step is to determine portfolio $\mathbf{x_t^*}(\beta(t))$ that maximizes $\pi_1(\mathbf{x_t}; t, \beta(t)|I_t)$. Finally, we substitute $\mathbf{x_t^*}(\beta(\mathbf{t}))$ into $\pi_1^*(t, \beta(t))$ in Equation (13.1) and prove that $\pi_1^*(t, \beta(t))$ is indeed concave in $\beta(t)$ and $\frac{\pi_1^*(t, \beta(t))}{d\beta(t)} \big|_{\beta(t)=0} = v_1(t) + w_1(t)$.

ASSUMPTION 3.1 $r_t > w_k(t) + v_k(t), k = 1, \cdots, n$ and $r_t \geq -h_t + \alpha(v_1(t+1) + w_1(t+1))$.

We start by presenting our first result in the following lemma: its proof is included in Appendix.

LEMMA 3.1 *With Assumption 3.1, for any $q_j(t), j \neq k$, $\pi_2(\mathbf{q_t}; t, \mathbf{x_t}|i)$ is concave in $q_k(t)$.*

We now develop the optimal option execution policy $\mathbf{q_t^*}$ for the given reserved capacity $\mathbf{x_t}$ with the updated demand information i. First, we present the following lemma.

LEMMA 3.2 $q_j^*(t) > 0 \Rightarrow q_k^*(t) = x_k(t), \forall k < j.$

Proof of this lemma can be found in Appendix.

Remark 3.1 Lemma 3.2 indicates that we can rank option contracts based on the execution cost. For such a list, a contract becomes active only if its preceding contract has been exhausted. In other words, if $q_i^*(t) < x_i(t)$, then $q_{i+1}(t) = q_{i+2}(t) = \cdots = q_n(t) = 0$.

With Lemma 3.2, it is also possible for us to construct an algorithm to find the optimal option execution policy. We start from the contract with cheapest execution price to determine $q_1(t)$. Then, it follows by determining $q_2(t)$. After figuring out $q_1^*(t), \cdots, q_{k-1}^*(t)$, we determine $q_k^*(t)$ between the following two cases. (1) If $q_k^*(t) < x_k(t)$, then let $q_j^*(t) = 0, \forall j > k$; (2) otherwise, $q_k^*(t) = x_k(t)$ and execute option contract $K + 1$ for $q_{k+1}^*(t)$, if $k + 1 \leq n$.

To facilitate this procedure of finding $\mathbf{q}_t^*$, let us define a base-stock level for contract k,

DEFINITION 3.1 $Q_k(i), k = 1, 2, \cdots, n$ *satisfies*

$$(r_t - w_k(t)) - (r_t + h_t)H(Q_k(i) \mid i) + \alpha \frac{d}{dq_k(t)} E_{D_t} \pi_1^*(t+1,$$

$$(\beta(t) + q_1(t) + \cdots + q_k(t) - D_t)^+) \mid_{\beta(t)+q_1(t)+\cdots+q_k(t)=Q_k(i)} = 0. \tag{13.3}$$

Let $Q_k(i)$ be the smallest one if there are multiple solutions of equation (13.3). To make sure that $Q_k(i)$ is well defined, we need the following lemma, the proof of which appears in Appendix.

LEMMA 3.3 *There exists a unique $Q_k(i), k = 1, 2, \cdots, n$. Moreover, $Q_1(i) > Q_2(i) \cdots > Q_n(i)$.*

We now demonstrate how $q_1^*(t)$ can be determined. Let $q_2(t) = q_3(t) = \cdots = q_n(t) = 0$. Lemma 3.1, $\pi_2(q_1(t), 0, \cdots, 0; t, \mathbf{x_t}, i)$ is concave in $q_1(t)$. Therefore, we choose $q_1^*(t)$ with the constraint of $q_t^1 \le x_1(t)$. To simplify the exposition, we use $\pi_2(\mathbf{q_t}; t)$ for $\pi_2(\mathbf{q_t}; t, \mathbf{x_t}, i)$ whenever there is no confusion. Rewrite Equation (13.2) as

$$\pi_2(\mathbf{q_t}; t) = -(r_t + h_t) \int_0^{\beta(t)+q_1(t)} (\beta(t) + q_1(t) - D_t) h(D \mid i) dD$$

$$+ r_t(q_1(t) + \beta(t)) - w_1(t) q_1(t)$$

$$+ \alpha E_{D_t} \pi_1^* \left(t+1, (\beta(t) + q_1(t) - D_t)^+ \right). \tag{13.4}$$

Then

$$\frac{d\pi_2(q_1(t), 0, \cdots, 0; t)}{dq_1(t)}$$

$$= (r_t - w_1(t)) - (r_t + h_t)H(q_1(t) + \beta(t) \mid i)$$

$$+ \alpha \frac{d}{dq_1(t)} E_{D_t} \pi_1^*(t+1, (q_1(t) + \beta(t) - D_t)^+). \tag{13.5}$$

With this expression and Lemma 3.1, we obtain the following lemma, which is required for proving Lemma 3.3. Proof of this lemma can be found in Appendix.

LEMMA 3.4 $-(r_t + h_t)H(\beta(t) \mid i) + \alpha \frac{d}{d\beta(t)} E_{D_t} \pi_1^*(t+1, (\beta(t) - D_t)^+)$ *is non-increasing in $\beta(t)$, i.e.*

$$-(r_t + h_t)h(\beta(t) \mid i) + \alpha \frac{d^2}{(d\beta(t))^2} E_{D_t} \pi_1^*(t+1, (\beta(t) - D_t)^+) \le 0. \tag{13.6}$$

With Definition 3.1, and Equation (13.5), the optimal execution of the supply contract 1 is

$$
q_1^*(t) = \begin{cases} 0, & \text{if } \beta(t) \geq Q_1(i); \\ Q_1(i), & \text{if } Q_1(i) - x_1(t) < \beta(t) < Q_1(i); \\ x_1(t), & \text{if } \beta(t) < Q_1(i) - x_1(t). \end{cases}
$$

We now move to determine $q_2^*(t)$. If $q_1^*(t) = x_1(t)$, then let $q_2(t)$ be the next decision variable. By Lemma 3.1, we know that $\pi_2(x_1(t), q_2(t), \cdots, 0; t)$ is concave in $q_2(t)$. Similarly we obtain

$$
\begin{aligned}
& \frac{d\pi_2(x_1(t), q_2(t), \cdots, 0; t)}{dq_2(t)} \\
= \; & (r_t - w_2(t)) - (r_t + h_t) H(q_2(t) + \beta(t) + x_1(t) \mid i) \\
& + \alpha \frac{d}{dq_2(t)} E_{D_t} \pi_1^*(t+1, (q_2(t) + \beta(t) + x_1(t) - D_t)^+).
\end{aligned}
$$

$$(13.7)$$

By Definition 3.1, $Q_2(i)$ satisfies

$$
\begin{aligned}
& (r_t - w_2(t)) - (r_t + h_t) H(Q_2(i)) + \alpha \frac{d}{dq_2(t)} E_{D_t} \pi_1^*(t+1, \\
& (\beta(t) + x_1(t) + q_2(t) - D_t)^+) \, |_{\beta(t) + x_1(t) + q_2(t) = Q_2(i)} \quad = \quad 0.
\end{aligned}
$$

$$(13.8)$$

The optimal option execution of supply contract 2 is

$$
q_2^*(t) = \begin{cases} 0, & \text{if } Q_2(i) - x_1(t) \leq \beta(t) < Q_1(i) - x_1(t); \\ Q_2(i) - \beta(t) - x_1(t), & \text{if } Q_2(i) - x_1(t) - x_2(t) \leq \beta(t) \\ & \qquad < Q_2(i) - x_1(t); \\ x_2(t), & \text{if } \beta(t) < Q_2(i) - x_1(t) - x_2(t). \end{cases}
$$

Following the same procedure, we can obtain the optimal option execution for other supply contracts. We summarize the optimal option execution process in the following Lemma.

LEMMA 3.5 *The optimal option execution* $(l = 1, \cdots, n)$ *is:*

case 1 *when* $Q_j(i) - x_{j-1}(t) - \cdots - x_1(t) \leq \beta(t) < Q_{j-1}(i) - x_{j-1}(t) - \cdots - x_1(t);$ $q_l^*(t) = 0, \forall l \geq j$ *and* $q_l^*(t) = x_l^*(t), \forall l < j;$

case 2 *when* $Q_j(i) - x_t^j - \cdots - x_1(t) \leq \beta(t) < Q_j(i) - x_{j-1}(t) - \cdots - x_1(t);$ $q_l^*(t) = 0, \forall l > j$, $q_j^*(t) = Q_j(i) - \beta(t) - \cdots - x_{j-1}(t)$ *and* $q_l^*(t) = x_l^*(t), \forall l < j.$

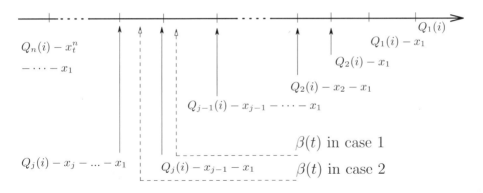

Figure 13.2. Replenishment policy illustration

The process of determining the optimal option execution is illustrated in Fig. 13.2.

To this end, we have solved the problem of option execution, and we are ready to deal with the problem of portfolio selection. Note that the portfolio selection problem involves an expectation over the demand information signal I. It would simplify the optimization process if we could connect the optimal option execution with the demand information signal. In what follows, we explore the relationship of the option execution and the demand information. We demonstrate that there is a one-to-one mapping between the order quantity on individual contracts and information intervals. To do that, we need to connect the option execution to the demand process. We assume that the demand is stochastically increasing, i.e. $\forall i_1 < i_2, H^{-1}(z|i_1) < H^{-1}(z|i_2)$.

Hence, $H(Q_k \mid i)$ is decreasing in i. By Lemma 3.4, the left-hand side of Equation (13.3) is non-increasing in $Q_k(i)$. Hence, $Q_k(i)$ is increasing in i, i.e.

$$Q_k(i_1) < Q_k(i_2), \forall i_1 < i_2. \tag{13.9}$$

Now, we define the critical demand information signals.

DEFINITION 3.2 *Define the critical information values* $(i_k, \hat{i}_k, k = 0, \cdots, n)$ *as follows, for* $x_j(t) > 0, \forall j = 1, \cdots, n,$

$$i_0 = \hat{i}_0 \quad \text{is s.t.} \quad Q_1(i_0) = \beta(t)$$

$$i_1 \quad \text{is s.t.} \quad Q_1(i_1) = \beta(t) + x_1(t)$$

$$\hat{i}_1 \quad \text{is s.t.} \quad Q_2(\hat{i}_1) = \beta(t) + x_1(t)$$

$$i_2 \quad \text{is s.t.} \quad Q_2(i_2) = \beta(t) + x_1(t) + x_2(t)$$

$$\vdots$$

$$\hat{i}_{n-1} \quad \text{is s.t.} \quad Q_n(\hat{i}_{n-1}) = \beta(t) + x_1(t) + \cdots + x_{n-1}(t)$$

$$i_n \quad \text{is s.t.} \quad Q_n(\hat{i}_{n-1}) = \beta(t) + x_1(t) + \cdots + x_n(t)$$

$$\hat{i}_n \quad \text{is} \quad +\infty$$

where $Q_k(i)$ *is defined in Definition 3.1.*

For some $k, k = 1, \cdots, n$, *s.t.* $x_t^k = 0$, *then* $\hat{i_{k-1}} = \hat{i_{k-2}}$ *and* $i_k = i_{k-1}$. *If* $\beta(t) = 0$, *then* $i_0 = -\infty$.

The option execution policy $\mathbf{q}_t^*$ of period t is as follows.

THEOREM 3.1 *Given the initial inventory* $\beta(t)$ *and reserved capacity* $\mathbf{x}_t^*$, *the option execution depends on the updated demand information* $I_t = i$. *Specifically, it follows the modified base-stock policy that is illustrated in Table 1.1.*

Proof. From Equation (13.9), we know that as the information i increases, the thresholds become larger and move to the righthand side of Fig. 13.2. Hence, the result is straightforward from the Definition 3.2 and Lemma 3.5. □

With Theorem 3.1, we are ready to deal with proving the concavity of $\pi_1(\mathbf{x_t}; t, \beta(t)|I_t)$ with respect to $\mathbf{x_t}$. We summarize our result in the following lemma, the proof of which appears in Appendix.

LEMMA 3.6 *Given initial inventory* $\beta(t)$, $\pi_1(\mathbf{x_t}; t, \beta(t)|I_t)$ *is concave in* $\mathbf{x_t}$.

To this end, we have demonstrated that for a given $\beta(t)$ with respect to $\mathbf{x_t}$, $\pi_1(\mathbf{x_t}; t, \beta(t)|I_t)$ is concave in the feasible region of the polyhedral

Table 13.1. Option execution: modified base-stock policy

Case	Information Revised i	option execution q_t^*
(0)	$(-\infty, i_0)$	$q_1^*(t) = q_2^*(t) = \cdots = 0$
(1)	$[i_0, i_1)$	$q_1^*(t) = Q_1(i) - \beta,\, q_2^*(t) = \cdots = q_n^*(t) = 0$
(2)	$[i_1, \hat{i_1})$	$q_1^*(t) = x_1(t),\, q_2^*(t) = \cdots = q_n^*(t) = 0$
$\vdots$	$\vdots$	$\vdots$
(2n-2)	$[i_{n-1}, \hat{i}_{n-1})$	$q_1^*(t) = x_1(t), \cdots, q_t^{n-1,*} = x_{n-1}(t),\, q_n^*(t) = 0$
(2n-1)	$[\hat{i}_{n-1}, i_n)$	$q_1^*(t) = x_1(t), \cdots,\, q_n^*(t) = Q_n(i) - \beta - x_1(t) - \cdots - x_{n-1}(t)$
(2n)	$[i_n, +\infty)$	$q_1^*(t) = x_1(t),\, q_2^*(t) = x_2(t), \cdots,\, q_n^*(t) = x_n(t)$

cone of nonempty interiors. This fact implies that K-K-T conditions are necessary and sufficient at optimality (Dimitri P. Bertsekas, (1995)). The portfolio selection $\mathbf{x_t^*}$ can be found by using K-K-T conditions.

Define the associate Lagrangian multiplier μ_j for each constraint $x_j(t) \geq 0, j = 1, \cdots, n$. Then, the K-K-T conditions are

$$i.e. \begin{cases} -\dfrac{\partial \pi_1(x_1^*(t), \cdots, x_n^*(t); t)}{\partial x_j(t)} - \mu_j^* = 0, \\ \mu_j^* \geq 0, \\ \mu_j^* = 0, \forall x_j(t) > 0, \\ \dfrac{\partial \pi_1(x_1^*(t), \cdots, x_n^*(t); t)}{\partial x_j(t)} = 0, \forall x_j^*(t) > 0, \\ \dfrac{\partial \pi_1(x_1^*(t), \cdots, x_n^*(t); t)}{\partial x_j(t)} \leq 0, \forall x_j^*(t) = 0, \end{cases} \quad j = 1, 2, \cdots, n.$$

DEFINITION 3.3 *Define* $y_n(t), \cdots, y_2(t), y_1(t)$, *such that*

$$\frac{\partial \pi_1(t)}{\partial x_n(t)} \Big|_{\beta(t) + \cdots + x_n(t) = y_n(t)} = 0,$$

$$\vdots$$

$$\frac{\partial \pi_1(t)}{\partial x_2(t)} \Big|_{\beta(t) + x_1(t) + x_2(t) = y_2(t), \cdots, \beta(t) + \cdots + x_n(t) = y_n(t)} = 0,$$

$$\frac{\partial \pi_1(t)}{\partial x_1(t)} \Big|_{\beta(t) + x_1(t) = y_1(t), \cdots, \beta(t) + \cdots + x_n(t) = y_n(t)} = 0. \tag{13.10}$$

If there are multiple values for $y_j(t), j = 1, \cdots, n$, then choose the smallest one.

To obtain $\mathbf{x}_t^*$, the following lemmas provide us with an algorithmic procedure. Proofs of Lemmas 3.7, 3.8, and 3.9 can be found in Appendix.

LEMMA 3.7 *There exist unique $y_n(t), \cdots, y_2(t), y_1(t)$. Moreover, $y_i(t)$ can be found one by one in the order of $y_n(t), \cdots, y_2(t), y_1(t)$.*

LEMMA 3.8 *If $y_k(t) = \beta(t) + x_1^*(t) + \cdots + x_k^*(t), k = j, \cdots, n$ and $y_j(t) \leq y_{j-1}(t)$, then $x_{j-1}^*(t) = 0$, where $j = 2, \cdots, n$.*

LEMMA 3.9 *If $y_k(t) = \beta(t) + x_1^*(t) + \cdots + x_k^*(t), k = j + 1, \cdots, n$ and $y_j(t) \leq \beta(t)$, then $x_j^*(t) = \cdots = x_1^*(t) = 0, j = 1, \cdots, n$.*

With the above lemmas, we now develop an algorithm in determining the optimal $\mathbf{x}_t^*$.

- If $y_n(t) \leq \beta(t)$, then by Lemma 3.9, $\mathbf{x_t^*} = \mathbf{0}$. Else go to the next step.

- If $y_{n-1}(t) \leq \beta(t)$, then by Lemma 3.9, $x_{n-1}^*(t) = x_{n-2}^*(t) = \cdots = x_1^*(t) = 0$, and then $x_n^*(t) = y_n(t) - \beta(t)$.

  Else if $y_{n-1}(t) > \beta(t)$ and $y_{n-1}(t) \geq y_n(t)$, then by Lemma 3.8, $x_{n-1}^*(t) = 0$, contract $n-1$ is *inferior* and go to the next step.

  Else if $y_{n-1}(t) > \beta(t)$ and $y_{n-1}(t) < y_n(t)$, then $y_{n-1}(t) = \beta(t) + x_1^*(t) + \cdots + x_{n-1}^*(t)$, $x_n^*(t) = y_n(t) - y_{n-1}(t)$ and go to the next step.

- $\vdots$

- If $y_1(t) \leq \beta(t)$, $x_1^*(t) = 0$, $\mathbf{x_t^*} = (0, y_2(t) - \beta(t), y_3(t) - y_2(t), \cdots, y_n(t) - y_{n-1}(t))$.

  Else if $y_1(t) > \beta(t)$ and $y_1(t) \geq y_2(t)$, then by Lemma 3.8, $x_1^*(t) = 0$, contract 1 is *inferior*.

  Else if $y_1(t) > \beta(t)$ and $y_1(t) < y_2(t)$, then $x_1^*(t) = y_1(t) - \beta(t)$, $x_2^*(t) = y_2(t) - y_1(t)$.

It can be seen from the above procedure that if $y_j(t) \leq y_{j-1}(t)$, then $x_{j-1}^*(t) = 0$, which is indifferent to the inventory level $\beta(t)$. Note that such a contract is known as an *inferior* contract. To simplify our procedure for finding $\mathbf{x}_t^*$, we remove all inferior contracts from further consideration. Hence, without loss of generality, we have the following assumption.

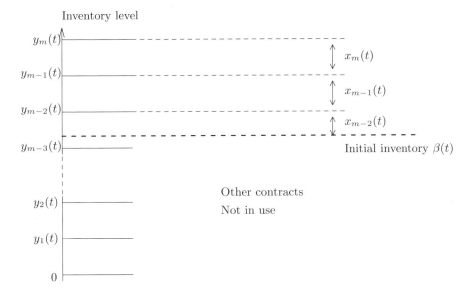

Figure 13.3. A demonstration of the optimal supply portfolio selection

ASSUMPTION 3.2 $y_1(t) < y_2(t) < \cdots < y_m(t)$, $m \le n$.

THEOREM 3.2 (i) *After eliminating all inferior contracts, for an initial inventory $\beta(t)$ the portfolio selection follows a base-stock policy. Specifically, if $\beta(t) > y_m(t)$, then $\mathbf{x_t^*} = \mathbf{0}$. Otherwise, we can find $1 \le k \le n$ such that $y_{k-1}(t) \le \beta(t) < y_k(t)$, $(y_0(t) = 0)$.*

$$x_j^*(t) = \begin{cases} y_j(t) - y_{j-1}(t), & j = k+1, \cdots, m; \\ y_k(t) - \beta(t), & j = k; \\ 0, & j = 1, \cdots, k-1. \end{cases}$$

(ii) *The base-stock levels $y_1(t), \cdots, y_m(t)$ are defined by Definition 3.3. Further, $y_k(t), k = 1, \cdots, m$ is decreasing in $v_k(t)$ or $w_k(t)$.*

Proof. From the above procedure for finding $\mathbf{x_t^*}$, we know that the optimal portfolio selection $\mathbf{x_t^*}$ follows the base-stock policy as illustrated in the first part of the theorem.

It is easy to see that the left-hand side of Equation (13.A.12) decreases as $v_n(t)$ or $w_n(t)$ increases. By Definition 3.3 and Lemma 3.6, the left-hand side of Equation (13.A.12) decreases as $y_n(t)$ is increasing. Hence, $y_n(t)$ is decreasing in $v_n(t)$ or $w_n(t)$. Proof for $y_j(t), j = n-1, \cdots, 1$ can be developed similarly. $\square$

We now use Fig 13.3 to illustrate the procedure of the optimal supply portfolio selection. Based on Definition 3.3 and Lemma 3.7, it is possible for us to calculate $y_i(t), i = 0, \cdots, m$. Note that $y_i(t)$ is independent of $\beta(t)$. As $y_i(t) > y_{i-1}(t)$, Figure 13.3 depicts $y_i(t)$ as different levels or known as base-stock levels on the left side, and the inventory level $\beta(t)$ as a dashed line. If there is a $y_i(t)$, such that $y_i(t) \geq \beta(t)$, all contracts $i, i + 1, \cdots, m$ are active. For $j > i$ the distance between $y_{j+1}(t)$ and $y_j(t)$ represents the capacity selected for contract $j + 1$. The distance between $y_i(t)$ and $\beta(t)$ is the capacity selected for contract i.

This result demonstrates that the selection process starts from the most flexible contracts. The less flexible suppliers are used only when the initial inventory is very low.

The follow lemmas show the concavity of $\pi_1^*(t, \beta(t))$ in $\beta(t)$, and their proofs appear in Appendix.

LEMMA 3.10 $\pi_1^*(t, \beta(t))$ *is concave in* $\beta(t), \forall \beta(t) \geq 0$.

LEMMA 3.11

$$\frac{d\pi_1^*(t, \beta(t))}{d\beta(t)} \mid_{\beta(t)=0} = v_t^1 + w_1(t), \ and \ \frac{d\pi_1^*(t, \beta(t))}{d\beta(t)} \mid_{\beta(t)=+\infty} < 0.$$

To this end, we complete the mathematical backward induction proof except for the last period.

When $t = T$, we use $+s$ instead of $-h_t$ and remove the last term of $\alpha \pi_1^*(t+1, (\beta(t) + \cdots + q_n(t) - D_t)^+)$ in the Equation (13.2). Then, in the same manner, it is straightforward to prove that $\pi_1^*(t, \beta(t))$ is concave in $\beta(T), \forall \beta(T) \geq 0$ and $\frac{d\pi_1^*(T,0)}{d\beta(T)} = v_T^1 + w_T^1$.

4. Conclusions and Research Directions

In this paper, we develop a model for supply portfolio selection and execution. We demonstrate the existence of an optimal selection and execution policy, and show that the selection process starts with the most flexible suppliers, and the execution process starts from the most less flexible suppliers.

It is worth noting that we only consider the decision process of the option contract selection and execution. The problem of designing the contract, which is the task on the supplier side, remains unsolved. In particular, the reservation and execution prices are treated as input parameters. How a supplier determines these prices remains an open question. These areas may be fruitful research directions in the future.

Acknowledgments

This paper is supported in part by RGC Competitive Earmarked Research Grants, CUHK4239/03E and CUHK4167/04E. The authors would like to thank Professor Qianchuan Zhao for technical comments on this paper.

References

Barnes-Schuster, D. Bassok, Y. and Anupindi, R. "Coordination and Flexibility in Supply Contracts with Options", *Manufacturing and Service Operations Management* Vol. 4, No. 3, (2002), 171-207.

Bertsekas, D. P. *Nonlinear Programming*. Athena Scientific, Belmont, Massachusetts (1995).

Brown, A.O. and Lee, H.L. "The impact of demand signal quality on optimal decisions in supply contracts", in *Stochastic Modeling and Optimization of Manufacturing Systems and Supply Chains*, eds. Shanthikumar, J.G., Yao, D.D. and Zijm, W.H.M., Kluwer Academic Publishers, Boston, MA, (2003), 299-328.

Cachon, G. P. "Supply Chain Coordination with Contracts", to appear as chapter 6 of *Handbooks in Operations Research and Management Science: Supply Chain Management*, edited by Steve Graves and Ton de Kok and published by North-Holland, (2003).

Cachon, G. P. and Lariviere, M. A. "Contracting to Assure Supply: How to Share Demand Forecasts in a Supply Chain", *Management Science*, Vol. 47, No. 5, (2001), 629-646.

Donohue K. L. "Efficient Supply Contracts for Fashion Goods with Forecast Updating and Two Production Modes", *Management Science*, Vol. 46, No. 11, (2000), 1397-1411.

Eppen, G. D. and Iyer, A. V. "Backup Agreements in Fashion Buying - The Value of Upstream Flexibility", *Management Science*, Vol. 43, No. 11, (1997).

Iyer, A. V. and Bergen, M. E. "Quick Response in Manufacturer-Retailer Channels", *Management Science*, Vol. 43, No. 4 (1997).

Li, C. and Kouvelis, P. "Flexible and Risk-Sharing Supply Contracts Under Price Uncertainty", *Management Science*, Vol. 45, No.10 (1999).

Martinez-de-Albeniz, V. and Simchi-Levi, D. "A Portfolio Approach to Procurement Contracts", Working Paper, Operations Research Center, MIT, U.S.A. (2003)

Sethi, S.P., Yan, H., and Zhang, H., "Quantity Flexible Contracts: Optimal Decisions with Information Updates", *Decision Sciences*, 35, 4, Fall (2004), 691-712.

Sethi, S.P., Yan, H., and Zhang, H., "Inventory and Supply Chain Management with Forecast Updates", in series International Series in Operations Research & Management Science, Springer, NY, (2005).

Tsay, A. A., "The Quantity Flexibility Contract and Supplier-Customer Incentives", *Management Science*, Vol. 45, No. 10, (1999).

Zhu, K. and Thonemann, U. W. "Modelling the Benefits of Sharing Future Demand Information", *Operations Research*, Vol. 52, No. 1 (2004), 136-147.

Appendix

Proof of Lemma 3.1. First, fixing other $q_j(t), j \neq k$, let

$$g_t(q_k(t)) = r_t(D_t \wedge (\beta(t) + \cdots + q_n(t))) - w_1(t)q_1(t) - \cdots - w_n(t)q_n(t)$$
$$-h_t(\beta(t) + \cdots + q_n(t) - D_t)^+ + \alpha\pi_{t+1}^{1,*} \ (\beta(t) + \cdots + q_n(t) - D_t)^+ \ . (13.A.1)$$

Hence by Equation (13.2), $\pi_2(\mathbf{q_t}; t, \mathbf{x_t}|i) = E_{D_t}(g_t(q_k(t)))$. As D_t is independent of $q_k(t)$, the concavity in $q_k(t)$ can be preserved after the expectation in D_t. Then we only need to show that $g_t(q_k(t))$ is concave in $q_k(t)$, given $q_j(t), j \neq k$.

Case 1 If $\beta(t) + \cdots + q_n(t) - D_t \leq 0$, $g_t(q_k(t)) = r_t(\beta(t) + \cdots + q_n(t)) - w_1(t)q_1(t) - \cdots - w_n(t)q_n(t) + \alpha\pi_{t+1}^{1,*}(0)$, and $\frac{dg_t(q_k(t))}{dq_k(t)} = r_t - w_k(t) > 0$;

Case 2 Otherwise, $\beta(t) + \cdots + q_n(t) - D_t \geq 0$, $g_t(q_k(t)) = r_t D_t - w_1(t)q_1(t) - \cdots - w_n(t)q_n(t) - h_t(\beta(t) + \cdots + q_n(t) - D_t) + \alpha\pi_{t+1}^{1,*}(\beta(t) + \cdots + q_n(t) - D_t)$ and $\frac{dg_t(q_k(t))}{dq_k(t)} = -w_k(t) - h_t + \alpha\frac{d\pi_1^*(t+1,\beta(t)+\cdots+q_n(t)-D_t)}{dq_k(t)}$. Recalling the supposition we know that $\frac{d\pi_1^*(t+1,\beta(t)+\cdots+q_n(t)-D_t)}{dq_k(t)}$ is non-increasing in $q_k(t)$ and consequently $\frac{dg_t(q_k(t))}{dq_k(t)}$ is non-increasing in $q_k(t)$. Additionally, when $\beta(t) + \cdots + q_n(t) - D_t = 0$, $\frac{dg_t(q_k(t))}{dq_k(t)} \mid_{q_k(t)=D_t-\sum_{j\neq k} q_j(t)} = -w_k(t) - h_t + \alpha(v_1(t+1) + w_1(t+1))$.

Recall Assumption 3.1, which guarantees that $r_t - w_k(t) \geq -w_k(t) - h_t + \alpha(v_1(t+1) + w_1(t+1))$, i.e. at the joint point of two cases, i.e. when $q_k(t) = D_t - \sum_{j\neq k} q_k(t)$, the left-hand derivative is no less than the right-hand derivative. Hence, $g_t(q_k(t))$ is concave in $q_k(t)$. The lemma is proved to be true.

Proof of Lemma 3.2 Given $q_j^*(t) > 0$, suppose that there is $k < j$, s.t. $q_k^*(t) < x_k(t)$. We can find $q_k^{'}(t) = q_k^*(t) + \varepsilon, \varepsilon > 0$ and $q_j^{'}(t) = q_j^*(t) - \varepsilon$. Then $-w_k(t)q_k^{'}(t) - w_j(t)q_j^{'}(t) = -w_k(t)q_k^*(t) - w_j(t)q_j^*(t) + (-w_k(t) + w_j(t))\varepsilon > -w_k(t)q_k^*(t) - w_j(t)q_j^*(t)$, since $w_k(t) < w_j(t)$ (Assumption 2.1). Hence we obtain

$$\begin{cases} -w_k(t)q_k^{'}(t) - w_j(t)q_j^{'}(t) > -w_k(t)q_k^*(t) - w_j(t)q_j^*(t) \\ \\ q_k^{'}(t) + q_j^{'}(t) = q_k^*(t) + q_j^*(t) \end{cases} .$$

Recall that $(q_k(t), q_j(t))$ appears in Equation (13.2) in terms of $-w_k(t)q_k(t) - w_j(t)q_j(t)$ and $q_k(t) + q_j(t)$, and D_t is independently distributed from $q_k(t)$ and $q_j(t)$ so that the expectation preserves the order of preference. Hence, to maximize Equation (13.2), it is better to choose $(q_k^{'}(t), q_j^{'}(t))$ than $(q_k^*(t), q_j^*(t))$, contradicting $(q_k^*(t), q_j^*(t))$ to be optimal.

Proof of Lemma 3.3. Let $Q_k(i), k = 1, 2, \cdots, n$ be the least solution to the following equation of u:

$$(r_t - w_k(t)) - (r_t + h_t)H(u \mid i) + \alpha \frac{d}{dq_k(t)} E_{D_t} \pi_1^*(t+1, (\beta(t) + q_1(t)$$

$$+ \cdots + q_k(t) - D_t)^+) \mid_{\beta(t)+q_1(t)+\cdots+q_k(t)=u} = 0. \tag{13.A.2}$$

Let us show that Equation (13.A.2) always has solutions so that $Q_k(i)$ is well defined for all k. First of all, by Lemma 3.4 the left-hand side of this equation is non-increasing in u. Secondly, when $u = 0$, the demand cumulative distribution function $H(u \mid i) = 0$. Then, the left-hand side is

$$r_t - w_k(t)$$

$$+ \alpha \frac{d}{d\beta(t)} E_{D_t} \pi_1^*(t+1, (\beta(t) + q_1(t) + \cdots + q_k(t) - D_t)^+) \mid_{\beta(t)+q_1(t)+\cdots+q_k(t)=0}$$

$$> r_t - w_k(t) > 0 \tag{13.A.3}$$

Thirdly, when $u \to +\infty$, the demand cumulative distribution function $H(u \mid i) = 1$. Then, the left-hand side of the above equation is

$$-w_k(t) - h_t + \alpha \frac{d}{d\beta(t)} E_{D_t} \pi_1^*(t+1, (\beta(t)$$

$$+ q_1(t) + \cdots + q_k(t) - D_t)^+) \mid_{\beta(t)+q_1(t)+\cdots+q_k(t)=+\infty}$$

$$= -w_k(t) - h_t + \alpha \int_0^{+\infty} \frac{d}{d\beta(t)} \pi_1^*(t+1, \beta(t)$$

$$+ q_1(t) + \cdots + q_k(t) - z) \mid_{\beta(t)+q_1(t)+\cdots+q_k(t)=+\infty} dH(z \mid i)$$

$$< -w_k(t) - h_t < 0.$$

The first inequality is by $\frac{d\pi_1^*(t+1,\beta(t+1))}{d\beta(t+1)} \mid_{\beta(t+1)=+\infty} < 0$, which is assumed at the beginning of Section 3. Hence, there is at least one real number u that satisfies Equation (13.A.2). By the last sentence of Definition 3.1, $Q_k(i)$ is unique.

As $w_k(t) < w_t^{k+1}$ (Assumption 2.1) and by Lemma 3.4 the left-hand side of Equation (13.3) is non-increasing in $Q_k(i)$, it is straightforward that $Q_k(i) > Q_{k+1}(i)$. Hence,

$$Q_1(i) > Q_2(i) \cdots > Q_n(i). \tag{13.A.4}$$

Proof of Lemma 3.4. By Lemma 3.1, i.e. the concavity of $\pi_2(q_1(t), 0, \cdots, 0; t)$ in $q_1(t)$, $\frac{d\pi_2(q_1(t),0,\cdots,0;t)}{dq_1(t)}$ is non-increasing in $q_1(t)$. Comparing the positions of $q_1(t)$ and $\beta(t)$ appearing in Equation (13.5), we know that given a specific $q_1(t)$, $(r_t - w_1(t)) - (r_t + h_t)H(q_1(t) + \beta(t) \mid i) + \frac{d}{dq_1(t)} E_{D_t} \pi_1^*(t+1, (q_1(t) + \beta(t) - D_t)^+)$ is non-increasing in $\beta(t)$. Hence, letting $q_1(t) = 0$, we obtain that $(r_t - w_1(t)) - (r_t + h_t)H(\beta(t) \mid i) + \frac{d}{d\beta(t)^\dagger} E_{D_t} \pi_1^*(t+1, (\beta(t) - D_t)^+)$ is non-increasing in $\beta(t)$.

Proof of Lemma 3.6. According to Theorem 3.1, rewrite Equation (13.1) as follows.

$$\pi_1(\mathbf{x_t}; t, \beta(t) \mid I_t)$$

$$= -v_1(t)x_1(t) - v_2(t)x_2(t) - \cdots - v_n(t)x_n(t) + \int_{-\infty}^{i_0} [\pi_2(0, \cdots, 0; t, i)] dF_I(i)$$

$$+ \int_{i_0}^{i_1} \left[\pi_2(Q_1(i) - \beta(t), 0, \cdots, 0; t, i) \right] dF_I(i)$$

$$+ \int_{i_1}^{\hat{i_1}} \left[\pi_2(x_1(t), 0, \cdots, 0; t, i) \right] dF_I(i) \tag{13.A.5}$$

$$\vdots$$

$$+ \int_{i_n}^{\infty} \left[\pi_2(x_1(t), \cdots, x_n(t); t, i) \right] dF_I(i) \tag{13.A.6}$$

s.t.

$$x_i \geq 0, i = 1, 2, \cdots, n. \tag{13.A.7}$$

Rewriting Equation (13.2) as

$$
\begin{aligned}
\pi_2(\mathbf{q_t}; t, i) = & -(r_t + h_t) \int_0^{\beta(t) + q_1(t) + \cdots + q_n(t)} H(D \mid i) dD \\
& + r_t(q_1(t) + \cdots + q_n(t) + \beta(t)) \\
& - w_1(t) - \cdots - w_n(t)q_n(t) + \alpha E_{D_t} \pi_{t+1}^{1,*} \ (\beta(t) + \cdots + q_n(t))^+ \ .
\end{aligned}
\tag{13.A.8}
$$

we then obtain

$$
\begin{aligned}
& \pi_1(\mathbf{x_t}; t, \beta(t) | I_t) \\
= \ & -v_1(t)x_1(t) - v_2(t)x_2(t) - \cdots - v_n(t)x_n(t) \\
& + \int_{-\infty}^{i_0} \left[-(r_t + h_t) \int_0^{\beta(t)} H(D \mid i) dD + r_t \beta(t) \right. \\
& \left. + \alpha E_{D_t} \pi_1^*(t+1, (\beta(t) - D_t)^+) \right] \ dF_I(i) \\
& + \int_{i_0}^{i_1} \left[-(r_t + h_t) \int_0^{Q_1(i)} H(D \mid i) dD + r_t Q_1(i) \right. \\
& \left. - w_1(t)(Q_1(i) - \beta(t)) + \alpha E_{D_t} \pi_1^*(t+1, (Q_1(i) - D_t)^+) \right] \ dF_I(i) \\
& + \int_{i_1}^{\hat{i_1}} \left[-(r_t + h_t) \int_0^{x_1(t)+\beta(t)} H(D \mid i) dD + r_t(x_1(t) + \beta(t)) \right. \\
& \left. - w_1(t)x_1(t) + \alpha E_{D_t} \pi_1^*(t+1, (\beta(t) + x_1(t) - D_t)^+) \right] \ dF_I(i) \ .
\end{aligned}
$$

$$\vdots$$

$$\vdots$$

$$
\begin{aligned}
& + \int_{i_{n-1}}^{i_n} \left[-(r_t + h_t) \int_0^{Q_n(i)} H(D \mid i) dD + r_t Q_n(i) \right. \\
& - w_1(t)x_1(t) - \cdots - w_t^{n-1} x_{n-1}(t) \\
& - w_n(t)Q_n(i) + w_n(t)(\beta(t) + \cdots + x_{n-1}(t)) \\
& \left. + \alpha E_{D_t} \pi_1^*(t+1, (Q_n(i) - D_t)^+) \right] \ dF_I(i) \\
& + \int_{i_n}^{\infty} \left[-(r_t + h_t) \int_0^{\beta(t)+x_1(t)+\cdots+x_n(t)} H(D \mid i) dD \right.
\end{aligned}
$$

$$+r_t(\beta(t) + x_1(t) + \cdots + x_n(t)) - w_1(t)x_1(t) - \cdots - w_n(t)x_n(t)$$

$$+\alpha E_{D_t} \pi_1^*(t+1, (\beta(t) + \cdots + x_n(t) - D_t)^+) \ dF_I(i). \qquad (13.A.9)$$

Take the partial derivatives with respect to $x_1(t), x_2(t), \cdots, x_n(t)$.

$$\frac{\partial \pi_1(t)}{\partial x_1(t)}$$

$$= \quad -v_1(t) - w_1(t)\bar{F}_I(i_1) + \int_{i_1}^{\hat{i}_1} \left[-(r_t + h_t)H(\beta(t) + x_1(t) \mid i) + r_t \right.$$

$$+\alpha \frac{d}{dx_1(t)} E_{D_t} \pi_1^*(t+1, (\beta(t) + x_1(t) - D_t)^+) \ dF_I(i) + \int_{\hat{i}_1}^{i_2} \left[w_2(t) \right] dF_I(i)$$

$$+ \int_{i_2}^{\hat{i}_2} \left[-(r_t + h_t)H(\beta(t) + x_1 + x_2 \mid i) + r_t \right.$$

$$+\alpha \frac{d}{dx_1(t)} E_{D_t} \pi_1^*(t+1, (\beta(t) + x_1(t) + x_2(t) - D_t)^+) \left] dF_I(i) \right.$$

$$+ \cdots \cdots$$

$$+ \int_{i_{n-1}}^{\hat{i}_{n-1}} \left[-(r_t + h_t)H(\beta(t) + \sum_{j=1}^{n-1} x_j(t) \mid i) + r_t \right.$$

$$+\alpha \frac{d}{dx_1(t)} E_{D_t} \pi_1^*(t+1, (\beta(t) + \sum_{j=1}^{n-1} x_j(t) - D_t)^+) \right] dF_I(i)$$

$$+ \int_{\hat{i}_{n-1}}^{i_n} \left[w_n(t) \right] dF_I(i)$$

$$+ \int_{i_n}^{+\infty} \left[-(r_t + h_t)H(\beta(t) + \sum_{j=1}^{n} x_j(t) \mid i) + r_t \right.$$

$$+\alpha \frac{d}{dx_1(t)} E_{D_t} \pi_1^*(t+1, (\beta(t) + \sum_{j=1}^{n} x_j(t) - D_t)^+) \right] dF_I(i)$$

$$+ \frac{di_1}{dx_1(t)} \pi_2(Q_1(i_1) - \beta(t), 0, \cdots, 0; t, i_1) f_I(i_1)$$

$$- \frac{di_1}{dx_1(t)} \pi_2(x_1(t), 0, \cdots, 0; t, i_1) f_I(i_1)$$

$$+ \frac{d\hat{i}_1}{dx_1(t)} \pi_2(x_1(t), 0, \cdots, 0; t, \hat{i}_1) f_I(\hat{i}_1)$$

$$- \frac{d\hat{i}_1}{dx_1(t)} \pi_2(x_1(t), Q_2(\hat{i}_1) - x_1(t) - \beta(t), \cdots, 0; t, \hat{i}_1) f_I(\hat{i}_1)$$

$$+ \cdots \cdots$$

$$+ \frac{di_n}{dx_1(t)} \pi_2(x_1(t), \cdots, x_{n-1}(t), Q_n(\hat{i_{n-1}}) - \beta(t) - \sum_{j=1}^{n-1} x_j(t); t, \hat{i}_1) f_I(i_n)$$

$$- \frac{di_n}{dx_1(t)} \pi_2(x_1(t), \cdots, x_{n-1}(t), x_n(t); t, \hat{i}_1) f_I(i_n)$$

$$= \quad -v_1(t) - w_1(t)\bar{F}_I(i_1) + \int_{i_1}^{\hat{i}_1} \left[-(r_t + h_t)H(\beta(t) + x_1(t) \mid i) + r_t \right.$$

$$+\alpha\frac{d}{dx_1(t)}E_{D_t}\pi_1^*(t+1,(\beta(t)+x_1(t)-D_t)^+)\ dF_I(i)$$

$$+\int_{\hat{i_1}}^{i_2}[w_2(t)]\,dF_I(i)$$

$$+\int_{i_2}^{\hat{i_2}}\left[-(r_t+h_t)H(\beta(t)+x_1+x_2\mid i)+r_t\right.$$

$$+\alpha\frac{d}{dx_1(t)}E_{D_t}\pi_1^*(t+1,(\beta(t)+x_1(t)+x_2(t)-D_t)^+)\ dF_I(i)$$

$$+\cdots\cdots$$

$$+\int_{i_{n-1}}^{\hat{i}_{n-1}}\left[-(r_t+h_t)H(\beta(t)+\sum_{j=1}^{n-1}x_j(t)\mid i)+r_t\right.$$

$$+\alpha\frac{d}{dx_1(t)}E_{D_t}\pi_1^*(t+1,(\beta(t)+\sum_{j=1}^{n-1}x_j(t)-D_t)^+)\right]dF_I(i)$$

$$+\int_{\hat{i}_{n-1}}^{i_n}[w_n(t)]\,dF_I(i)$$

$$+\int_{i_n}^{+\infty}\left[-(r_t+h_t)H(\beta(t)+\sum_{j=1}^{n}x_j(t)\mid i)+\right.$$

$$r_t+\alpha\frac{d}{dx_1(t)}E_{D_t}\pi_1^*(t+1,(\beta(t)+\sum_{j=1}^{n}x_j(t)-D_t)^+)\right]dF_I(i).\qquad(13.A.10)$$

The last equality holds by the definitions (see Definition 3.2) of $Q_k(i)$ when the information $i=i_{\hat{k}-1},i_k$ respectively. Similarly,

$$\frac{\partial\pi_1(t)}{\partial x_2(t)}$$

$$=\quad -v_2(t)-w_2(t)\bar{F}_I(i_2)$$

$$+\int_{i_2}^{\hat{i_2}}\left[-(r_t+h_t)H(\beta(t)+x_1+x_2\mid i)+r_t\right.$$

$$+\alpha\frac{d}{dx_2(t)}E_{D_t}\pi_1^*(t+1,(\beta(t)+x_1(t)+x_2(t)-D_t)^+)\ dF_I(i)$$

$$+\cdots\cdots$$

$$+\int_{i_{n-1}}^{\hat{i}_{n-1}}\left[-(r_t+h_t)H(\beta(t)+\sum_{j=1}^{n-1}x_j(t)\mid i)+r_t\right.$$

$$+\alpha\frac{d}{dx_2(t)}E_{D_t}\pi_1^*(t+1,(\beta(t)+\sum_{j=1}^{n-1}x_j(t)-D_t)^+)\right]dF_I(i)$$

$$+\int_{\hat{i}_{n-1}}^{i_n}[w_n(t)]\,dF_I(i)$$

$$+\int_{i_n}^{+\infty}\left[-(r_t+h_t)H(\beta(t)+\sum_{j=1}^{n}x_j(t)\mid i)+r_t\right.$$

$$+\alpha \frac{d}{dx_2(t)} E_{D_t}\pi_1^*(t+1, (\beta(t) + \sum_{j=1}^{n} x_j(t) - D_t)^+)\Bigg] dF_I(i), \quad (13.\text{A}.11)$$

$$\vdots$$
$$\vdots$$

$$\frac{\partial \pi_1}{\partial x_n} = -v_n(t) - w_n(t)\bar{F}_I(i_n)$$

$$+ \int_{i_n}^{+\infty} \Bigg[-(r_t + h_t)H(\beta(t) + \sum_{j=1}^{n} x_j(t) \mid i) + r_t$$

$$+\alpha \frac{d}{dx_n(t)} E_{D_t}\pi_1^*(t+1, (\beta(t) + \sum_{j=1}^{n} x_j(t) - D_t)^+)\Bigg] dF_I(i).$$

$$(13.\text{A}.12)$$

To ascertain whether it is concave in the $\mathbf{x_t}$, a vector, we need to obtain the second-order derivatives as follows:

$$\frac{\partial^2 \pi_1(t)}{(\partial x_1(t))^2}$$

$$= \int_{i_1}^{\hat{i_1}} [-(r_t + h_t)h(\beta(t) + x_1(t) \mid i)$$

$$+\alpha \frac{d^2}{(dx_1(t))^2} E_{D_t}\pi_1^*(t+1, (\beta(t) + x_1(t) - D_t)^+) \ dF_I(i)$$

$$+ \int_{i_2}^{\hat{i_2}} [-(r_t + h_t)h(\beta(t) + x_1 + x_2 \mid i)$$

$$+\alpha \frac{d^2}{(dx_1(t))^2} E_{D_t}\pi_1^*(t+1, (\beta(t) + x_1(t) + x_2(t) - D_t)^+) \ dF_I(i)$$

$$+ \cdots\cdots$$

$$+ \int_{i_{n-1}}^{\hat{i}_{n-1}} \Bigg[-(r_t + h_t)h(\beta(t) + \sum_{j=1}^{n-1} x_j(t) \mid i)$$

$$+\alpha \frac{d^2}{(dx_1(t))^2} E_{D_t}\pi_1^*(t+1, (\beta(t) + \sum_{j=1}^{n-1} x_j(t) - D_t)^+)\Bigg] dF_I(i)$$

$$+ \int_{i_n}^{+\infty} \Bigg[-(r_t + h_t)h(\beta(t) + \sum_{j=1}^{n} x_j(t) \mid i)$$

$$+\alpha \frac{d^2}{(dx_1(t))^2} E_{D_t}\pi_1^*(t+1, (\beta(t) + \sum_{j=1}^{n} x_j(t) - D_t)^+)\Bigg] dF_I(i)$$

$$-\frac{di_1}{dx_1(t)}[-(r_t + h_t)H(\beta(t) + x_1(t) \mid i_1) + r_t$$

$$+\alpha \frac{d}{dx_1(t)} E_{D_t} \pi_1^*(t+1, (\beta(t) + x_1(t) - D_t)^+) - w_1(t)]f_I(i_1)$$

$$+\frac{d\hat{i_1}}{dx_1(t)}[-(r_t + h_t)H(\beta(t) + x_1(t) \mid \hat{i_1}) + r_t$$

$$+\alpha \frac{d}{dx_1(t)} E_{D_t} \pi_1^*(t+1, (\beta(t) + x_1(t) - D_t)^+) - w_1(t)]f_I(\hat{i_1})$$

$$-\frac{d\hat{i_1}}{dx_1(t)}[w_2(t) - w_1(t)]f_I(\hat{i_1})$$

$$+\cdots\cdots$$

$$+\frac{di_n}{dx_1(t)}[w_n(t) - w_1(t)]f_I(i_n)$$

$$-\frac{di_n}{dx_1(t)}\left[-(r_t + h_t)H(\beta(t) + \sum_{j=1}^{n} x_j(t) \mid i_n) + r_t\right.$$

$$+\alpha \frac{d}{dx_1(t)} E_{D_t} \pi_1^*(t+1, (\beta(t) + \sum_{j=1}^{n} x_j(t) - D_t)^+) - w_1(t)]f_I(i_n)$$

$$= \int_{i_1}^{\hat{i_1}} \left[-(r_t + h_t)h(\beta(t) + x_1(t) \mid i)\right.$$

$$+\alpha \frac{d^2}{(dx_1(t))^2} E_{D_t} \pi_1^*(t+1, (\beta(t) + x_1(t) - D_t)^+) \; dF_I(i)$$

$$+\int_{i_2}^{\hat{i_2}} \left[-(r_t + h_t)h(\beta(t) + x_1 + x_2 \mid i)\right.$$

$$+\alpha \frac{d^2}{(dx_1(t))^2} E_{D_t} \pi_1^*(t+1, (\beta(t) + x_1(t) + x_2(t) - D_t)^+) \; dF_I(i)$$

$$+\cdots\cdots$$

$$+\int_{i_{n-1}}^{\hat{i_{n-1}}} \left[-(r_t + h_t)h(\beta(t) + \sum_{j=1}^{n-1} x_j(t) \mid i)\right.$$

$$+\alpha \frac{d^2}{(dx_1(t))^2} E_{D_t} \pi_1^*(t+1, (\beta(t) + \sum_{j=1}^{n-1} x_j(t) - D_t)^+) \right] dF_I(i)$$

$$+\int_{i_n}^{+\infty} \left[-(r_t + h_t)h(\beta(t) + \sum_{j=1}^{n} x_j(t) \mid i)\right.$$

$$+\alpha \frac{d^2}{(dx_1(t))^2} E_{D_t} \pi_1^*(t+1, (\beta(t) + \sum_{j=1}^{n} x_j(t) - D_t)^+) \right] dF_I(i).$$

$$(13.A.13)$$

The last equality holds by recalling the definition of $Q_t^k(i_k)$ and $Q_t^k(\hat{i_{k-1}})$. Specifically, by Definition 3.1 and 3.2,

$$-(r_t + h_t)H(\beta(t) + x_1(t) \mid i_1) + r_t + \alpha \frac{d}{dx_1(t)} E_{D_t} \pi_1^*(t+1, (\beta(t) + x_1(t) - D_t^1)^+) - w_1(t) = 0,$$

$$(13.A.14)$$

$$-(r_t + h_t)H(\beta(t) + x_1(t) \mid \hat{i}_1) + r_t + \alpha \frac{d}{dx_1(t)} E_{D_t} \pi_1^*(t + 1, (\beta(t) + x_1(t) - D_t^1)^+)$$

$$-w_1(t) = w_2(t) - w_1(t), \qquad (13.A.15)$$

$$\vdots$$

$$-(r_t + h_t)H(\beta(t) + \sum_{j=1}^{n} x_j(t) \mid i_n) + r_t$$

$$+\alpha \frac{d}{dx_1(t)} E_{D_t} \pi_1^*(t + 1, (\beta(t) + \sum_{j=1}^{n} x_j(t) - D_t^1)^+) - w_1(t)$$

$$= w_n(t) - w_1(t). \qquad (13.A.16)$$

Moreover, by Equation (13.5)and Lemma 3.1, $\frac{d\pi_2(q_1(t),0,\cdots,0;t)}{dq_1(t)}$ is non-increasing in $q_1(t)$. That is

$$\frac{d^2 \pi_2(q_1(t), 0, \cdots, 0; t)}{(dq_1(t))^2} \leq 0; \qquad (13.A.17)$$

i.e.

$$-(r_t + h_t)h(q_1(t) + \beta(t) \mid i) + \alpha \frac{d^2}{(dq_1(t))^2} E_{D_t} \pi_1^*(t + 1, (q_1(t) + \beta(t) - D_t^1)^+) \leq 0. \qquad (13.A.18)$$

Similarly, we can obtain $\frac{d^2 \pi_2(x_1(t),q_2(t),\cdots,0;t)}{(dq_2(t))^2} \leq 0, \cdots, \frac{d^2 \pi_2(x_1(t),x_2(t),\cdots,q_n(t);t)}{(dq_n(t))^2} \leq 0$, and strictly less than 0. Substitute these inequalities into Equation (13.A.13) and we can obtain

$$\frac{\partial^2 \pi_1(t)}{(\partial x_1(t))^2} = a_1 \leq 0.$$

Further, where $k = 2, \cdots, n$

$$\frac{d^2}{(dx_1(t))^2} E_{D_t} \pi_1^*(t + 1, (\beta(t) + \sum_{j=1}^{k} x_j(t) - D_t)^+)$$

$$= \frac{d^2}{(dx_1(t))(dx_2(t))} E_{D_t} \pi_1^*(t + 1, (\beta(t) + \sum_{j=1}^{k} x_j(t) - D_t)^+)$$

$$= \frac{d^2}{(dx_2(t))^2} E_{D_t} \pi_1^*(t + 1, (\beta(t) + \sum_{j=1}^{k} x_j(t) - D_t)^+). \qquad (13.A.19)$$

Similarly,

$$\frac{\partial^2 \pi_1(t)}{(\partial x_2(t))^2}$$

$$= + \int_{i_2}^{\hat{i}_2} [-(r_t + h_t)h(\beta(t) + x_1 + x_2 \mid i)$$

$$+\alpha \frac{d^2}{(dx_1(t))^2} E_{D_t} \pi_1^*(t + 1, (\beta(t) + x_1(t) + x_2(t) - D_t)^+) \ dF_I(i)$$

$$+ \cdots \cdots$$

$$+ \int_{i_{n-1}}^{\hat{i_{n-1}}} \Bigg[-(r_t + h_t)h(\beta(t) + \sum_{j=1}^{n-1} x_j(t) \mid i)$$

$$+ \alpha \frac{d^2}{(dx_1(t))^2} E_{D_t} \pi_1^*(t+1, (\beta(t) + \sum_{j=1}^{n-1} x_j(t) - D_t)^+) \Bigg] dF_I(i)$$

$$+ \int_{i_n}^{+\infty} \Bigg[-(r_t + h_t)h(\beta(t) + \sum_{j=1}^{n} x_j(t) \mid i)$$

$$+ \alpha \frac{d^2}{(dx_1(t))^2} E_{D_t} \pi_1^*(t+1, (\beta(t) + \sum_{j=1}^{n} x_j(t) - D_t)^+) \Bigg] dF_I(i)$$

$$= \frac{\partial^2 \pi_1(t)}{(\partial x_1(t))(\partial x_2(t))} = a_2 \le 0. \qquad (13.A.20)$$

and

$$a_1 \le a_2$$

$$\vdots$$

$$\frac{\partial^2 \pi_1(t)}{(\partial x_n(t))^2}$$

$$= + \int_{i_n}^{+\infty} \Bigg[-(r_t + h_t)h(\beta(t) + \sum_{j=1}^{n} x_j(t) \mid i)$$

$$+ \alpha \frac{d^2}{(dx_1(t))^2} E_{D_t} \pi_1^*(t+1, (\beta(t) + \sum_{j=1}^{n} x_j(t) - D_t)^+) \Bigg] dF_I(i)$$

$$= \frac{\partial^2 \pi_1(t)}{(\partial x_n(t))(\partial x_{n-1}(t))} = \cdots = \frac{\partial^2 \pi_1(t)}{(\partial x_n(t))(\partial x_1(t))} = a_n \le 0, \quad (13.A.21)$$

and

$$a_{n-1} \le a_n.$$

Hence, the Hessian Matrix

$$A_n = \begin{pmatrix} a_1 & a_2 & \cdots & a_n \\ a_2 & a_2 & \cdots & a_n \\ \vdots & \vdots & \ddots & \vdots \\ a_n & a_n & \cdots & a_n \end{pmatrix}.$$

where $a_1 \le a_2 \le \cdots \le a_n \le 0$.
   Let, $k = 1, 2, \cdots, n$,

$$B_k = A_k \begin{pmatrix} 1 & -1 & & & \\ & 1 & -1 & & \\ & & \ddots & \ddots & \\ & & & 1 & -1 \\ & & & & 1 \end{pmatrix}$$

$$
= \begin{pmatrix}
a_1 & -a_1 + a_2 & -a_2 + a_3 & \cdots & -a_{k-1} + a_k \\
a_2 & 0 & -a_2 + a_3 & \cdots & -a_{k-1} + a_k \\
a_3 & 0 & 0 & \cdots & \vdots \\
\vdots & \vdots & \vdots & \ddots & -a_{k-1} + a_k \\
a_k & 0 & 0 & \cdots & 0
\end{pmatrix} . \quad (13.A.22)
$$

Hence, principal minors of A_n

$$
|A_k| = |B_k| = (-1)^k (-a_k)(-a_1 + a_2)(-a_2 + a_3) \cdots (-a_{k-1} + a_k); \quad (13.A.23)
$$

$$
i.e. \quad \begin{array}{l} |A_k| \geq 0, \quad \text{if } k \text{ is even}; \\ |A_k| \leq 0, \quad \text{if } k \text{ is odd}. \end{array} \quad (13.A.24)
$$

which means A_n is negative semi-definite.

Hence, given $\beta(t)$, $\pi_1(\mathbf{x_t}; t, \beta(t) | I_t)$ is concave in the options portfolio vector $\mathbf{x_t}$.

Proof of Lemma 3.7. First we develop the following algorithm to find out the values of $y_n(t), \cdots, y_2(t), y_1(t)$, if they exist. We start with finding out the value of $y_n(t)$ by its definition $\frac{\partial \pi_1(t)}{\partial x_n(t)} \big|_{\beta(t) + \cdots + x_n(t) = y_n(t)} = 0$, where $\frac{\partial \pi_1(t)}{\partial x_n(t)}$ can be found in Equation (13.A.12). After finding out $y_n(t), \cdots, y_{j+1}(t)$, we can find out $y_j(t)$ by the definition of $\frac{\partial \pi_1(t)}{\partial x_j(t)} \big|_{\beta(t) + \cdots + x_j(t) = y_j(t), \cdots, \beta(t) + \cdots + x_n(t) = y_n(t)} = 0$ similarly. After determining $y_n(t), \cdots, y_2(t)$, the last step is to use equation

$$
\frac{\partial \pi_1(t)}{\partial x_1(t)} \big|_{\beta(t) + x_1(t) = y_1(t), \cdots, \beta(t) + \cdots + x_n(t) = y_n(t)} = 0
$$

to find out the value of $y_1(t)$, where $\frac{\partial \pi_1(t)}{\partial x_1(t)}$ can be found in Equation (13.A.10). To finish the proof of existence, we only need to show the existence of $y_j(t), j = 1, \cdots, n$ solved by the algorithm mentioned above. Taking $y_n(t)$ as an example, the right-hand side of Equation (13.A.12) can be considered as a function of $\beta(t) + \cdots + x_n(t) = y_n(t)$. It is non-increasing in $y_n(t)$ by its concavity in $x_n(t)$. It is easy to check that the right-hand side of Equation (13.A.12) is positive when $y_n(t) = 0$ and negative when $y_n(t) = +\infty$. Hence, there is a $y_n(t)$, which ensures that expression (13.A.12) equals 0.

The uniqueness of this is proved by the last sentence of Definition 3.3.

Proof. of Lemma 3.8 We have two choices of $x_{j-1}^*(t)$. The first case is $x_{j-1}^*(t) > 0$ and $\frac{\partial \pi_1(\mathbf{x_t^*}; t)}{\partial x_{j-1}(t)} = 0$. Then, $y_{j-1}(t) \geq y_j(t) = \beta(t) + x_1^*(t) + \cdots + x_j^*(t) > \beta(t) + x_1^*(t) + \cdots + x_{j-1}^*(t)$. Hence, recalling Lemma 3.6 and Definition 3.3 for $y_{j-1}(t)$, we can obtain $\frac{\partial \pi_1(\mathbf{x_t^*}; t)}{\partial x_{j-1}(t)} > \frac{\partial \pi_1(t)}{\partial x_{j-1}(t)} \big|_{\beta(t) + x_1(t) + \cdots + x_{j-1}(t) = y_{j-1}(t)} = 0$, which contradicts the K-K-T conditions. The second case is $x_{j-1}^*(t) = 0$ and $\frac{\partial \pi_1(\mathbf{x_t^*}; t)}{\partial x_{j-1}} \leq 0$, where there is no contradiction.

Proof of Lemma 3.9. If $y_j(t) \leq \beta(t)$, as $x_1^*(t), \cdots, x_j^*(t) \geq 0$, then $\beta(t) + x_1^*(t) + \cdots + x_j^*(t) \geq y_j(t)$. Hence, by Lemma 3.6 and Definition 3.3, $\frac{\partial \pi_1(\mathbf{x_t^*}; t)}{\partial x_j(t)} \leq 0$, which results in $x_j^*(t) = 0$. As we have $y_1(t), \cdots, y_{j-1}(t) \leq y_j(t) \leq \beta(t)$, we can obtain $x_1^*(t), \cdots, x_{j-1}^*(t) = 0$ similarly.

Proof of Lemma 3.10. According to Theorem 3.2, the feasible interval of $\beta(t)$ is divided into $n + 1$ subintervals, i.e. $[0, y_1(t)], [y_1(t), y_2(t)], \cdots, [y_{n-1}(t), y_n(t)], [y_n(t), +\infty)$. We will show that $\pi_1^*(t, \beta(t))$ is concave in $\beta(t)$ in each subintervals and the right-hand and left-hand derivatives are equal at the joint points.

Case 1: $\beta(t) \in [y_n(t), +\infty)$ In this case, according to Theorem 3.2, $\mathbf{x_t^*} = \mathbf{0}$. Then, by Definition 3.2 $i_0 = \cdots = i_n$ and by Theorem 3.1 $\mathbf{q_t^*} = \mathbf{0}$. Hence, by Equation (13.1)

$$\pi_1^*(t, \beta(t)) = \pi_1(\mathbf{0}; t, \beta(t)) = \int_{-\infty}^{+\infty} [\pi_2(\mathbf{0}; t, i)] dF_I(i), \qquad (13.\text{A}.25)$$

where, by Equation (13.2),

$$\pi_2(\mathbf{0}; t, i) = E_{D_t}[r_t(D_t \wedge \beta(t)) - h_t(\beta(t) - D_t)^+ + \alpha \pi_1^*(t+1, (\beta(t) - D_t)^+)]. \quad (13.\text{A}.26)$$

From Equation (13.A.8), we have

$$
\frac{d\pi_1^*(t, \beta(t))}{d\beta(t)} = w_n(t) + \int_{-\infty}^{+\infty} \{(r_t - w_n(t))
$$
$$
- (r_t + h_t)H(\beta(t) \mid i) + \alpha \frac{d}{d\beta(t)} E_{D_t} \; \pi_1^*(t+1, (\beta(t) - D_t)^+) \; \} dF_I(i).
$$
$$(13.\text{A}.27)$$

By Lemma 3.4, we know that Equation (13.A.27) is non-increasing in $\beta(t)$. Hence, $\pi_1^*(t, \beta(t))$ is concave in $\beta(t)$ for Case 1.

When $\beta(t) = y_n(t)$, $i_0 = i_n$, then we have

$$
\frac{d\pi_1^*(t, \beta(t))}{d\beta(t)} \Big|_{\beta(t) = y_n(t)}
$$
$$
= w_n(t) + \int_{-\infty}^{+i_0} \{(r_t - w_n(t)) - (r_t + h_t)H(y_n(t) \mid i)
$$
$$
+ \alpha \frac{d}{d\beta(t)} E_{D_t} \; \pi_1^*(t+1, (y_n(t) - D_t)^+) \quad dF_I(i)
$$
$$
+ \int_{i_n}^{+\infty} \{(r_t - w_n(t)) - (r_t + h_t)H(y_n(t) \mid i)
$$
$$
+ \alpha \frac{d}{d\beta(t)} E_{D_t} \; \pi_1^*(t+1, (y_n(t) - D_t)^+) \; \} dF_I(i). \qquad (13.\text{A}.28)
$$

By Definition 3.3 for $y_n(t)$, the last term equals $v_n(t)$ and we have

$$
\frac{d\pi_1^*(t, \beta(t))}{d\beta(t)} \Big|_{\beta(t) = y_n(t)}
$$
$$
= v_n(t) + w_n(t) + \int_{-\infty}^{+i_0} \{(r_t - w_n(t)) - (r_t + h_t)H(y_n(t) \mid i)
$$
$$
+ \alpha \frac{d}{d\beta(t)} E_{D_t} \; \pi_1^*(t+1, (y_n(t) - D_t)^+) \; \} dF_I(i). \qquad (13.\text{A}.29)
$$

Case 2: $\beta(t) \in [y_{n-1}(t), y_n(t)]$ In this case, according to Theorem 3.2, $\mathbf{x_t^*} = (0, \cdots, 0, y_n(t) - \beta(t))$ and $\pi_1^*(t, \beta(t)) = \pi_1(0, \cdots, 0, y_n(t) - \beta(t); t, \beta(t))$. By Definition 3.2, $i_0 = \cdots = i_{n-1}$. Hence,

$$
\pi_1^*(t, \beta(t)) = \pi_1(0, \cdots, 0, y_n(t) - \beta(t); t, \beta(t))
$$
$$
= -v_n(t)(y_n(t) - \beta(t)) + \int_{-\infty}^{i_0} \pi_2(0, \cdots, 0; t, i) dF_I(i)
$$

$$+ \int_{i_0}^{i_n} \pi_2(0, \cdots, Q_t^n - \beta(t); t, i) dF_I(i)$$

$$+ \int_{i_n}^{+\infty} \pi_2(0, \cdots, y_n(t) - \beta(t); t, i) dF_I(i)$$

$$= -v_n(t)(y_n(t) - \beta(t)) + \int_{-\infty}^{i_0} [-(r_t + h_t) \int_0^{\beta(t)} H(D \mid i) dD + r_t \beta(t) +$$

$$\alpha E_{D_t} \pi_1^*(t+1, (\beta(t) - D_t)^+)] dF_I(i)$$

$$+ \int_{i_0}^{i_n} [-(r_t + h_t) \int_0^{Q_t^n} H(D \mid i) dD + r_t Q_t^n - w_n(t)(Q_t^n - \beta(t))$$

$$+\alpha E_{D_t} \pi_1^*(t+1, (Q_t^n - D_t)^+)] dF_I(i)$$

$$+ \int_{i_n}^{+\infty} [-(r_t + h_t) \int_0^{y_n(t)} H(D \mid i) dD + r_t y_n(t) - w_n(t)(y_n(t) - \beta(t))$$

$$+\alpha E_{D_t} \pi_1^*(t+1, (y_n(t) - D_t)^+)] dF_I(i). \qquad (13.A.30)$$

The first equality holds with Equation (13.A.7) and the second equality holds with Equation (13.A.8). Moreover,

$$\frac{d\pi_1^*(t, \beta(t))}{d\beta(t)}$$

$$= v_n(t) + w_n(t) + \int_{-\infty}^{+i_0} \{(r_t - w_n(t)) - (r_t + h_t)H(\beta(t) \mid i)$$

$$+\alpha \frac{d}{d\beta(t)} E_{D_t} \ \pi_1^*(t+1, (\beta(t) - D_t)^+) \ \} dF_I(i)$$

$$+\frac{di_0}{d\beta(t)} [-(r_t + h_t) \int_0^{\beta(t)} H(D \mid i_0) dD + r_t \beta(t)$$

$$+\alpha E_{D_t} \pi_1^*(t+1, (\beta(t) - D_t)^+)] f_I(i_0)$$

$$-\frac{di_0}{d\beta(t)} [-(r_t + h_t) \int_0^{Q_t^n(i_0)} H(D \mid i_0) dD + r_t Q_t^n(i_0)$$

$$-w_n(t)(Q_t^n(i_0) - \beta(t))$$

$$+\alpha E_{D_t} \pi_1^*(t+1, (Q_t^n(i_0) - D_t)^+)] f_I(i_0). \qquad (13.A.31)$$

Note that, in this case, $Q_t^n(i_0) = Q_n(\hat{i_{n-1}}) = \beta(t) + \cdots + x_{n-1}(t) = \beta(t)$ (see Definition 3.2), hence

$$\frac{d\pi_1^*(t, \beta(t))}{d\beta(t)}$$

$$= v_n(t) + w_n(t) + \int_{-\infty}^{+i_0} \{(r_t - w_n(t)) - (r_t + h_t)H(\beta(t) \mid i)$$

$$+\alpha \frac{d}{d\beta(t)} E_{D_t} \ \pi_1^*(t+1, (\beta(t) - D_t)^+) \ \} dF_I(i). \qquad (13.A.32)$$

Take the second-order derivative and we can obtain,

$$\frac{d^2 \pi_1^*(t, \beta(t))}{(d\beta(t))^2}$$

$$= \int_{-\infty}^{+i_0} \{-(r_t + h_t)h(\beta(t) \mid i)$$

$$+\alpha \frac{d^2}{(d\beta(t))^2} E_{D_t} \ \pi_1^*(t+1, (\beta(t) - D_t)^+) \ \} dF_I(i)$$

$$+\frac{di_0}{d\beta(t)}[(r_t - w_n(t)) - (r_t + h_t)H(\beta(t) \mid i_0)$$

$$+\alpha \frac{d}{d\beta(t)} E_{D_t} \ \pi_1^*(t+1, (\beta(t) - D_t)^+) \ ]f_I(i_0)$$

$$= \int_{-\infty}^{+i_0} \{-(r_t + h_t)h(\beta(t) \mid i)$$

$$+\alpha \frac{d^2}{(d\beta(t))^2} E_{D_t} \ \pi_1^*(t+1, (\beta(t) - D_t)^+) \ \} dF_I(i)$$

$$\leq \quad 0. \tag{13.A.33}$$

Because $i_0 = \hat{i_{n-1}}$, $\beta(t) = \beta(t) + \cdots + x_{n-1}(t) = Q_t^n(\hat{i_n})$ in this case. Hence, by Definition 3.1 ($k = n$), the second equality holds. The last inequality is by Lemma 3.4. Hence, $\pi_1^*(t, \beta(t))$ is proved to be concave in $\beta(t)$ in case 2.

When $\beta(t) = y_n(t)$,

$$\frac{d\pi_1^*(t, \beta(t))}{d\beta(t)} \mid_{\beta(t)=y_n(t)}$$

$$= \quad v_n(t) + w_n(t) + \int_{-\infty}^{+i_0} \{(r_t - w_n(t)) - (r_t + h_t)H(y_n(t) \mid i)$$

$$+\alpha \frac{d}{d\beta(t)} E_{D_t} \ \pi_1^*(t+1, (y_n(t) - D_t)^+) \ \} dF_I(i). \tag{13.A.34}$$

We can see that equations (13.A.29) and (13.A.34) are equal, i.e. the joint point of case 1 and 2 has the equivalent left-hand and right-hand derivatives. Hence, $\pi_1^*(t, \beta(t))$ is concave in $\beta(t)$ in $[y_{n-1}(t), +\infty)$.

When $\beta(t) = y_{n-1}(t)$,

$$\frac{d\pi_1^*(t, \beta(t))}{d\beta(t)} \mid_{\beta(t)=y_{n-1}(t)}$$

$$= \quad v_n(t) + w_n(t) + \int_{-\infty}^{+i_0} \{(r_t - w_n(t)) - (r_t + h_t)H(y_{n-1}(t) \mid i)$$

$$+\alpha \frac{d}{d\beta(t)} E_{D_t} \ \pi_1^*(t+1, (y_{n-1}(t) - D_t)^+) \ \} dF_I(i). \tag{13.A.35}$$

Case 3: $\beta(t) \in [y_{n-2}(t), y_{n-1}(t)]$ In this case, according to Theorem 3.2, $\mathbf{x_t^*} = (0, \cdots, 0, y_{n-1}(t) - \beta(t), y_n(t) - y_{n-1}(t))$ and $\pi_1^*(t, \beta(t)) = \pi_1(0, \cdots, 0, y_{n-1}(t) - \beta(t), y_n(t) - y_{n-1}(t); t, \beta(t))$. Then, by Definition 3.2 $i_0 = \cdots = \hat{i}_{n-2}$.

Similarly, we have

$$\frac{d\pi_1^*(t, \beta(t))}{d\beta(t)}$$

$$= \quad v_t^{n-1} + w_t^{n-1} + \int_{-\infty}^{+i_0} (r_t - w_t^{n-1}) - (r_t + h_t)H(\beta(t) \mid i)$$

$$+\alpha \frac{d}{d\beta(t)} E_{D_t} \ \pi_1^*(t+1, (\beta(t) - D_t)^+) \ \} dF_I(i), \tag{13.A.36}$$

which is similar to Equation (13.A.32). Take the second-order derivative and we obtain,

$$\frac{d^2\pi_1^*(t,\beta(t))}{(d\beta(t))^2}$$

$$= \int_{-\infty}^{+i_0} \{-(r_t+h_t)h(\beta(t)\mid i)+$$

$$\alpha\frac{d^2}{(d\beta(t))^2}E_{D_t}\ \pi_1^*(t+1,(\beta(t)-D_t)^+)\ \}\,dF_I(i)$$

$$+\frac{di_0}{d\beta(t)}[(r_t-w_t^{n-1})-(r_t+h_t)H(\beta(t)\mid i_0)$$

$$+\alpha\frac{d}{d\beta(t)}E_{D_t}\ \pi_1^*(t+1,(\beta(t)-D_t)^+)\ ]f_I(i_0)$$

$$= \int_{-\infty}^{+i_0} \{-(r_t+h_t)h(\beta(t)\mid i)$$

$$+\alpha\frac{d^2}{(d\beta(t))^2}E_{D_t}\ \pi_1^*(t+1,(\beta(t)-D_t)^+)\ \}\,dF_I(i)$$

$$\leq 0. \tag{13.A.37}$$

Note that $i_0 = \hat{i}_{n-2}$, $\beta(t) = \beta(t) + \cdots + x_t^{n-2} = Q_t^{n-1}(\hat{i}_{n-2})$ in this case. Hence, by Definition 3.1 ($k = n-1$), the second equality holds. The last inequality holds by Lemma 3.4. Hence, $\pi_1^*(t,\beta(t))$ is proved to be concave in $\beta(t)$ in Case 3. When $\beta(t) = y_{n-1}(t)$,

$$\frac{d\pi_1^*(t,\beta(t))}{d\beta(t)}\Big|_{\beta(t)=y_{n-1}(t)}$$

$$= v_t^{n-1}+w_t^{n-1}$$

$$+\int_{-\infty}^{+i_0}(r_t-w_t^{n-1})-(r_t+h_t)H(y_{n-1}(t)\mid i)$$

$$+\alpha\frac{d}{d\beta(t)}E_{D_t}\ \pi_1^*(t+1,(y_{n-1}(t)-D_t)^+)\ \}\,dF_I(i). \tag{13.A.38}$$

By Definition 3.3 for $y_n(t)$ and $y_{n-1}(t)$, after some algebraic calculation it is easy to obtain

$$-v_t^{n-1}+\int_{\hat{i}_{n-1}}^{\hat{i}_{n-1}}[-(r_t+h_t)H(\beta(t)+\cdots+x_{n-1}(t)\mid i)$$

$$+r_t-w_t^{n-1}+\alpha\frac{d}{dx_{n-1}(t)}\pi_1^*(t,(\beta(t)+\cdots+x_{n-1}(t)-D_t)^+)]\,dF_I(i)$$

$$+\int_{\hat{i}_{n-1}}^{+\infty}(w_n(t)-w_t^{n-1})dF_I(i)+v_n(t)=0. \tag{13.A.39}$$

When $\beta(t) = y_{n-1}(t)$, $\mathbf{x_t^*} = (0,\cdots,0,0,y_n(t)-y_{n-1}(t))$. Then, by Definition 3.2, we have $i_0 = \cdots = \hat{i}_{n-1}$. Then, it can be reduced to be

$$-v_t^{n-1}+\int_{i_0}^{+\infty}[w_n(t)-w_t^{n-1}]dF_I(i)+v_n(t)=0. \tag{13.A.40}$$

Substitute Equation (13.A.40) into Equation (13.A.38), and we obtain that Equation (13.A.38) and Equation (13.A.35) are equal, i.e. the joint point of case 2 and 3 has equivalent left-hand and right-hand derivatives. Hence, $\pi_1^*(t, \beta(t))$ is concave in $\beta(t)$ in $[y_{n-2}(t), +\infty)$.

Similarly, we can obtain that $\pi_1^*(t, \beta(t))$ is concave in the remaining subintervals and the right-hand and left-hand derivatives are equal at joint points.

Proof of Lemma 3.11. In Case 1, when $\beta(t) = +\infty$ such that $H(\beta(t) \mid i) = 1$, by Equation (13.A.27), it becomes

$$
\frac{d\pi_1^*(t, \beta(t))}{d\beta(t)} \Big|_{\beta(t)=+\infty}
$$

$$
= w_n(t) + \int_{-\infty}^{+\infty} \{(r_t - w_n(t)) - (r_t + h_t)
$$

$$
+ \alpha \frac{d}{d\beta(t)} E_{D_t} \pi_1^*(t+1, (\beta(t) - D_t)^+) \Big|_{\beta(t)=+\infty} \} dF_I(i)
$$

$$
= -h_t + \int_{-\infty}^{+\infty} \alpha \frac{d}{d\beta(t)} E_{D_t} \pi_1^*(t+1, (\beta(t) - D_t)^+) \Big|_{\beta(t)=+\infty} dF_I(i)
$$

$$
= -h_t + \int_{-\infty}^{+\infty} \alpha \int_0^{+\infty} \frac{d}{d\beta(t)} [\pi_1^*(t+1, \beta(t) - z)] \Big|_{\beta(t)=+\infty} dH(z \mid i) \ dF_I(i)
$$

$$
< 0. \tag{13.A.41}
$$

The last inequality holds by our early assumption of $\frac{d\pi_1^*(t+1, \beta(t+1))}{d\beta(t+1)} \Big|_{\beta(t+1)=+\infty} < 0$.

In case $n+1$: $\beta(t) \in [0, y_1(t)]$, according to Theorem 3.2, $\mathbf{x}_t^* = (y_1(t) - \beta(t), y_2(t) - y_1(t), \cdots, y_n(t) - y_{n-1}(t))$ and $\pi^*(t, \beta(t)) = \pi_1(y_1(t) - \beta(t), y_2(t) - y_1(t), \cdots, y_n(t) - y_{n-1}(t); t, \beta(t))$. Hence, by equations (13.A.7) and (13.A.8),

$$
\pi_1^*(t, \beta(t))
$$

$$
= -v_1(t)(y_1(t) - \beta(t)) - v_2(t)(y_2(t) - y_1(t)) - \cdots - v_n(t)(y_n(t) - y_{n-1}(t))
$$

$$
+ \int_{-\infty}^{i_0} \left[-(r_t + h_t) \int_0^{\beta(t)} H(D \mid i)dD + r_t \beta(t) \right.
$$

$$
+ \alpha E_{D_t} \pi_1^*(t+1, (\beta(t) - D_t)^+) \right] dF_I(i)
$$

$$
+ \int_{i_0}^{i_1} \left[-(r_t + h_t) \int_0^{Q_1(i)} H(D \mid i)dD + r_t Q_1(i) \right.
$$

$$
- w_1(t)(Q_1(i) - \beta(t)) + \alpha E_{D_t} \pi_1^*(t+1, (Q_1(i) - D_t)^+) \right] dF_I(i)
$$

$$
+ \int_{i_1}^{\hat{i_1}} \left[-(r_t + h_t) \int_0^{y_1(t)} H(D \mid i)dD + r_t(y_1(t)) - w_1(t)(y_1(t) - \beta(t)) \right.
$$

$$
+ \alpha E_{D_t} \pi_1^*(t+1, (y_1(t) - D_t)^+) \right] dF_I(i)
$$

$$
\vdots
$$

$$
\vdots
$$

$$
+ \int_{\hat{i}_{n-1}}^{i_n} \left[-(r_t + h_t) \int_0^{Q_n(i)} H(D \mid i)dD + r_t Q_n(i) \right.
$$

$$
- w_1(t)(y_1(t) - \beta(t)) - \cdots - w_t^{n-1}(y_{n-1}(t) - y_{n-2}(t))
$$

$$-w_n(t)Q_n(i) + w_n(t)y_{n-1}(t) + \alpha E_{D_t}\pi_1^*(t+1,(Q_n(i)-D_t)^+) \; dF_I(i)$$

$$+\int_{i_n}^{\infty}\left[-(r_t+h_t)\int_0^{y_n(t)}H(D\mid i)dD + r_t y_n(t)\right.$$

$$-w_1(t)(y_1(t)-\beta(t)) - \cdots - w_n(t)(y_n(t)-y_{n-1}(t))$$

$$\left.+\alpha E_{D_t}\pi_1^*(t+1,(y_n(t)-D_t)^+)\,\right]dF_I(i). \tag{13.A.42}$$

Hence, similarly, we obtain

$$\frac{d\pi_1^*(t,\beta(t))}{d\beta(t)}$$

$$= \quad v_t^1 + w_1(t) + \int_{-\infty}^{+i_0}\{(r_t-w_1(t)) - (r_t+h_t)H(\beta(t)\mid i)$$

$$+\alpha\frac{d}{d\beta(t)}E_{D_t}\;\pi_1^*(t+1,(\beta(t)-D_t)^+)\;\}dF_I(i). \tag{13.A.43}$$

When $\beta(t)=0$, by Definition 3.2, $i_0 = -\infty$ and

$$\frac{d\pi_1^*(t,0)}{d\beta(t)} = v_1(t) + w_1(t). \tag{13.A.44}$$

Chapter 14

A REGIME-SWITCHING MODEL FOR EUROPEAN OPTIONS

David D. Yao

Department of Industrial Engineering and Operations Research
Columbia University, New York, NY 10027[*]
yao@columbia.edu

Qing Zhang

Department of Mathematics, University of Georgia, Athens, GA 30602[†]
qingz@math.uga.edu

Xun Yu Zhou

Department of Systems Engineering and Engineering Management
The Chinese University of Hong Kong, Shatin, Hong Kong[‡]
xyzhou@se.cuhk.edu.hk

Abstract   We study the pricing of European-style options, with the rate of return and the volatility of the underlying asset depending on the market mode or regime that switches among a finite number of states. This regime-switching model is formulated as a geometric Brownian motion modulated by a finite-state Markov chain. With a Girsanov-like change of measure, we derive the option price using risk-neutral valuation. We also develop a numerical approach to compute the pricing formula, using a successive approximation scheme with a geometric rate of convergence. Using numerical examples of simple, two- or three-state Markov chain

[*]Research undertaken while on leave at the Department of Systems Engineering and Engineering Management, The Chinese University of Hong Kong. Supported in part by NSF under Grant DMI-00-85124, and by RGC Earmarked Grant CUHK4175/00E.
[†]Supported in part by USAF Grant F30602-99-2-0548.
[‡]Supported in part by RGC Earmarked Grants CUHK4175/00E and CUHK4234/01E.

models, we are able to demonstrate the presence of the volatility smile and volatility term structure.

Keywords: Regime switching, option pricing, successive approximations, volatility smile and term structure.

1.   Introduction

The classical Black-Scholes formula for option pricing uses a geometric Brownian motion model to capture the price dynamics of the underlying security. The model involves two parameters, the expected rate of return and the volatility, both assumed to be deterministic constants. It is well known, however, that the stochastic variability in the market parameters is not reflected in the Black-Scholes model. Another widely acknowledged shortfall of the model is its failure to capture what is known as "volatility smile." That is, the implied volatility of the underlying security (implied by the market price of the option on the underlying via the Black-Scholes formula), rather than being a constant, should change with respect to the maturity and the exercise price of the option.

Emerging interests in this area have focused on the so-called regime-switching model, which stems from the need of more realistic models that better reflect random market environment. Since a major factor that governs the movement of an individual stock is the trend of the general market, it is necessary to allow the key parameters of the stock to respond to the general market movements. The regime-switching model is one of such formulations, where the stock parameters depend on the market mode (or, "regime") that switches among a finite number of states. The market regime could reflect the state of the underlying economy, the general mood of investors in the market, and other economic factors. The regime-switching model was first introduced by Hamilton (1989) to describe a regime-switching time series. Di Masi et al. (1994) discuss mean-variance hedging for regime-switching European option pricing. To price regime-switching American and European options, Bollen (1998) employs lattice method and simulation, whereas Buffington and Elliott (2002) use risk-neutral pricing and derive a set of partial differential equations for option price. Guo (1999) and Shepp (2002) use regime-switching to model option pricing with inside information. Duan et al. (2002) establish a class of GARCH option models under regime switching. For the important issue of fitting the regime-switching model parameters, Hardy (2001) develops maximum likelihood estimation using real data from the S&P 500 and TSE 300 indices. In addition to option pricing, regime-switching models have also been formulated and investigated for other problems; see Zhang (2001) for the development of an optimal stock selling rule, Zhang and Yin (2004) for applications in portfolio management, and Zhou and Yin (2003) for a dynamic Markowitz problem.

In the regime-switching model, one typically "modulates" the rate of return and the volatility by a finite-state Markov chain $\alpha(\cdot) = \{\alpha(t) : t \geq 0\}$, which represents the market regime. For example, $\alpha(t) \in \{-1, 1\}$ with 1 representing the bullish (up-trend) market and -1 the bearish (down-trend) one. In general, we can take $\mathcal{M} = \{1, 2, \ldots, m\}$. More specifically, let $X(t)$, the price of a stock at time t, be governed by the following equation:

$$(14.1) \quad dX(t) = X(t)[\mu(\alpha(t))dt + \sigma(\alpha(t))dw(t)], \quad 0 \leq t \leq T;$$

where $X(0) = X_0$ is the stock price at $t = 0$; $\mu(i)$ and $\sigma(i)$, for each $i \in \mathcal{M}$, represent the expected rate of return and the volatility of the stock price at regime i; and $w(\cdot)$ denotes the standard (one-dimensional) Brownian motion. Equation (14.1) is also called a *hybrid* model, where randomness is characterized by the pair $(\alpha(t), w(t))$, with $w(\cdot)$ corresponding to the usual noise involved in the classical geometric Brownian motion model while $\alpha(t)$ capturing the higher-level noise associated with infrequent yet extremal events. For example, it is known that the up-trend volatility of a stock tends to be smaller than its down-trend volatility. When the market trends up, investors are often cautious and move slowly, which leads to a smaller volatility. On the other hand, during a sharp market downturn when investors get panic, the volatility tends to be much higher. (This observation is supported by an initial numerical study reported in Zhang (2001) in which the average historical volatility of the NASDAQ Composite is substantially greater when its price trends down than when it moves up.) Furthermore, when a market moves sideways the corresponding volatility appears to be even smaller. As an example, suppose we take the value of $\alpha(t)$ to be

$$-2 = \text{severe downtrend ('crash')}, \quad -1 = \text{downtrend},$$
$$0 = \text{sideways}, \quad 1 = \text{rally}, \quad 2 = \text{strong rally}.$$

Then, the sample paths of $\alpha(\cdot)$ and $X(\cdot)$ are given in Fig. 14.1. The 'crash' state is reached between sessions 70-80, which simulates a steep downward movement in price (namely the daily high is lower than last session's low).

Our main objective in this paper is to price regime-switching European options. We develop a successive approximation procedure, based

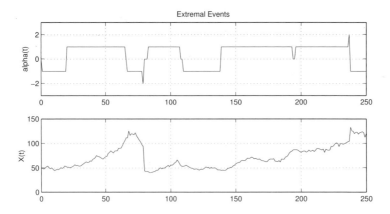

Figure 14.1. Sample Paths of $\alpha(\cdot)$ and $X(\cdot)$.

on the fixed-point of a certain integral operator with a Gaussian kernel. The procedure is easy to implement without having to solve the differential equations; moreover, it has a geometric rate of convergence. Using this numerical procedure, we demonstrate that our regime-switching model does generate the desired volatility smile and term structure.

We now briefly review other related literature. For derivative pricing in general, we refer the reader to the books by Duffie (1996) and Karatzas and Shreve (1998). In recent years there has been extensive research effort in enhancing the classical geometric Brownian motion model. Merton (1976) introduces additive Poisson jumps into the geometric Brownian motion, aiming to capture discontinuities in the price trajectory. The drawback, however, is the difficulty in handling the associated dynamic programming equations, which take the form of quasi-variational inequalities. Hull and White (1987) develop stochastic volatility models, which price the European options as the expected value of the Black-Scholes price with respect to the distribution of the stochastic volatility. These models are revisited by Fouque et al. (2000) using a singular perturbation approach. Albanese et al. (preprint) study a model that is a composition of a Brownian motion and a gamma process, with the latter used to rescale the time. In addition, the parameters of the gamma process are allowed to evolve according to a two-state Markov chain. While the model does capture the volatility smile, the resulting pricing formula appears to be quite involved and difficult to implement. Renault and Touzi (1996) demonstrate volatility smile using a model with pure diffusion, which is on one hand more complex in structure than regime switching models, and on the other hand does not in any case specialize to the latter.

Compared with diffusion type volatility models, regime-switching models have two more advantages. First, the discrete jump Markov process captures more directly the dynamics of events that are less frequent (occasional) but nevertheless more significant to longer-term system behavior. For example, the Markov chain can represent discrete events such as market trends and other economic factors that are difficult to be incorporated into a diffusion model. Second, regime-switching models require very limited data input, essentially, the parameters $\mu(i)$, $\sigma(i)$ for each state i, and the Q matrix. For a two-state Markov chain, a simple procedure to estimate these parameters is given in Zhang (2001). For more general Markov chains, refer to the approach in Yin et al. (2003) based on stochastic approximation.

The rest of the paper is organized as follows. Our starting point is the hybrid model in (14.1) for the price dynamics of the underlying security, detailed in §2. We establish a Girsanov-like theorem (Lemma 1), which leads to an equivalent martingale measure, and therefore a risk-neutral pricing scheme. Based on this pricing scheme, we develop in §3 a numerical approach to compute the option price, which is a successive approximation procedure with a geometric rate of convergence. In §4, we demonstrate via numerical examples that with a simple, two- or three-state Markov chain modulating the volatility we can produce the anticipated volatility smile and volatility term structure.

2. Risk-Neutral Pricing

A standard approach in derivative pricing is risk-neutral valuation. The idea is to derive a suitable probability space upon which the expected rate of return of all securities is equal to the risk-free interest rate. Mathematically, this requires that the discounted asset price be a martingale; and the associated probability space is referred to as the risk-neutral world. The price of the option on the asset is then the expected value, with respect to this martingale measure, of the discounted option payoff. In a nutshell, this is also the route we are taking here, with the martingale measure identified in Lemma 1 below, and related computational issues deferred to §3.

Let $(\Omega, \mathcal{F}, \mathsf{P})$ denote the probability space, upon which all the processes below are defined. Let $\{\alpha(t)\}$ denote a continuous-time Markov chain with state space $\mathcal{M} = \{1, 2, \ldots, m\}$. Note that for simplicity, we use each element in $\mathcal{M}$ as an index, which can be associated with a more elaborate state description, for instance, a vector. Let $Q = (q_{ij})_{m \times m}$ be the generator of α with $q_{ij} \geq 0$ for $i \neq j$ and $\sum_{j=1}^{m} q_{ij} = 0$ for each $i \in \mathcal{M}$. Moreover, for any function f on $\mathcal{M}$ we denote $Qf(\cdot)(i) := \sum_{j=1}^{m} q_{ij} f(j)$.

Let $X(t)$ denote the price of a stock at time t which satisfies (14.1). We assume that X_0, $\alpha(\cdot)$, and $w(\cdot)$ are mutually independent; and $\sigma^2(i) > 0$, for all $i \in \mathcal{M}$.

Let $\mathcal{F}_t$ denote the sigma field generated by $\{(\alpha(u), w(u)) : 0 \le u \le t\}$. Throughout, all (local) martingales concerned are with respect to the filtration $\mathcal{F}_t$. Therefore, in the sequel we shall omit reference to the filtration when a (local) martingale is mentioned. Clearly, $\{w(t)\}$ and $\{w^2(t) - t\}$ are both martingales (since $\alpha(\cdot)$ and $w(\cdot)$ are independent).

Let $r > 0$ denote the risk-free rate. For $0 \le t \le T$, let

$$Z_t := \exp\left(\int_0^t \beta(u)dw(u) - \frac{1}{2}\int_0^t \beta^2(u)du\right),$$

where

(14.2) $$\beta(u) := \frac{r - \mu(\alpha(u))}{\sigma(\alpha(u))}.$$

Then, applying Itô's rule, we have

$$\frac{dZ_t}{Z_t} = \beta(t)dw(t);$$

and Z_t is a local martingale, with $\mathsf{E}Z_t = 1$, $0 \le t \le T$. Define an equivalent measure $\widetilde{\mathsf{P}}$ via the following:

(14.3) $$\frac{d\widetilde{\mathsf{P}}}{d\mathsf{P}} = Z_T.$$

The lemma below is essentially a generalized Girsanov's theorem for Markov-modulated processes. (While results of this type are generally known, a proof is included since a specific reference is not readily available.)

LEMMA 1 (1) Let $\widetilde{w}(t) := w(t) - \int_0^t \beta(u)du$. Then, $\widetilde{w}(\cdot)$ is a $\widetilde{\mathsf{P}}$-Brownian motion.

(2) $X(0)$, $\alpha(\cdot)$ and $\widetilde{w}(\cdot)$ are mutually independent under $\widetilde{\mathsf{P}}$;

(3) (Dynkin's formula) For any smooth function $F(t, x, i)$, we have

$$F(t, X(t), \alpha(t)) = F(s, X(s), \alpha(s)) + \int_s^t \mathcal{A}F(u, X(u), \alpha(u))du + M(t) - M(s),$$

where $M(\cdot)$ is a $\widetilde{\mathsf{P}}$-martingale and $\mathcal{A}$ is an generator given by

$$\mathcal{A}F = \frac{\partial}{\partial t}F(t, x, i) + \frac{1}{2}x^2\sigma^2(i)\frac{\partial^2}{\partial x^2}F(t, x, i) + rx\frac{\partial}{\partial x}F(t, x, i) + QF(t, x, \cdot)(i).$$

This implies that $(X(t), \alpha(t))$ is a Markov process with generator $\mathcal{A}$.

Proof. Define a row vector $\Psi(t) = \left(I_{\{\alpha(t)=1\}}, \ldots, I_{\{\alpha(t)=m\}} \right)$, where I_A is the indicator function of a set A. Let

$$z(t) = \Psi(t) - \Psi(0) - \int_0^t \Psi(u) Q du.$$

Note that both $\{(z(u), w(u)) : u \leq t\}$ and $\{(\alpha(u), w(u)) : u \leq t\}$ generate the same sigma field $\mathcal{F}_t$. Thus, $(z(t), w(t))$ is a P-martingale. Let Θ denote a column vector and θ a scaler. Define $V(t) = z(t)\Theta + w(t)\theta$. Then, $V(t)$ is a P-martingale. Let

$$\eta(t) = \int_0^t \beta(u) dw(u) - \frac{1}{2} \int_0^t \beta(u)^2 du.$$

Then, for each θ and Θ, $\widetilde{V}(t) = V(t) - \langle V, \eta \rangle_t$ is a $\widetilde{\mathsf{P}}$-martingale, where $\langle V, \eta \rangle_t = \theta \int_0^t \beta(u) du$. Thus, $\widetilde{V}(t) = z(t)\Theta + \widetilde{w}(t)\theta$. Hence, in view of Elliott (1982) Thm. 13.19 $(z(t), \widetilde{w}(t))$ is a $\widetilde{\mathsf{P}}$-martingale. Moreover, since $\widetilde{w}(\cdot)$ and $(\widetilde{w}^2(t) - t)$ are both $\widetilde{\mathsf{P}}$-martingales, $\widetilde{w}(\cdot)$ is a $\widetilde{\mathsf{P}}$-Brownian motion (see Elliott (1982) Cor 13.25).

We next show the mutual independence of X_0, $\alpha(\cdot)$ and $\widetilde{w}(\cdot)$ under $\widetilde{\mathsf{P}}$. Note that Z_T is $\mathcal{F}_T$ measurable, and $\mathsf{E}Z_T = 1$. Let ζ_1 denote a random variable, measurable with respect to X_0. Then, making use of (14.3), we have

$$\widetilde{\mathsf{E}}\zeta_1 = \mathsf{E}(Z_T \zeta_1) = (\mathsf{E}Z_T)(\mathsf{E}\zeta_1) = \mathsf{E}\zeta_1.$$

Furthermore, for any $\mathcal{F}_T$ measurable random variable ζ_2, we have

$$\widetilde{\mathsf{E}}(\zeta_1 \zeta_2) = \mathsf{E}(Z_T \zeta_1 \zeta_2) = (\mathsf{E}\zeta_1)\mathsf{E}(Z_T \zeta_2) = (\widetilde{\mathsf{E}}\zeta_1)(\widetilde{\mathsf{E}}\zeta_2).$$

This implies the independence between X_0 and $(\alpha(\cdot), \widetilde{w}(\cdot))$ up to time T. To show the independence between $\alpha(\cdot)$ and $\widetilde{w}(\cdot)$, for a given $f(x, i)$, let

$$\mathcal{A}^0 f(x, i) = \frac{1}{2} \frac{\partial^2}{\partial x^2} f(x, i) + Q f(x, \cdot)(i).$$

Then, the associated martingale problem has a unique solution (see, e.g., Yin and Zhang (1998) p. 199). Using Itô's rule, we can show that $(\alpha(\cdot), \widetilde{w}(\cdot))$ is a solution to the martingale problem under $\widetilde{\mathsf{P}}$. Since $(\alpha(\cdot), w(\cdot))$ is also a solution to the same martingale problem under P, it must be equal in distribution to $(\alpha(\cdot), \widetilde{w}(\cdot))$. The independence between $\alpha(\cdot)$ and $\widetilde{w}(\cdot)$ then follows from the independence between $\alpha(\cdot)$ and $w(\cdot)$.

Under $\widetilde{\mathsf{P}}$, (14.1) becomes

$$dX(t) = X(t)[rdt + \sigma(\alpha(t))d\widetilde{w}(t)], \ 0 \leq t \leq T; \qquad X(0) = X_0.$$

We now prove the Dynkin's formula. First, write

$$F(X(t), \alpha(t)) = \Psi(t)\overline{F}(X(t)),$$

where $\overline{F}(X(t)) = (F(X(t), 1), \ldots, F(X(t), m))'$. Applying Elliott (1982) Cor. 12.22, we have

$$dF(X(t), \alpha(t)) = \Psi(t)d(\overline{F}(X(t)) + (d\Psi(t))\overline{F}(X(t)) + d[\Psi, F]_t.$$

Since Ψ is a pure jump process, we have $[\Psi, F]_t = 0$. In addition, we have

$$\Psi(t)Q(F(X(t), 1), \ldots, F(X(t), m))' = QF(X(t), \cdot)(\alpha(t)).$$

Hence, Dynkin's formula follows.

Finally, the Markov property of $(X(t), \alpha(t))$ under $\widetilde{P}$ can be established following the same argument as in Ghosh et al. (1993). $\square$

Therefore, following the above lemma and in view of Hull (2000) and Fouque et al. (2000), $(\Omega, \mathcal{F}, \{\mathcal{F}_t\}, \widetilde{P})$ defines a risk-neutral world. Moreover, $e^{-rt}X(t)$ is a $\widetilde{P}$-martingale. Note that the risk-neutral martingale measure may not be unique. The market model under consideration has two types of random sources, $w(\cdot)$ and $\alpha(\cdot)$. The inclusion of $\alpha(\cdot)$ makes the underlying market incomplete. Nevertheless, the market can be made complete by introducing switching-cost securities such as those of the Arrow-Debreu type; refer to Guo (1999) for related discussions.

Consider a European-style call option with strike price K and maturity T. Let

$$h(x) = (x - K)^+ := \max\{x - K, 0\}.$$

The call option premium at time s, given the stock price $X(s) = x$ and the state of the Markov chain $\alpha(s) = i$, can be expressed as follows:

$$(14.4) \quad c(s, x, i) = \widetilde{E}[e^{-r(T-s)}h(X(T))|X(s) = x, \alpha(s) = i].$$

Throughout, we shall focus on call options only. For European put options, the analysis is similar, with the h function changed to $h(x) = (K - x)^+$.

3. Successive Approximations

We develop a numerical technique that is directly based on the risk-neutral valuation in §2. Let

$$H(s, t) := \int_s^t [r - \frac{1}{2}\sigma^2(\alpha(u))]du + \int_s^t \sigma(\alpha(u))d\widetilde{w}(u)$$

and
$$Y(t) := y + H(s,t), \qquad \text{with} \quad y = \log x.$$

Itô's rule implies
$$X(t) = \exp[Y(t)].$$

Let
$$\psi(s,y,i) = e^{-y}c(s,e^y,i).$$

Then, combining the above with (14.4) and taking into account $h(x) \leq |x|$, we have
$$\psi(s,y,i) \leq e^{-y-r(T-s)}\widetilde{\mathsf{E}}[e^{y+H(s,T)}|Y(s) = y, \alpha(s) = i] \leq C,$$

for some constant C. Let
$$\psi^0(s,y,i) = e^{-y}\widetilde{\mathsf{E}}[e^{-r(T-s)}h(e^{y+H(s,T)})|Y(s) = y, \alpha(u) = i, s \leq u \leq T],$$

which corresponds to the case when $\alpha(\cdot)$ has no jump in $[s,T]$. Then,

$$(14.5) \qquad
\begin{aligned}
\psi^0(s,y,i) &= e^{-y-r(T-s)} \\
&\quad \int_{-\infty}^{\infty} h(e^{y+u})N(u, m(T-s,i), \Sigma^2(T-s,i))du,
\end{aligned}$$

where N is the Gaussian density function with mean
$$m(t,i) = [r - \frac{1}{2}\sigma^2(i)]t,$$

and variance $\Sigma^2(t,i) = \sigma^2(i)t$. Note that, for each i, $e^y\psi^0(s,y,i)$ gives the standard Black-Scholes price, as expected.

If $q_{ii} = 0$, then $\psi(s,y,i) = \psi^0(s,y,i)$. For $q_{ii} \neq 0$, let
$$\tau = \inf\{t \geq s: \alpha(t) \neq \alpha(s)\},$$

i.e., τ is the first jump epoch of $\alpha(\cdot)$. Then,
$$\mathsf{P}(\tau > u|\alpha(s) = i) = e^{q_{ii}(u-s)}.$$

For any bounded measurable function f on $[0,T] \times \mathbb{R} \times \mathcal{M}$, define its norm as follow:
$$\|f\| = \sup_{s,y,i} |f(s,y,i)|.$$

This induces a Banach space $\mathcal{S}$ of all the bounded measurable functions on $[0,T] \times \mathbb{R} \times \mathcal{M}$. Also, define a mapping on $\mathcal{S}$:

$$(\mathcal{T}f)(s,y,i) = \int_s^T \sum_{j \neq i} e^{-r(t-s)}$$
$$\times \left(\int_{-\infty}^{\infty} e^u f(t, y+u, j)N(u, m(t-s,i), \Sigma^2(t-s,i))du \right) q_{ij}e^{q_{ii}(t-s)}dt.$$

THEOREM 1. (1) ψ is a unique solution to the equation

(14.6) $\qquad \psi(s,y,i) = \mathcal{T}\psi(s,y,i) + e^{q_{ii}(T-s)}\psi^0(s,y,i).$

(2) Let $\psi_0 = \psi^0$; and define $\{\psi_n\}$, for $n = 1, 2, \ldots$, recursively as follows:

(14.7) $\qquad \psi_{n+1}(s,y,i) = \mathcal{T}\psi_n(s,y,i) + e^{q_{ii}(T-s)}\psi^0(s,y,i).$

Then, the sequence $\{\psi_n\}$ converges to the solution ψ.

Proof. First, we note that

$$e^{-y}\widetilde{\mathsf{E}}_{s,y,i}[e^{-r(T-s)}h(e^{Y(T)})I_{\{\tau>T\}}] = e^{q_{ii}(T-s)}\psi^0(s,y,i),$$

where $\widetilde{\mathsf{E}}_{s,y,i}[\cdot] := \widetilde{\mathsf{E}}[\cdot|Y(s) = y, \alpha(s) = i]$. Therefore,

(14.8) $\qquad \begin{aligned} \psi(s,y,i) &= e^{-y}\widetilde{\mathsf{E}}_{s,y,i}[e^{-r(T-s)}h(e^{Y(T)})I_{\{\tau\le T\}}] \\ &\quad + e^{q_{ii}(T-s)}\psi^0(s,y,i). \end{aligned}$

By conditioning on $\tau = t$, we write its first term as follows:

$$\int_s^T e^{-y-r(T-s)}\widetilde{\mathsf{E}}_{s,y,i}[h(e^{Y(T)})|\tau = t](-q_{ii}e^{q_{ii}(t-s)})dt.$$

Recall that τ is the first jump time of $\alpha(\cdot)$. Therefore, given $\{\tau = t\}$, the post-jump distribution of $\alpha(t)$ is equal to $q_{ij}/|q_{ii}|$, $j \in \mathcal{M}$. Moreover,

(14.9) $\quad Y(t) = y + [r - \frac{1}{2}\sigma^2(i)](t-s) + \sigma(i)[\widetilde{w}(t) - \widetilde{w}(s)],$

which has a Gaussian distribution and is independent of $\alpha(\cdot)$. In view of these, it follows that, for $s \le t \le T$,

$$\begin{aligned} &\widetilde{\mathsf{E}}_{s,y,i}[h(e^{Y(T)})|\tau = t] \\ =\ &\widetilde{\mathsf{E}}_{s,y,i}[\widetilde{\mathsf{E}}_{s,y,i}[h(e^{Y(T)})|Y(t), \alpha(t)]|\tau = t] \\ =\ &\widetilde{\mathsf{E}}_{s,y,i}[e^{Y(t)}e^{r(T-t)}\psi(t, Y(t), \alpha(t))|\tau = t] \\ =\ &\sum_{j\ne i}\frac{q_{ij}}{-q_{ii}}\int_{-\infty}^{\infty}e^{y+u}e^{r(T-t)}\psi(t, y+u, j)N(u, m(t-s,i), \Sigma^2(t-s,i))du. \end{aligned}$$

Thus, the first term in (14.8) is equal to

$$\begin{aligned} \int_s^T &\bigg(\sum_{j\ne i}e^{-r(t-s)} \\ &\times\int_{-\infty}^{\infty}e^u\psi(t, y+u, j)N(u, m(t-s,i), \Sigma^2(t-s,i))du\bigg)q_{ij}e^{q_{ii}(t-s)}dt. \end{aligned}$$

Let

$$\rho(i) = \int_s^T \sum_{j \neq i} e^{-r(t-s)}$$

$$\times \left(\int_{-\infty}^{\infty} e^u N(u, m(t-s,i), \Sigma^2(t-s,i)) du \right) q_{ij} e^{q_{ii}(t-s)} dt.$$

We want to show that $0 \leq \rho(i) < 1$, for $i = 1, 2, \ldots, m$. In fact, let

$$A(u,i) = \int_{-\infty}^{\infty} e^u N(u, m(t-s,i), \Sigma^2(t-s,i)) du.$$

Then it is readily verified that

$$A(u,i) = \exp[m(u,i) + \Sigma^2(u,i)] = \exp(ru).$$

Thus,

$$\text{(14.10)} \qquad \begin{aligned} \rho(i) &= \int_s^T \sum_{j \neq i} e^{-r(t-s)} A(t-s,i) q_{ij} e^{q_{ii}(t-s)} dt \\ &= 1 - e^{q_{ii}(T-s)} < 1, \end{aligned}$$

when $q_{ii} \neq 0$. Let $\rho = \max\{\rho(i) : i \in \mathcal{M}\}$. Then, $0 \leq \rho < 1$ and

$$\|\mathcal{T}f\| \leq \rho\|f\|,$$

i.e., $\mathcal{T}$ is a contraction mapping on $\mathcal{S}$. Therefore, in view of the contraction mapping fixed point theorem, we know equation (14.6) has a unique solution. This implies also the convergence of the sequence $\{\psi_n\}$ to ψ. $\square$

Note that the convergence of ψ_n to ψ is geometric. In fact, it is easy to see that $\psi_{n+1} - \psi = \mathcal{T}(\psi_n - \psi)$, which implies $\|\psi_{n+1} - \psi\| \leq \rho\|\psi_n - \psi\|$. Therefore,

$$\|\psi_n - \psi\| \leq C\rho^n,$$

for some constant C. In addition, the convergence rate ρ depends on the jump rates of $\alpha(\cdot)$. By and large, the less frequent the jumps, the faster is the convergence. This can be seen from (14.10).

Therefore, to evaluate ψ, we solve equation (14.6) via the successive approximations in (14.7). Finally, the call option price is as follows. (Recall $y = \log x$.)

$$\text{(14.11)} \qquad c(s,x,i) = x\psi(s, \log x, i).$$

4.   Volatility Smile and Term Structure

We now illustrate the volatility smile and volatility term structure implied in our model. We first consider a case in which the volatility (as well as the return rate) is modulated by a two-state Markov chain, i.e., $\mathcal{M} = \{1, 2\}$ and $Q = \begin{pmatrix} -\lambda & \lambda \\ 0 & 0 \end{pmatrix}$. That is, state 2 is an absorbing state. As it will be demonstrated the volatility smile presents in even such a simple case. If we take $\alpha(s) = 1$, then there exists a stopping time τ such that $(\tau - s)$ is exponentially distributed with parameter λ and

$$(14.12) \qquad \alpha(t) = \begin{cases} 1 & \text{if } t < \tau \\ 2 & \text{if } t \geq \tau. \end{cases}$$

Therefore, the volatility process $\sigma(\alpha(t))$ jumps at most once at time $t = \tau$. Its jump size is given by $\sigma(2) - \sigma(1)$, and the average sojourn time in state 1 (before jumping to state 2) is $1/\lambda$. For instance, if the time unit is one year and $\lambda = 6$, then it means that the expected time for the volatility to jump from $\sigma(1)$ to $\sigma(2)$ is two months. In this case, the volatility is characterized by a vector $(\sigma(1), \sigma(2), \lambda)$.

Applying the successive approximation in §3, we have

$$
\begin{aligned}
(14.13) \quad \psi(s, y, 1) = &\int_s^T e^{-r(t-s)} \left(\int_{-\infty}^\infty e^u \psi(t, y + u, 2) \right. \\
& \left. \times N(u, m(t - s, 1), \Sigma^2(t - s, 1)) du \right) \lambda e^{-\lambda(t-s)} dt \\
& + e^{-\lambda(T-s)} \psi^0(s, y, 1),
\end{aligned}
$$

and

$$(14.14) \qquad \psi(s, y, 2) = \psi^0(s, y, 2),$$

where $\psi^0(s, y, i)$, $i = 1, 2$, are defined in (14.5).

Let $s = 0$. Given the risk-free rate r, the current stock price x, the maturity T, the strike price K, and the volatility vector $(\sigma(1), \sigma(2), \lambda)$, in view of (14.11), (14.13) and (14.14), the call option can be priced as follows:

$$
\begin{aligned}
(14.15) \quad c(0, x, 1) = &x \int_0^T e^{-rt} \left(\int_{-\infty}^\infty e^u \psi^0(t, u + \log x, 2) \right. \\
& \left. \times N(u, m(t, 1), \Sigma^2(t, 1)) du \right) \lambda e^{-\lambda t} dt + x e^{-\lambda T} \psi^0(0, \log x, 1).
\end{aligned}
$$

This pricing formula consists of two parts: the classical Black-Scholes part (with no jump) and a correction part. In addition, it is a natural extension to the classical Black-Scholes formula by incorporating a possible volatility jump, which is usually adequate for near-term options. Moreover, both of these two parts are given in analytic form which is very helpful for evaluating option Greeks and making various numerical comparisons.

Given the option price $c(0, x, 1)$, we can derive the so-called implied volatility using the standard Black-Scholes formula as in Hull (2000).

The following numerical cases illustrate the volatility smile and volatility term structure implied in our model. In all cases, we fix $r = 0.04$ and $x = 50$, while varying the other parameters. First we consider the cases with $\sigma(1) \leq \sigma(2)$. Let

$$\Gamma_\lambda = \{1, 2, \ldots, 20\},$$
$$\Gamma_K = \{30, 35, \ldots, 70\},$$
$$\Gamma_T = \{20/252, 40/252, \ldots, 240/252\},$$
$$\Gamma_\sigma = \{0, 0.1, \ldots, 2\}.$$

Case (1a): Here, we fix $T = 60/252$ (three months to maturity), $\sigma(1) = 0.3$, $\sigma(2) = 0.8$, $\lambda \in \Gamma_\lambda$, and $K \in \Gamma_K$. We plot the implied volatility against the strike price (K) and the jump rate (λ) in Fig. 14.2 (a).

As can be observed from Fig. 14.2 (a), for each fixed $\lambda \in \Gamma_\lambda$, the implied volatility reaches its minimum at $K = 50$ (at money) and increases as K moves away from $K = 50$. This is the well-known volatility smile phenomenon in stock options Hull (2000). In addition, for fixed $K \in \Gamma_K$, the implied volatility is increasing in λ, corresponding to a sooner jump from $\sigma(1)$ to $\sigma(2)$.

Case (1b): In this case, we take $\sigma(1) = 0.3$ and fix $\lambda = 1$, and replace the λ-axis in Fig. 14.2 (a) by the volatility jump size $\sigma(2) - \sigma(1) \in \Gamma_\sigma$. As can be observed from Fig. 14.2 (b), the smile increases in the jump size. In addition, the implied volatility is an increasing function of $\sigma(2) - \sigma(1)$, for each fixed $K \in \Gamma_K$.

Case (1c): In this case, we take $\sigma(1) = 0.3$, $\sigma(2) = 0.8$ and $K = 50$, and replace the strike price in Fig. 14.2 (a) by the maturity (T). Then, Fig. 14.2 (c) shows that for fixed λ, the implied volatility increases in T. Similarly, for fixed T, the implied volatility also increases in λ.

Case (1d): Here, we fix $\sigma(1) = 0.3$, $\lambda = 1$, $K = 50$, and continue with Case (1c), but replace λ by the increase in volatility. Then, the implied volatility is also increasing in $\sigma(2) - \sigma(1) \in \Gamma_\sigma$.

To get a better view of the volatility smile, we plotted in Fig. 14.3 the two dimensional truncation of Case (1a) with fixed $\lambda = 1$ and $\lambda = 5$. It

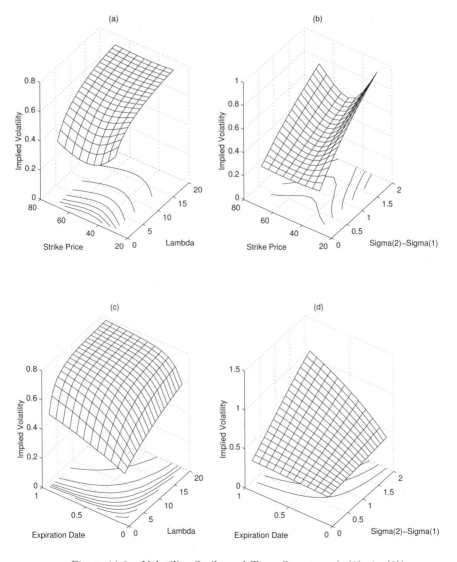

Figure 14.2.   Volatility Smile and Term Structure $(\sigma(1) \leq \sigma(2))$

is clear from this picture that the volatility smile reached the minimum at $x = 50$ and it is asymmetric with respect to strike prices.

Whereas in the above cases we have $\sigma(1) < \sigma(2)$, in the next set of cases we consider $\sigma(1) \geq \sigma(2)$. In these cases, the market anticipates a decline in volatility.

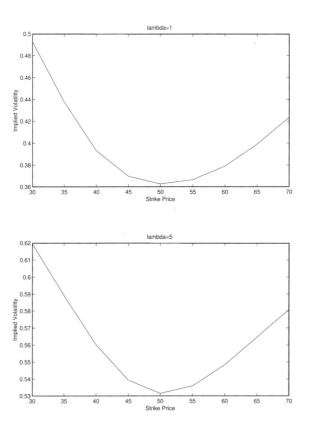

Figure 14.3.   Volatility Smile in Case (1a) with $\lambda = 1$ and $\lambda = 5$, resp.

Case (2a): We take $\sigma(1) = 0.8$ and $\sigma(2) = 0.3$, with $T = 60/252$, $\lambda \in \Gamma_\lambda$, and $K \in \Gamma_K$. The implied volatility against the strike price (K) and the jump rate (λ) is plotted in Fig. 14.4 (a). This case is similar to Case (a).

Case (2b): Here, we let $\sigma(1) = 2.3$, fix $\lambda = 1$, and replace the λ-axis in Fig. 14.4 (a) by $\sigma(2) - \sigma(1) \in \Gamma_\sigma^1$, where $\Gamma_\sigma^1 = \{0, -0.1, \ldots, -2\}$. As can be seen from Fig. 14.4 (b), the smile increases in the jump size $|\sigma(2) - \sigma(1)|$.

Case (2c): In this case, we take $\sigma(1) = 0.8$, $\sigma(2) = 0.3$ and $K = 50$. In contrast to the earlier Case (c), Fig. 14.4 (c) shows that for fixed λ, the implied volatility decreases in (T) (and also in (λ) with fixed T).

Case (2d): In this last case, we fix $\sigma(1) = 2.3$, $\lambda = 1$, and $K = 50$. The implied volatility decreases in (T) and also in $|\sigma(2) - \sigma(1)|$.

Next we consider a three-state model without any absorbing state. Let $T = 0.5$, $X_0 = 50$, $r = 0.04$, $\sigma(1) = 0.2$, $\sigma(2) = 0.5$, $\sigma(3) = 0.3$, and let the generator be

$$Q = \begin{pmatrix} -1.0 & 1.0 & 0.0 \\ 0.5 & -1.0 & 0.5 \\ 0.0 & 1.0 & -1.0 \end{pmatrix}.$$

Let $c(0, x, i)$, $i = 1, 2, 3$, denote the call prices. The corresponding implied volatilities are plotted in Fig. 14.5, which depicts a "grimace" curve typical in equity markets (e.g. SP500); see Hull (2000). In addition, note that in this case $\sigma(2) > \sigma(3) > \sigma(1)$, and the implied volatility exhibits a similar order: $IV(\alpha = 2) > IV(\alpha = 3) > IV(\alpha = 1)$.

The above examples clearly illustrate the advantage of the Markov-chain modulated volatility model, in particular, its striking simplicity — it requires fewer parameters than most stochastic volatility models.

Finally, we compare our model with the diffusion-type volatility model of Hull-White (1987) (also refer to Hull (2000) pp. 458-459). These are two very different models as explained in the Introduction. The comparison below aims to investigate whether the Hull-White model can be adapted, via taking expectation with respect to the probability law of the switching mechanism, to price regime-switching options. For simplicity, consider the two-state Markov chain introduced at the beginning of this section. Recall τ is the switchover time (from state 1 to state 2).

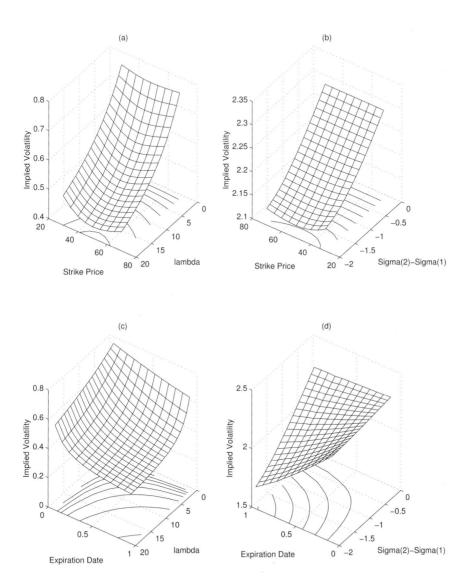

Figure 14.4. Volatility Smile and Term Structure $(\sigma(1) \geq \sigma(2))$

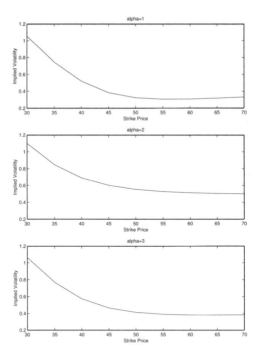

Figure 14.5. Volatility Grimace.

Given τ, we first characterize the average volatility rate needed in the Hull-White model as follows:

$$(14.16) \quad \overline{\sigma} = \begin{cases} \sqrt{(\tau/T)\sigma^2(1) + (1 - \tau/T)\sigma^2(2)}, & \text{if } \tau < T \\ \sigma(1), & \text{if } \tau \geq T. \end{cases}$$

Let $c_{\text{BS}}(\sigma)$ denote the Black-Scholes call price with constant volatility σ; let $c(\tau) = c_{\text{BS}}(\overline{\sigma}(\tau))$. Then, based on the Hull-White model, the option is priced as $c_{\text{HW}} = \mathsf{E}[c(\tau)]$.

Consider a set of parameters with $K = 50$, $X_0 = 50$, $r = 0.04$, $\sigma(1) = 0.2$, $\sigma(2) = 1$, $\lambda = 0.5$, and the maturity T varies from 0.1 to 1. From the results summarized in Table 1, it is evident that the adapted H-W model fails to match the *exact* prices (computed using our formula (14.15)), especially for options with a short maturity.

T	0.10	0.12	0.14	0.20	0.25	0.33	0.50	1.00
H-W	1.0108	1.3040	1.5086	2.1295	2.6316	3.4037	4.8165	8.7926
Exact	1.5114	1.7438	1.9041	2.3975	2.8157	3.4976	4.8389	8.7929

Table 1. Comparison against Conditional H-W.

We have also tried alternative ways to characterize the average volatility rate, such as

$$\overline{\sigma} = \sqrt{(\tau/2)\sigma^2(1) + (1 - \tau/2)\sigma^2(2)},$$

with τ following an exponential distribution confined to $[0, 2]$, i.e., with the density $\lambda \exp(-\lambda t)/(1 - \exp(-2\lambda))$, $t \in [0, 2]$. (Truncating the distribution at $t = 2$ is based on the fact that the longest maturity considered here is $T = 1$.) These alternatives all seem to perform worse than the one in (14.16). It is clear that while the adapted H-W model can be used as an approximation to price regime-switching options, it is not a substitute for our exact model in general.

References

C. Albanese, S. Jaimungal, and D.H. Rubisov, A jump model with binomial volatility, preprint.

N. P. B. Bollen, Valuing options in regime-switching models, *Journal of Derivatives*, vol. 6, pp. 38-49, (1998).

J. Buffington and R. J. Elliott, American options with regime switching, *International Journal of Theoretical and Applied Finance*, vol. 5, pp. 497-514, (2002).

G. B. Di Masi, Y. M. Kabanov and W. J. Runggaldier, Mean variance hedging of options on stocks with Markov volatility, *Theory of Probability and Applications*, vol. 39, pp. 173-181, (1994).

J.-C. Duan, I. Popova and P. Ritchken, Option pricing under regime switching, *Quantitative Finance*, vol. 2, pp.116-132, (2002).

D. Duffie, *Dynamic Asset Pricing Theory*, 2nd Ed., Princeton University Press, Princeton, NJ, 1996.

R. J. Elliott, *Stochastic Calculus and Applications*, Springer-Verlag, New York, 1982.

J. P. Fouque, G. Papanicolaou, and K. R. Sircar, *Derivatives in Financial Markets with Stochastic Volatility*, Cambridge University Press, 2000.

M.K. Ghosh, A. Aropostathis, and S.I. Marcus, Optimal control of switching diffusions with application to flexible manufacturing systems, *SIAM J. Contr. Optim.*, vol. 31, pp. 1183-1204, (1993).

X. Guo, *Inside Information and Stock Fluctuations*, Ph.D. thesis, Rutgers University, 1999.

J. D. Hamilton, A new approach to the economic analysis of non-stationary time series, *Econometrica*, vol. 57, pp. 357-384, (1989).

M. R. Hardy, A regime-switching model of long-term stock returns, *North American Actuarial Journal*, vol. 5, pp. 41-53, (2001).

J. C. Hull, *Options, Futures, and Other Derivatives*, 4th Ed., Prentice Hall, Upper Saddle River, NJ, 2000.

J. C. Hull and A. White, The pricing of options on assets with stochastic volatilities, *Journal of Finance,* vol. 42, pp. 281-300, (1987).

I. Karatzas and S. E. Shreve, *Methods of Mathematical Finance*, Springer, New York, 1998.

R. C. Merton, Option pricing when underlying stock returns are discontinuous, *Journal of Financial Economics*, vol. 3, pp. 125-144, (1976).

M. Musiela and M. Rutkowski, *Martingale Methods in Financial Modeling*, Springer, New York, 1997.

E. Renault and N. Touzi, Option hedging and implied volatilities in a stochastic volatility model, *Mathematical Finance*, vol. 6, pp. 279-302, (1996).

A. Quarteroni and A. Valli, *Domain Decomposition Methods for Partial Differential Equations*, Oxford Science Publications 1999.

L. Shepp, A model for stock price fluctuations based on information, *IEEE Transactions on Information Theory*, vol. 48, pp.1372-1378, (2002).

G. Yin and Q. Zhang, *Continuous-Time Markov Chains and Applications: A Singular Perturbation Approach*, Springer-Verlag, New York, 1998.

G. Yin, Q. Zhang, and K. Yin, Constrained stochastic estimation algorithms for a class of hybrid stock market models, *J. Optim. Theory Appl.*, vol. 118, No. 1, pp. 157-182, (2003).

J. Yong and X.Y. Zhou, *Stochastic Controls: Hamiltonian Systems and HJB Equations*, Springer-Verlag, New York, 1999.

Q. Zhang, Stock trading: An optimal selling rule, *SIAM J. Contr. Optim.*, vol. 40, pp. 64-87, (2001).

Q. Zhang and G. Yin, Nearly optimal asset allocation in hybrid stock-investment models, *J. Optim. Theory Appl.*, vol. 121, pp. 419-444, (2004).

X.Y. Zhou and G. Yin, Markowitz's mean-variance portfolio selection with regime switching: A continuous-time model, *SIAM Journal on Control and Optimization*, vol. 42, pp. 1466-1482, (2003).

Chapter 15

PRICING AMERICAN PUT OPTIONS USING STOCHASTIC OPTIMIZATION METHODS

G. Yin

Department of Mathematics
Wayne State University
Detroit, MI 48202

gyin@math.wayne.edu

J.W. Wang

Asset Finance Group-New Products
Moodys Investors Service
99 Church St., New York, NY 10007

jianwu.wang@moodys.com

Q. Zhang

Department of Mathematics
Boyd GSRC, The University of Georgia
Athens, GA 30602-7403

qingz@math.uga.edu

Y.J. Liu

Department of Mathematics
Missouri Southern State University
Joplin, MO 64801-1595

liu-y@mssu.edu

R.H. Liu

Department of Mathematics
University of Dayton
300 College Park, Dayton, OH 45469-2316

ruihua.liu@notes.udayton.edu

Abstract This work develops a class of stochastic optimization algorithms for pricing American put options. The stock model is a regime-switching geometric Brownian motion. The switching process represents macro market states such as market trends, interest rates, etc. The solutions of pricing American options may be characterized by certain threshold values. Here, we show how one can use a stochastic approximation (SA) method to determine the optimal threshold levels. For option pricing in a finite horizon, a SA procedure is carried for a fixed time T. As T varies, the optimal threshold values obtained using stochastic approximation trace out a curve, called the threshold frontier. Convergence and rates of convergence are obtained using weak convergence methods and martingale averaging techniques. The proposed approach provides us with a viable computational approach, and has advantage in terms of the reduced computational complexity compared with the variational or quasi-variational inequality approach for optimal stopping.

Keywords: Stochastic approximation, stochastic optimization, weak convergence, geometric Brownian motion, regime switching, American put option.

1. Introduction

This paper complements Yin, Wang, Zhang and Liu (2005), in which we proposed a class of recursive stochastic approximation algorithms for pricing American put options. Asymptotic properties of the algorithms such as convergence and rates of convergence were stated together with simulation results. Due to page limitation, the proofs of the results were omitted in the aforementioned paper, however. Here, we provide the detailed proofs and developments of the results. The stock model is a hybrid switching diffusion process, in which a number of diffusions are modulated by a finite-state Markov chain. The premise of this model is that the financial markets are sometimes quite calm and at other times much more volatile. To describe the volatility changes over time, we use a Markov chain to capture discrete shifts such as market trends and interest rates etc. For example, a two-state Markov chain can be used to characterize the up and down trends of a market. It has been well recognized that the market volatility has a close correlation with market trends (e.g., volatility associated with a bear market is much greater than that of a bull market). In the finance literature, the coexistence of continuous dynamics and discrete events is also referred to as regime switching. There is a substantial research devoted to such models. For example, regime-switching time series was treated in Hamilton (1989); American options were considered in Barone-Adesi and Whaley (1987); mean-variance hedging of European options was studied in Di Masi, Kabanov and Runggaldier (1994). A successive approximation scheme for pricing European options was developed in Yao, Zhang and Zhou (2006), in which the analytic solution was derived and moreover, a regime-switching model was shown to generate the desired volatility smile and term structure. In addition to option pricing, regime-switching models have also been used for such problems as optimal stock selling rules (Zhang (2001), Yin, Liu and Zhang (2002)), portfolio management (Zhang and Yin (2004)), and dynamic Markowitz's problems (Zhou and Yin (2003)).

We remark that there has been extensive effort for extending classical geometric Brownian motion models. Additive Poisson jumps were introduced in Merton (1969) together with the geometric Brownian motion for capturing discontinuities in the price trajectories. Stochastic volatility models were introduced in Hull and White (1987), in which it was proven that the European option can be priced as the expected value of the Black-Scholes price with respect to the distribution of the stochastic volatility when the volatility is uncorrelated with the asset price. These models were revisited in Fouque, Papanicolaou, and Sircar (2000) using a singular perturbation approach.

For Monte Carlo methods used in financial engineering, see Glasserman (2003) and the references therein. Recently, perpetual American options were treated in Guo and Zhang (2004), where it was shown that in many cases the optimal solutions can be represented by threshold levels (see also Buffington and Elliott (2002)), each of which corresponds to a given system mode. Although the solution provides insight and eases difficulty, the computational effort for solving the associated system of equations proved to be extensive, especially for Markov chains with many states, in which closed-form solution may be virtually impossible to obtain. Thus it is of practical value to seek alternatives.

Our work is motivated by the results in Guo and Zhang (2004). Here we consider a more challenging problem–pricing American options with finite expiration time in which the modulating Markov chain has possibly more than two states. We reformulate the optimal stopping problem as a stochastic optimization problem (in the sense of Kiefer and Wolfowitz; see Kushner and Yin (2003)) with the objective of finding the optimum within the class of threshold-value-dependent stopping rules. Using gradient estimates and stochastic approximation methods, we carry out decision making task, construct recursive algorithms, and establish their convergence and rates of convergence. For recent work on using stochastic optimization type algorithms without Markovian switching, see Fu, Wu, Gürkan and Demir (2000) and Fu, Laprise, Madan, Su and Wu (2001) and references therein.

The premise of our approach is to concentrate on the class of stopping rules depending on some threshold values. We make no attempt to solve the corresponding variational inequalities or partial differential equations, but rather treat the underlying problem perimetrically. For option pricing in a finite horizon, we develop procedures with a fixed expiration time T and obtain recursive estimates for the optimal threshold value associated with this fixed T. By varying T within a certain range, we obtain a curve of the threshold points as a function of T. We call this curve a *threshold frontier*. For demonstration purpose, we provide a simple example for illustration in the numerical section. However, as far as the development of the stochastic approximation algorithm is concerned, it suffices to work with a fixed T, so we focus on such a case.

Owing to the appearance of the Markov chain, the stochastic approximation algorithm is not of the usual form. It may be viewed as a Markov modulated stochastic function optimization problem, for which care must be taken. We demonstrate that the stochastic approximation/optimization approach provides an efficient and systematic computation scheme. In the proposed algorithm, since the noise varies much faster than that of the parameter, certain averaging takes place and the noise is averaged out resulting in a projected ordinary differential equation whose stationary point is the optimum we are searching for. After establishing the convergence of the algorithm, we reveal how a suitably scaled and centered estimation error sequence evolve dynamically. It is shown that a stochastic differential equation is obtained via martingale averaging techniques under proper normalization. The scaling factor together with the stationary covariance of this diffusion process gives us the rates of convergence. For recent development and up-to-dated account on stochastic approximation methods, we refer the reader to Kushner and Yin (2003) and the references therein.

The rest of the paper is arranged as follows. Section 2 begins with the precise formulation of the problem. Focusing on the class of stopping times depending on threshold values, we use stochastic approximation methods to resolve the decision making problem by searching for the optimal threshold values. Finite-difference type gradient estimates are designed in conjunction with stochastic approximation algorithms. Projections are used to ensure the iterates to remain in a bounded domain together with the constraints required by the threshold values. Although the recursive formula is in discrete time, it depends on the values of a continuous-time Markov chain describing the regime changes. Thus it may be considered as a "mixed-time" formulation. Section 3 proceeds with the convergence and rate of convergence analysis, in which weak convergence methods are used. Further remarks are made in Section 4.

Finally, the proofs of results are relegated to Section 5 for preserving the flow of presentation.

2. Formulation

Hybrid Geometric Brownian Motion Model

Suppose that $\alpha(t)$ is a finite-state, continuous-time Markov chain with state space $\mathcal{M} = \{1, \ldots, m\}$, which represents market trends and other economic factors. As a simple example, when $m = 2$, $\alpha(t) = 1$ denotes a bullish market, whereas $\alpha(t) = 2$ represents a bearish market. We adopt the risk-neutral valuation setup. In fact, beginning with a regime-switching model, one may derive a suitable probability space upon which the expected rate of return of all securities is equal to the risk-free interest rate; see Yao, Zhang and Zhou (2006).

Let $S(t)$ be the stock price at time t. We consider a hybrid geometric Brownian motion model (or geometric Brownian motion with regime switching). Given a finite horizon $T > 0$, suppose that $S(t)$ satisfies the stochastic differential equation

$$(1) \qquad \frac{dS(t)}{S(t)} = \mu dt + \sigma(\alpha(t)) dw(t), \quad 0 \le t \le T, \quad S(0) = S_0,$$

where μ is the risk-free interest rate under a risk-neutral setup, $\sigma(i)$ is the volatility when the Markov chain $\alpha(t)$ takes the value i, and $w(\cdot)$ is a real-valued standard Brownian motion that is independent of $\alpha(\cdot)$. Note that in (1), the volatility depends on the Markov chain $\alpha(t)$. Define another process

$$(2) \qquad X(t) = \int_0^t r(\alpha(s)) ds + \int_0^t \sigma(\alpha(s)) dw(s),$$

where

$$(3) \qquad r(i) = \mu - \frac{\sigma^2(i)}{2} \quad \text{for each } i = 1, \ldots, m.$$

Using $X(t)$, we can write the solution of (1) as

$$(4) \qquad S(t) = S_0 \exp(X(t)).$$

Let $\mathcal{F}_t$ be the σ-algebra generated by $\{w(s), \alpha(s) : s \le t\}$ and $\mathcal{A}_T$ be the class of $\mathcal{F}_t$-stopping times that are bounded by T, i.e., $\mathcal{A}_T = \{\tau : \tau$ is an $\mathcal{F}_t$-stopping time and $\tau \le T$ w.p.1$\}$. Consider the American put option with strike price K and a fixed expiration time T. The objective is to find the maximum value of a discounted payoff over a class of stopping times. The value (optimal payoff) function takes the form: for each $i \in \mathcal{M}$,

(5) $v(S_0, i) = \sup_{\tau \in \mathcal{A}_T} E[\exp(-\mu\tau)(K - S(\tau))^+ | S(0) = S_0, \alpha(0) = i].$

In Guo and Zhang (2004), assuming that the modulating Markov chain has only two states, perpetual American put options are considered. It is shown that the optimal stopping rule is given in terms of two threshold levels (θ^1, θ^2) such that the optimal exit time is

(6) $\tau^* = \inf\{t \geq 0 : (S(t), \alpha(t)) \notin D^*\}$

with $D^* = \{(\theta^{1,*}, \infty) \times \{1\}\} \cup \{(\theta^{2,*}, \infty) \times \{2\}\}$, and that the threshold pair $(\theta^{1,*}, \theta^{2,*})$ can be obtained by solving a set of algebraic equations.

In general, the situation is much more involved and difficult when we treat American put options with a finite expiration date and allow the state space $\mathcal{M}$ to have more than two elements. It is shown in Buffington and Elliott (2002) that the continuation regions under the optimal stopping setting can be written in terms of threshold pairs $(\theta^{1,*}, \theta^{2,*})$. In this paper, we consider the case when the optimal stopping rule is of the threshold form (6). We propose an alternative approach using stochastic approximation method aiming to provide a systematic treatment for more general situation for a finite T. At any finite time T, we obtain an approximation to the optimal threshold value. For different T, we obtain a collection of associated optimal threshold estimates. The trajectory of threshold points, as a function of T, will be defined as a threshold frontier in what follows.

A Stochastic Approximation Approach

Owing to the dependence of optimal solution on the threshold values in connection with option pricing, we focus on a class of stopping times, which depend on a vector-valued parameter θ. The problem is converted to a stochastic approximation problem. The basic premise stems from a twist of the optimal stopping rules. The rational is to concentrate on the class of stopping times depending on threshold values in lieu of finding the optimal stopping time among all stopping rules. It will be seen that such an approach provides us with a viable computational approach, and has distinct advantage in terms of the reduced computational complexity compared with the optimal stopping approach. Within the class of threshold-type solutions, let τ be a stopping time depending on θ defined by
(7) $\tau = \tau(\theta) = \inf\{t > 0 : (X(t), \alpha(t)) \notin D(\theta)\} \wedge T,$

where $\theta = (\theta^1, \ldots, \theta^m)' \in \mathbb{R}^{m \times 1}$ with

$$(8) \qquad D(\theta) = \{(\theta^1, \infty) \times \{1\}\} \cup \cdots \cup \{(\theta^m, \infty) \times \{m\}\}.$$

We aim at finding the optimal threshold level $\theta_* = (\theta_*^1, \ldots, \theta_*^m)'$ so that the expected return is maximized. The problem can be rewritten as:

$$\text{Problem } \mathcal{P} : \begin{cases} \text{Find argmax } \varphi(\theta), \\ \varphi(\theta) = E\left\{\exp(-\mu\tau(\theta))(K - S(\tau(\theta)))^+\right\}. \end{cases}$$

Note that the expectation above depends on the initial Markov state α, so does the objective function. For notational simplicity, we suppress the α dependence in what follows.

SA for Fixed T**.** Consider the problem of pricing a finite horizon American put option. For each fixed $T > 0$, we develop a stochastic recursive procedure to resolve the problem by constructing a sequence of estimates of the optimal threshold value θ_* using

$$\theta_{n+1} = \theta_n + \varepsilon_n\{\text{noisy gradient estimate of } \varphi(\theta_n)\},$$

where $\{\varepsilon_n\}$, representing the step size of the algorithm, is a sequence of real numbers satisfying $\varepsilon_n > 0$, $\varepsilon_n \to 0$, and $\sum_n \varepsilon_n = \infty$.

Threshold Trajectory. For different T, we obtain a collection of associated optimal threshold estimates. The trajectory of threshold points is a function of T, which we call it a threshold frontier or threshold trajectory. As far as the stochastic recursions are concerned, the algorithms are the same for each T. Thus, in what follows, we will concentrate on a fixed T. Nevertheless, in the numerical demonstration, we will depict the threshold evolution with respect to the time T.

Recursive Algorithms

We use a simple noisy finite-difference scheme for the gradient estimates of $\varphi(\theta)$. Note (1) and use $X(t)$ given by (2). The algorithm can be outlined as follows:

1 Initialization: Choose an arbitrary θ_0.

2 Computer θ_1

- Determine $\tau(\theta_0)$.
- Construct gradient estimate.
- Carry out one step SA to get θ_1.

3 Iteration: Assuming θ_n has been constructed, repeat Step 2 above with θ_1 replaced by θ_{n+1} and θ_0 replaced by θ_n.

Now, we provide more detailed description of the algorithm as follows. Initially, choose an arbitrary estimate $\theta_0 = (\theta_0^1, \ldots, \theta_0^m)' \in \mathbb{R}^{m \times 1}$. Determine $\tau(\theta_0)$, the first time that $(X(t), \alpha(t))$ escapes from $D(\theta_0)$ defined in (8). That is,

$$\tau(\theta_0) = \inf\{t > 0 : (X(t), \alpha(t)) \notin D(\theta_0)\} \wedge T.$$

Depending on if we use simulation (or observe the market data), $\exp(-\mu\tau(\theta_0))(K - S(\tau(\theta_0)))^+$ can be simulated (or observed) through

$$\widehat{O}(\theta_0, \widetilde{\xi}_0) = \widetilde{\varphi}(\theta_0) + \chi(\theta_0, \widetilde{\xi}_0), \quad \text{with}$$
$$\widetilde{\varphi}(\theta_0) = \exp(-\mu\tau(\theta_0))(K - S(\tau(\theta_0)))^+,$$

where $\chi(\theta_0, \widetilde{\xi}_0)$ is the simulation error (or observation noise). Here ξ_0 (and in what follows, ξ_n) is a combined process that includes the random effects from $X(t)$ and the stopping time at designated stopping time. [In what follows, we call $\{\xi_n\}$ the sequence of collective noise or simply refer to it as the noise throughout the rest of the paper.] Construct the difference quotient

$$(D\widehat{\varphi}_0)^i$$
$$= \frac{\widehat{O}(\theta_0^1, \ldots, \theta_0^i + \delta_0, \ldots, \theta_0^m, \xi_{0,i}^+) - \widehat{O}(\theta_0^1, \ldots, \theta_0^i - \delta_0, \ldots, \theta_0^m, \xi_{0,i}^-)}{2\delta_0},$$

where $\xi_{0,i}^{\pm}$ are two different noise vectors used, and $\{\delta_n\}$ is a sequence of real numbers satisfying $\delta_n \geq 0$ and $\delta_n \to 0$ in a suitable way (e.g., if we choose $\varepsilon_n = 1/(n+1)$, we can use $\delta_n = 1/(n+1)^{1/6}$; see Kushner and Yin (2003)). Now, $(D\widehat{\varphi}_0)^i I_{\{\alpha(\tau(\theta_0))=i\}}$ denotes the ith component of the gradient estimate. With the θ_0 and the above gradient estimate, we compute $\theta_1 = (\theta_1^1, \theta_1^2, \ldots, \theta_1^m)' \in \mathbb{R}^{m \times 1}$ according to

$$\theta_1^i = \theta_0^i + \varepsilon_0(D\widehat{\varphi}_0)^i I_{\{\alpha(\tau(\theta_0))=i\}},$$

where I_A is the indicator of the set A.

Suppose that θ_n has been computed. Choose

$$\tau(\theta_n) = \inf\{t \geq 0 : (X(t), \alpha(t)) \notin D(\theta_n)\} \wedge T,$$

and observe

$$\widehat{O}(\theta_n, \widetilde{\xi}_n) = \widetilde{\varphi}(\theta_n) + \chi(\theta_n, \widetilde{\xi}_n), \quad \text{where}$$
$$\widetilde{\varphi}(\theta_n) = \exp(-\mu\tau(\theta_n))(K - S(\tau(\theta_n)))^+,$$

and $\chi(\theta_n, \tilde{\xi}_n)$ is the simulation error or observation noise. Construct

$$(D\widehat{\varphi}_n)^i$$
$$= \frac{\widehat{O}(\theta_n^1, \ldots, \theta_n^i + \delta_n, \ldots, \theta_n^m, \xi_{n,i}^+) - \widehat{O}(\theta_n^1, \ldots, \theta_n^i - \delta_n, \ldots, \theta_n^m, \xi_{n,i}^-)}{2\delta_n}.$$

Then the stochastic approximation algorithm takes the form

$$(9) \qquad \theta_{n+1}^i = \theta_n^i + \varepsilon_n (D\widehat{\varphi}_n)^i I_{\{\alpha(\tau(\theta_n))=i\}}, \quad \text{for} \quad i = 1, \ldots, m.$$

Before proceeding further, we would like to comment on the above algorithm. Owing to the presence of the Markov chain, (9) is not a standard stochastic approximation algorithm. Care must be taken to deal with the added complexity.

To ensure the boundedness of the iterates, we use a projection algorithm and write

$$(10) \quad \theta_{n+1}^i = \Pi_{[\theta_l^i, \theta_u^i]}[\theta_n^i + \varepsilon_n (D\widehat{\varphi}_n)^i I_{\{\alpha(\tau(\theta_n))=i\}}], \quad \text{for} \quad i = 1, 2, \ldots, m,$$

where for each real-valued θ^i,

$$\Pi_{[\theta_l^i, \theta_u^i]}\theta^i = \begin{cases} \theta_l^i, & \text{if } \theta^i < \theta_l^i, \\ \theta_u^i, & \text{if } \theta^i > \theta_u^i, \\ \theta^i, & \text{otherwise.} \end{cases}$$

The projection can be explained as follows. For component i, after $\theta_n^i + \varepsilon_n (D\widehat{\varphi}_n)^i I_{\{\alpha(\tau(\theta_n))=i\}}$ is computed, compare its value with the bounds θ_l^i and θ_u^i. If the update is smaller than the lower value θ_l^i, reset the value to θ_l^i, if it is larger than the upper value θ_u^i, reset its value to θ_u^i, otherwise keep its value as it was.

3.   Asymptotic Properties

This section presents asymptotic properties of the recursive algorithm. It consists of two parts, convergence and rates of convergence. To analyze the recursive algorithm proposed in the last section, we use weak convergence method and martingale averaging. The results are presented, and the proofs are deferred to Section 5.

Convergence

The basic idea lies in using an approach, known as ODE method (see Kushner and Yin (2003)) in the literature, to connect discrete-time iterates with continuous-time dynamic systems. Instead of dealing with the discrete iterates directly, we take a continuous-time interpolation and examine its asymptotic properties. To be more specific, let

$$(11) \qquad t_n = \sum_{j=0}^{n-1} \varepsilon_j, \quad \text{and} \quad m(t) = \max\{n : t_n \leq t\},$$

and define

$$\theta^0(t) = \theta_n \quad \text{for} \quad t \in [t_n, t_{n+1}), \quad \text{and} \quad \theta^n(t) = \theta^0(t + t_n).$$

Thus $\theta^0(\cdot)$ is a piecewise constant process and $\theta^n(\cdot)$ is its shift. The shift is used to bring the asymptotics to the foreground. It is readily seen that $\theta^n(\cdot)$ lives in $D([0, T] : \mathbb{R}^m)$, the space of functions defined on $[0, T]$ taking values in $\mathbb{R}^m$ such that the functions are right continuous, have left limits, and are endowed with the Skorohod topology (see Kushner and Yin (2003)).

To proceed, introduce the notation

$$D\widehat{\varphi}_n = ((D\widehat{\varphi}_n)^1, \ldots, (D\widehat{\varphi}_n)^m)' \in \mathbb{R}^{m \times 1}.$$

Then (10) can be written as

$$(12) \qquad \theta_{n+1} = \Pi[\theta_n + \varepsilon_n 1_{\alpha(\tau(\theta_n))} D\widehat{\varphi}_n].$$

Moreover, (12) can be further written as

$$(13) \qquad \theta_{n+1} = \theta_n + \varepsilon_n 1_{\alpha(\tau(\theta_n))} D\widehat{\varphi}_n + \varepsilon_n z_n,$$

where $\varepsilon_n z_n = \theta_{n+1} - \theta_n - \varepsilon_n 1_{\alpha(\tau(\theta_n))} D\widehat{\varphi}_n$ is known as a "reflection" term, the minimal force needed to bring the iterates back to the projection region if they ever escape from the constraint set (18). Using the notation in (13), it is readily seen that

$$(14) \qquad \theta^n(t) = \theta_n + \sum_{j=n}^{m(t_n+t)-1} \varepsilon_j 1_{\alpha(\tau(\theta_j))} D\widehat{\varphi}_j + \sum_{j=n}^{m(t_n+t)-1} \varepsilon_j z_j.$$

Set

$$(15) \qquad Z^n(t) = \sum_{j=n}^{m(t_n+t)-1} \varepsilon_j z_j.$$

In analyzing stochastic recursive algorithms, one often wishes to separate the effect of bias and noise. This can be done as follows:

$$b_n^i = \frac{\widetilde{\varphi}(\theta_n^1, \ldots, \theta_n^i + \delta_n, \ldots, \theta_n^m) - \widetilde{\varphi}(\theta_n^1, \ldots, \theta_n^i - \delta_n, \ldots, \theta_n^m)}{2\delta_n}$$

$$\times I_{\alpha(\tau(\theta_n))=i\}} - \varphi_{\theta^i}(\theta_n),$$

$$b_n = (b_n^1, \ldots, b_n^m)',$$

$$\psi^i(\theta, \xi_{n,i}) = \chi(\theta, \xi_{n,i}^+) - \chi(\theta, \xi_{n,i}^-),$$

$$\psi(\theta_n, \xi_n) = (\psi^1(\theta_n, \xi_{n,1}), \ldots, \psi^m(\theta_n, \xi_{n,m}))',$$

where $\varphi_{\theta^i}(\theta) = (\partial/\partial\theta^i)\varphi(\theta)$. In the above, b_n is known as a bias term and $\psi(\theta, \xi_n)$ as a "noise" term. Note that in fact b_n is also θ_n dependent. For future use, using b_n and $\psi(\cdot)$, we may also write the recursive formula in an expansive form as

$$(16) \quad \theta_{n+1}^i = \Pi_{[\theta_l^i, \theta_u^i]}[\theta_n^i + \varepsilon_n[\varphi_{\theta^i}(\theta_n) + b_n^i + \frac{\psi^i(\theta_n, \xi_{n,i})}{2\delta_n} I_{\{\alpha(\tau(\theta_n))=i\}}]],$$

or its equivalent vector notation

$$(17) \quad \theta_{n+1} = \Pi\left[\theta_n + \varepsilon_n[\varphi_\theta(\theta_n) + b_n + 1_{\alpha(\tau(\theta_n))} \frac{\psi(\theta_n, \xi_n)}{2\delta_n}]\right],$$

where Π is the projection onto the box constraint set

$$(18) \qquad [\theta_l^1, \theta_u^1] \times [\theta_l^2, \theta_u^2] \times \cdots \times [\theta_l^m, \theta_u^m],$$

$$1_{\alpha(\tau(\theta))} = \mathrm{diag}(I_{\{\alpha(\tau(\theta))=1\}}, \ldots, I_{\{\alpha(\tau(\theta))=m\}}) \in \mathbb{R}^{m \times m},$$

and $\mathrm{diag}(A_1, \ldots, A_l)$ denotes a block diagonal matrix such that each A_i has an appropriate dimension. Using an expansive form (17), we have

$$(19) \quad \theta_{n+1} = \theta_n + \varepsilon_n\left[\varphi_\theta(\theta_n) + b_n + 1_{\alpha(\tau(\theta_n))}\psi(\theta_n, \xi_n)\right] + \varepsilon_n z_n.$$

To establish the convergence of the algorithm, We will use the following conditions.

(A1) The sequences $\{\varepsilon_n\}$ and $\{\delta_n\}$ are chosen so that $0 < \delta_n$, $\varepsilon_n \to 0$, $\varepsilon_n/\delta_n^2 \to 0$, as $n \to \infty$, and that $\sum_n \varepsilon_n = \infty$. Moreover,

$$\limsup_n (\varepsilon_{n+k}/\varepsilon_n) < \infty, \quad \limsup_n (\delta_{n+k}/\delta_n) < \infty,$$

$$\limsup_n [(\varepsilon_{n+k}/\delta_{n+k}^2)/(\varepsilon_n/\delta_n^2)] < \infty.$$

In addition, there is a sequence of positive integers $\{\gamma_n\}$ satisfying $\gamma_n \to \infty$ sufficiently slowly such that

(20)
$$\sup_{0 \le i \le \gamma_n} \left| \frac{\varepsilon_{n+i}}{\varepsilon_n} - 1 \right| \to 0 \ \text{ as } \ n \to \infty.$$

(A2) For each ξ, the function $\psi(\cdot, \xi)$ is continuous; for each $i = 1, \ldots, m$, the sequences $\{\xi_{n,i}^\pm\}$ are stationary such that

$$E|\chi(\theta, \xi_{n,i}^\pm)|^{2+\gamma} < \infty \text{ for some } \gamma > 0 \text{ and } E\chi(\theta, \xi_{n,i}^\pm) = 0 \text{ for each } \theta.$$

Moreover, as $k \to \infty$,

(21)
$$\frac{1}{k} \sum_{j=\ell}^{\ell+k} E_\ell \frac{\psi(\theta, \xi_j)}{2\delta_j} \to 0 \ \text{ in probability,}$$

where E_ℓ denotes the conditional expectation with respect to $\mathcal{F}_\ell$, the σ-algebra generated by $\{\alpha(u), u \le t_\ell; \theta_0, \xi_{j,i}^\pm : j < \ell, \ i = 1, \ldots, m\}$.

REMARK 3.1 Condition (A1) is for convenience. It is not a restriction since we can choose $\{\delta_n\}$ and $\{\varepsilon_n\}$ at our will. For example, we may choose $\varepsilon_n = O(1/n)$ and $\delta_n = O(1/n^{1/6})$; in this case, (A1) is readily verified. In view of the definition of $\psi(\theta, \xi_n)$, it is easily seen that $E|\psi(\theta, \xi_n)|^{2+\gamma} < \infty$ and $E\psi(\theta, \xi_n) = 0$ for each θ. Note that (21) is an averaging condition of law of large numbers type. In fact, we only require the law of large numbers hold in the sense of in probability. It indicates that the observation noise is averaged out. In the simulation, one often uses uncorrelated sequences. In such a case, the averaging condition is readily verified. If $\{\xi_n\}$ is a sequence of ϕ-mixing processes (see Billingsley (1968)), which indicates the remote past and distant future being asymptotically independently, then (21) is also verified. The continuity of $\psi(\cdot, \xi)$ is for convenience. In fact, only weak continuity (continuity in the sense of in expectation) is needed. Thus, indicator type of functions can be treated. To proceed, we state the convergence result next.

THEOREM 3.2. *Assume* (A1) *and* (A2). *Suppose, in addition, the differential equation*
(22)
$$\dot{\theta}(t) = \varphi_\theta(\theta(t)) + z(t)$$

has a unique solution for each initial condition $\theta(0)$. *Then* $\theta^n(\cdot)$ *converges weakly to* $\theta(\cdot)$ *such that* $\theta(\cdot)$ *is a solution of* (22).

COROLLARY 3.3. *Under the conditions of Theorem 3.2, suppose that θ_* is a unique asymptotically stable point of (22) being interior to the constraint set and that $\{\widehat{T}_n\}$ is a sequence of real numbers satisfying $\widehat{T}_n \to \infty$ as $n \to \infty$. Then $\theta^n(\widehat{T}_n + \cdot)$ converges weakly to θ_*.*

REMARK 3.4. Note that (22) is known as a projected ODE; see Chapter 4 of Kushner and Yin (2003). The term $z(t)$ is due to the reflection or projection constraint. To prove Theorem 3.2, our plan is as follows. We first show that $\{\theta^n(\cdot)\}$ is tight and then we characterize the limit process by use of the martingale averaging techniques. To prove Theorem 3.2, we proceed by establishing a sequence of lemmas.

Tightness of $\theta^n(\cdot)$.

LEMMA 3.5. *Under the conditions of Theorem 3.2, $\theta^n(\cdot)$ is tight in $D([0,T] : \mathbb{R}^m)$.*

In fact, we obtain the tightness of a pair of processes $(\theta^n(\cdot), Z^n(\cdot))$. Therefore, by Prohorov's theorem, we can extract a weakly convergent subsequence. Select such a subsequence and still denote it by $(\theta^n(\cdot), Z^n(\cdot))$ without loss of generality and for notational simplicity. Denote the limit by $(\theta(\cdot), Z(\cdot))$. By the Skorohod representation (without changing notation), we may assume that $(\theta^n(\cdot), Z^n(\cdot)) \to (\theta(\cdot), Z(\cdot))$ w.p.1 and the convergence is uniform on any compact time interval.

ODE Limit. To characterize the limit process, choose a sequence of positive integers $\{\beta_n\}$ satisfying $\beta_n \le \gamma_n$ (with γ_n given in (A1)) and $\beta_n \to \infty$. For $j \ge n$, define

$$(23) \qquad q_j^n = \sum_{i=n}^{j-1} \beta_i, \quad \text{and} \quad t_j^n = \sum_{i=n}^{n+q_j^n-1} \varepsilon_i.$$

Using the sequence t_j^n defined in (23), in lieu of (14), $Z^n(t)$ may be rewritten as

$$Z^n(t) = \sum_{j: t_j^n < t} \varepsilon_j z_j.$$

Owing to (23), $\beta_j = q_{j+1}^n - q_j^n$, and

$$\Delta_j^n = t_{j+1}^n - t_j^n = \sum_{i=n+q_j^n}^{n+q_{j+1}^n-1} \varepsilon_i$$

$$= \sum_{i=n+q_j^n}^{n+q_{j+1}^n-1} \left(\frac{\varepsilon_i}{\varepsilon_{n+q_j^n}} - 1 \right) \varepsilon_{n+q_j^n} + \sum_{i=n+q_j^n}^{n+q_{j+1}^n-1} \varepsilon_{n+q_j^n}.$$

By virtue of (A1) and (23),

$$(24) \qquad \beta_j \varepsilon_{n+q_j^n} = \Delta_j^n + \eta(j,n),$$

where $\eta(j,n) \to 0$ as $j, n \to \infty$. In other words, we can approximate $\beta_j \varepsilon_{n+q_j^n}$ by Δ_j^n with an approximation error going to 0 as $j, n \to \infty$.

Using the approximation (24), in view of the continuous-time interpolation, we have

$$
\theta^n(t+s) - \theta^n(t)
$$

$$
= \sum_{j:t_j^n \in [t,t+s)} (\theta_{j+1} - \theta_j)
$$

$$
= \sum_{j:t_j^n \in [t,t+s)} \varepsilon_j 1_{\alpha(\tau(\theta_j))} D\widehat{\varphi}_j + \sum_{j:t_j^n \in [t,t+s)} \varepsilon_j z_j
$$

$$
= \sum_{j:t_j^n \in [t,t+s)} \sum_{i=n+q_j^n}^{n+q_{j+1}^n-1} \varepsilon_i 1_{\alpha(\tau(\theta_i))} D\widehat{\varphi}_i + \sum_{j:t_j^n \in [t,t+s)} \sum_{i=n+q_j^n}^{n+q_{j+1}^n-1} \varepsilon_i z_i
$$

$$
= \sum_{j:t_j^n \in [t,t+s)} \beta_j \varepsilon_{n+q_j^n} \frac{1}{\beta_j} \sum_{i=n+q_j^n}^{n+q_{j+1}^n-1} 1_{\alpha(\tau(\theta_i))} D\widehat{\varphi}_i
$$

$$
+ \sum_{j:t_j^n \in [t,t+s)} \beta_j \varepsilon_{n+q_j^n} \frac{1}{\beta_j} \sum_{i=n+q_j^n}^{n+q_{j+1}^n-1} z_i
$$

$$
+ \sum_{j:t_j^n \in [t,t+s)} \frac{1}{\beta_j} \sum_{i=n+q_j^n}^{n+q_{j+1}^n-1} [\beta_j(\varepsilon_i - \varepsilon_{n+q_j^n})] 1_{\alpha(\tau(\theta_i))} D\varphi_i
$$

$$
+ \sum_{j:t_j^n \in [t,t+s)} \frac{1}{\beta_j} \sum_{i=n+q_j^n}^{n+q_{j+1}^n-1} [\beta_j(\varepsilon_i - \varepsilon_{n+q_j^n})] z_i,
$$

as a result,

$$
\theta^n(t+s) - \theta^n(t) = \sum_{j:t_j^n \in [t,t+s)} \Delta_j^n \frac{1}{\beta_j} \sum_{i=n+q_j^n}^{n+q_{j+1}^n-1} 1_{\alpha(\tau(\theta_i))} D\widehat{\varphi}_i
$$

$$
(25)
$$

$$
+ \sum_{j:t_j^n \in [t,t+s)} \Delta_j^n \frac{1}{\beta_j} \sum_{i=n+q_j^n}^{n+q_{j+1}^n-1} z_i + o(1),
$$

where $o(1) \to 0$ in probability uniformly in t.

Compared with the usual stochastic approximation algorithms, the iterations and the interpolated processes all involve a Markov chain. To take into account of this additional Markov chain, we state two lemmas first. The first lemma is on the weak continuity of $\tau(\theta)$, and the second one reveals that the matrix-valued indicator $1_{\alpha(\tau(\theta_i))}$ can be replaced by $1_{\alpha(\tau(\theta_{n+q_j^n}))}$ with an error term goes to 0 in probability. These lemmas will help us to derive the limit results Theorem 3.2 and Corollary 3.3. The proofs are postponed to Section 5.

LEMMA 3.6.   *Assume that* (A1) *and* (A2) *hold. Then* $\tau(\theta)$ *is weakly continuous in the sense that* $\theta \to \widetilde{\theta}$ *implies* $E\tau(\theta) \to E\tau(\widetilde{\theta})$.

LEMMA 3.7.   *Assume that* (A1) *and* (A2) *hold. Then*

$$
\zeta^n = \sum_{j:t_j^n \in [t,t+s]} \Delta_j^n \widetilde{\zeta}_j^n \overset{\text{def}}{=} \sum_{j:t_j^n \in [t,t+s]} \Delta_j^n \frac{1}{\beta_j} \sum_{i=n+q_j^n}^{n+q_{j+1}^n - 1} 1_{\alpha(\tau(\theta_i))} D\widehat{\varphi}_i
$$

$$
= \sum_{j:t_j^n \in [t,t+s]} \Delta_j^n \frac{1}{\beta_j} \sum_{i=n+q_j^n}^{n+q_{j+1}^n - 1} 1_{\alpha(\tau(\theta_{n+q_j^n}))} D\widehat{\varphi}_i + o(1),
$$

where $o(1) \to 0$ *in probability uniformly in* t.

Rates of Convergence

We begin with the examination of scaled sequences of the centered estimation error $\theta_n - \theta_*$. Suppose that $\varepsilon_n = O(1/n^{\gamma_1})$, $\delta_n = O(1/n^{\gamma_2})$, with $0 < \gamma_1, \gamma_2 \le 1$. Rates of convergence of the recursive algorithm are concerned with the scaling of $n^{\gamma_3}(\theta_n - \theta_*)$, leading a nontrivial limit, and the corresponding asymptotic covariance. It turns out, γ_i satisfy $\gamma_3 - 2\gamma_2 \le 0$ and $\gamma_3 + \gamma_2 - \gamma_1/2 \le 0$, and the largest γ_3 is obtained with equality replacing the inequalities.

This subsection is divided into two parts. In the first part, using perturbed Liapunov function argument, we establish the tightness of the sequence $\{n^{(1/3)\gamma_1}(\theta_n - \theta_*)\}$. This is essentially a stability result. It reveals how fast the estimation error varies and exploits the dependence of the error $\theta_n - \theta_*$ on the iteration number n. In the second part, we further examine the scaled sequence through its continuous-time interpolation, and show that the scaled sequence converges weakly to a diffusion process. The scaling factor together with the stationary covariance of the diffusion process gives us the desired rate result.

Tightness of Scaled and Centered Iterates.   Let us assume that the following conditions are satisfied.

(A3) Suppose that there is a Liapunov function $V(\cdot) : \mathbb{R}^m \to \mathbb{R}$ such that the second order mixed partial derivatives of $V(\cdot)$ are continuous, and $V_\theta(\theta)\varphi_\theta(\theta) \leq -\lambda V(\theta)$ for some $\lambda > 0$.

(A4) The second mixed partial derivatives of $\varphi(\cdot)$ are continuous. For each ξ,

$$\psi(\theta, \xi) = \psi(\theta_*, \xi) + \int_0^1 [\psi(\theta_* + s(\theta - \theta_*), \xi) - \psi(\theta_*, \xi)]ds.$$

For each θ,

$$\sum_{j=n}^{\infty} E|E_n\psi(\theta, \xi_j)| < \infty.$$

REMARK 3.8. Condition (A3) requires the existence of a Liapunov function for the limit ODE. The precise form of $V(\cdot)$ need not be known. Condition (A4) poses certain smoothness like conditions on $\psi(\theta, \xi)$. If $\psi(\theta, \xi)$ is independent of θ, then this condition is not needed. The summability of $\sum_j E|E_n\psi(\theta, \xi_j)|$ is satisfied if $\psi(\theta, \xi_j)$ is a uniform mixing process with a summable mixing rate (see Billingsley (1968)), which is a typical assumption.

THEOREM 3.9. *Assume* (A1)–(A4). *Use* $\varepsilon_n = 1/(n+1)^{\gamma_1}$, $\delta_n = \delta/(n+1)^{\gamma_1/6}$, *and* $\gamma_3 = 1/(n+1)^{\gamma_1/3}$. *Then* $EV(\theta_n) = O(1/n^{(2/3)\gamma_1})$.

COROLLARY 3.10. *Take* $\gamma_1 = 1$. *If the Liapunov function is locally quadratic, i.e.,*

$$V(\theta) = (\theta - \theta_*)'A(\theta - \theta_*) + o(|\theta - \theta_*|^2),$$

where A is a symmetric positive definite matrix, then $\{n^{1/3}(\theta_n - \theta_*)\}$ is tight.

Diffusion Limit. This section is devoted to the limit of the scaled estimation error $\{n^{1/3}(\theta_n - \theta_*)\}$. Define $u_n = n^{1/3}(\theta_n - \theta_*)$. Since θ_* is interior to the constraint set, we may drop the term z_n without loss of generality. We will also work with interpolated process, and define

$$u^n(t) = u_{n+i} \text{ for } t \in [t_{n+i} - t_n, t_{n+i+1} - t_n) \text{ for } i \geq 0,$$

where t_n is defined in (11). In what follows, for simplicity, we work with the case $\varepsilon_n = O(1/n)$. There are also apparent analog of the cases $\varepsilon_n = O(1/n^{\gamma_1})$ although we will not dwell on it here. To proceed, we state the following condition.

(A5) The following conditions hold:

    (a) The matrix $\varphi_{\theta\theta}(\theta_*) + I/3$ is stable (i.e., all of its eigenvalues have negative real parts).

    (b) The sequence

$$\sum_{j=m(t_n)}^{m(t_n+t)} \frac{1}{j^{1/2}} \psi(\theta_*, \xi_j)$$

    converges weakly to a Brownian motion $\widetilde{w}(t)$ with covariance $\Sigma(\theta_*)t$.

    (c) For each ξ, $\psi_\theta(\cdot, \xi)$ exists and is continuous.

In view of the scaling and (A5), we have

$$u_{n+1} = \left(\frac{n+1}{n}\right)^{1/3} u_n + \left(\frac{n+1}{n}\right)^{1/3} \frac{1}{n}\Big[\varphi_{\theta\theta}(\widetilde{\theta}_{n,*})u_n + n^{1/3}b_n$$
$$+ n^{1/3}1_{\alpha(\tau(\theta_n))}\Big[\frac{\psi(\theta_*, \xi_n)}{2\delta_n} + \frac{\psi_\theta(\widehat{\theta}_{n,*}, \xi_n)}{2\delta_n}u_n\Big]\Big],$$

where $\widetilde{\theta}_{n,*}$ and $\widehat{\theta}_{n,*}$ denote points on the line segments joining θ_n and θ_*. Noting $((n+1)/n)^{1/3} = 1 + 1/(3n) + o(1/n)$, we need only consider another sequence $\{v_n\}$ defined by

$$v_{n+1} = v_n + \frac{1}{n}\left(\varphi_{\theta\theta}(\theta_*) + \frac{I}{3n}\right)v_n$$
$$+ \frac{1}{n}[\varphi_{\theta\theta}(\widetilde{\theta}_{n,*}) - \varphi_{\theta\theta}(\theta_*)]v_n + \frac{1}{n}(n^{1/3}b_n)$$
$$+ \frac{1}{n^{1/2}}1_{\alpha(\tau(\theta_*))}\psi(\theta_*, \xi_n) + \frac{1}{n^{1/2}}1_{\alpha(\tau(\theta_*))}\psi_\theta(\widehat{\theta}_{n,*}, \xi_n)v_n.$$

Define the corresponding interpolation $v^n(t)$ as in that of $u^n(\cdot)$ with the replacement of u_n by v_n. It can be shown that the $v^n(\cdot)$ so defined will have the same limit as that of $u^n(\cdot)$. In fact $b_n = b_n(\theta)$. An expansion of the term $b_n(\theta_n)$ yields that

$$n^{1/3}b_n^i = n^{1/3}b_n^i(\theta_n) = \frac{1}{3!}\varphi_{\theta^i\theta^i\theta^i}(\theta_*)\delta^2 + o(1), \quad i \in \mathcal{M}.$$

In the above, i indices the ith component. Using Lemma 3.6, it can be proved that

$$
\begin{aligned}
v^{n,i}(t+s) = v^{n,i}(t) &+ \sum_{j=m(t_n+t)}^{m(t_n+t+s)-1} \frac{1}{j} [(\varphi_{\theta\theta}(\theta_*) + \frac{I}{3})v_j]^i \\
&+ \sum_{j=m(t_n+t)}^{m(t_n+t+s)-1} \frac{1}{j} \Big[[\varphi_{\theta\theta}(\widetilde{\theta}_{n,*}) - \varphi_{\theta\theta}(\theta_*)]v_j \Big]^i \\
&+ \sum_{j=m(t_n+t)}^{m(t_n+t+s)-1} \frac{1}{3!} \frac{1}{j} \varphi_{\theta^i\theta^i\theta^i}(\theta_*)\delta^2 \\
&+ \sum_{j=m(t_n+t)}^{m(t_n+t+s)-1} \frac{1}{\sqrt{j}} [1_{\alpha(\tau(\theta_*))}\psi(\theta_*, \xi_j)]^i + o(1),
\end{aligned}
$$

where $o(1) \to 0$ in probability uniformly in t. Using an argument similar to the proof of Theorem 3.2 for this Markov modulated stochastic approximation together with the techniques in Kushner and Yin (2003), we derive the following theorem although the details are omitted.

THEOREM 3.11. *Under conditions* (A1)–(A5), $u^n(\cdot)$ *converges weakly to* $u(\cdot)$ *such that* $u(\cdot)$ *satisfies the stochastic differential equation*

$$
(26) \qquad
\begin{aligned}
du = (\varphi_{\theta\theta}(\theta_*) + I/3)u\,dt &+ \begin{pmatrix} \varphi_{\theta^1\theta^1\theta^1}(\theta_*) \\ \vdots \\ \varphi_{\theta^m\theta^m\theta^m}(\theta_*) \end{pmatrix} \frac{\delta^2}{3!} dt \\
&+ \frac{1_{\alpha(\tau(\theta_*))}\Sigma(\theta_*)}{2\delta} dw,
\end{aligned}
$$

where $w(\cdot)$ *is a standard Brownian motion.*

REMARK 3.12. Note that in Theorem 3.11, we have replaced the Brownian motion $\widetilde{w}(\cdot)$ by a standard Brownian motion together with a diffusion matrix $1_{\alpha(\tau(\theta_*))}\Sigma(\theta_*)$. If in lieu of $\varepsilon_n = O(1/n)$, $\varepsilon_n = O(1/n^{\gamma_1})$ is used with $0 < \gamma_1 < 1$, then u_n is changed to $u_n = n^{(1/3)\gamma_1}(\theta_n - \theta_*)$. Redefine $u^n(\cdot)$. Under similar conditions as in Theorem 3.11 with (A5) (a) replaced by $\varphi_{\theta\theta}(\theta_*)$ being stable, Theorem 3.11 still holds. However, (26) is changed to

$$
(27) \qquad du = \varphi_{\theta\theta}(\theta_*)u\,dt + \begin{pmatrix} \varphi_{\theta^1\theta^1\theta^1}(\theta_*) \\ \vdots \\ \varphi_{\theta^m\theta^m\theta^m}(\theta_*) \end{pmatrix} \frac{\delta^2}{3!} dt + \frac{1_{\alpha(\tau(\theta_*))}\Sigma(\theta_*)}{2\delta} dw.
$$

4.   Further Remarks

In this paper, we developed stochastic approximation algorithms for pricing American put options. The recursive algorithms enable us to determine the optimal threshold levels in a systematic way. Only finite-difference gradient estimates are considered; variants of such algorithms can be designed. Simple finite difference gradient estimates are used in this paper. To construct the gradient estimates, the so-called infinites-imal perturbation analysis in Ho and Cao (1991) can be used; see also Fu, Wu, Gurkan and Demir (2000).

5.   Proofs of Results

Proof of Lemma 3.5. For any $\delta > 0$, $t, s > 0$ with $0 < s \leq \delta$,

$$E\left|\theta^n(t+s) - \theta^n(t)\right|^2 \leq KE\left|\sum_{j=m(t_n+t)}^{m(t_n+t+s)-1} \varepsilon_j 1_{\alpha(\tau(\theta_j))} D\widehat{\varphi}_j\right|^2$$

$$+ KE\left|\sum_{j=m(t_n+t)}^{m(t_n+t+s)-1} \varepsilon_j z_j\right|^2.$$

Owing to the truncation, $\{\theta_n\}$ is bounded, so

$$\lim_{\delta \to 0}\limsup_{n \to \infty} E\left|\sum_{j=m(t_n+t)}^{m(t_n+t+s)-1} \varepsilon_j 1_{\alpha(\tau(\theta_j))} D\widehat{\varphi}_j\right|^2 = \lim_{\delta \to 0}\limsup_{n \to \infty} O(\delta^2) = 0.$$

Moreover, we also obtain

$$\lim_{\delta \to 0}\limsup_{n \to \infty} E\left|\sum_{j=m(t_n+t)}^{m(t_n+t+s)-1} \varepsilon_j z_j\right|^2$$

$$\leq K \lim_{\delta \to 0}\limsup_{n \to \infty} E\left[\sum_{j=m(t_n+t)}^{m(t_n+t+s)-1} \varepsilon_j |D\widehat{\varphi}_j|\right]^2$$

$$\leq K \lim_{\delta \to 0}\limsup_{n \to \infty} E\left[\sum_{j=m(t_n+t)}^{m(t_n+t+s)-1} \varepsilon_j |\widehat{O}(\theta_j, \widetilde{\xi}_j)|\right]^2$$

$$= \lim_{\delta \to 0}\limsup_{n \to \infty} O(\delta^2) = 0.$$

The estimate above implies that $\{Z^n(\cdot)\}$ is tight. In the above, we have used (A2) to get the boundedness of the second moment. Combining the estimates above, an application of Theorem 3 in page 47 of Kushner (1984) yields the tightness of $\{\theta^n(\cdot)\}$. The lemma is concluded. $\square$

Proof of Lemma 3.6. Suppose not. That is, $\tau(\theta)$ is not weakly continuous, namely,

$$(28) \qquad \theta - \widetilde{\theta} \to 0 \quad \text{but} \quad E\tau(\theta) - E\tau(\widetilde{\theta}) \not\to 0.$$

Without loss of generality, we may assume that $E\tau(\theta) - E\tau(\widetilde{\theta}) > 0$. By virtue of the definition of $D(\theta)$, we have $|\mathrm{diam}(D(\theta)) - \mathrm{diam}(D(\widetilde{\theta}))| \to 0$, where $\mathrm{diam}(A)$ denotes the diameter of the set A. Since

$$(X(\tau(\theta)), \alpha(\tau(\theta))) \in D(\theta) \quad \text{and} \quad (X(\tau(\widetilde{\theta})), \alpha(\tau(\widetilde{\theta}))) \in D(\widetilde{\theta}),$$

$$(29) \qquad EX(\tau(\theta)) - EX(\tau(\widetilde{\theta})) \to 0.$$

On the other hand, using (2),

$$(30) \qquad EX(\tau(\theta)) - EX(\tau(\widetilde{\theta})) = E \int_{\tau(\widetilde{\theta})}^{\tau(\theta)} r(\alpha(s)) ds.$$

Suppose that there is no jump between $\tau(\widetilde{\theta})$ and $\tau(\theta)$. Then by virtue of the piecewise constant behavior of $\alpha(t)$ and (29), it is easily shown that $|E\tau(\theta) - E\tau(\widetilde{\theta})| \to 0$, which is a contradiction to (28). If there is one jump between $\tau(\widetilde{\theta})$ and $\tau(\theta)$, denote the jump time by $\widetilde{\tau}_1$. Then $\left| E \int_{\tau(\widetilde{\theta})}^{\tau(\theta)} r(\alpha(s)) ds \right| = \left| E[r_0(\widetilde{\tau}_1 - \tau(\widetilde{\theta})) + r_1(\tau(\theta) - \widetilde{\tau}_1)] \right|$, so

$$(31) \qquad \left| E \int_{\tau(\widetilde{\theta})}^{\tau(\theta)} r(\alpha(s)) ds \right| \geq E[|r_0|(\widetilde{\tau}_1 - \tau(\widetilde{\theta})) - |r_1|(\tau(\theta) - \widetilde{\tau}_1)],$$

where r_0 and r_1 are the constant values taken by $r(\alpha(t))$ at $\tau(\widetilde{\theta})$ and $\widetilde{\tau}_1$, respectively, and where without loss of generality, we have assumed that $|r_0|E(\widetilde{\tau}_1 - \tau(\widetilde{\theta})) - |r_1|E(\tau(\theta) - \widetilde{\tau}_1) > 0$. Let $0 < c_1 < 1$ be a constant satisfying

$$(32) \qquad c_1 E|r_0|(\widetilde{\tau}_1 - \tau(\widetilde{\theta})) > E|r_1|(\tau(\theta) - \widetilde{\tau}_1).$$

Then it follows from (31),

$$(33) \qquad \left| \int_{\tau(\widetilde{\theta})}^{\tau(\theta)} r(\alpha(s)) ds \right| \geq (1 - c_1)|r_0|E(\widetilde{\tau}_1 - \tau(\widetilde{\theta})).$$

This implies that $E[\widetilde{\tau}_1 - \tau(\widetilde{\theta})] \to 0$. Using (32) again, we also have $E[\tau(\theta) - \widetilde{\tau}_1] \to 0$. Therefore,

$$E[\tau(\theta) - \tau(\widetilde{\theta})] = E[\tau(\theta) - \widetilde{\tau}_1] + E[\widetilde{\tau}_1 - \tau(\widetilde{\theta})] \to 0,$$

which is a contradiction to (28). Suppose that more than one jumps occur between $\tau(\widetilde{\theta})$ and $\tau(\theta)$. We denote the jump times by $\widetilde{\tau}_j$ with $\tau(\widetilde{\theta}) < \widetilde{\tau}_1 < \widetilde{\tau}_2 < \cdots < \widetilde{\tau}_{k_1} < \tau(\theta)$. In view of (31),

$$
\left| E \int_{\tau(\widetilde{\theta})}^{\tau(\theta)} r(\alpha(s)) ds \right|
$$

$$
\geq E \left[|r_0|(\widetilde{\tau}_1 - \tau(\widetilde{\tau})) - \left| \sum_{j=1}^{k_1-1} r_j(\widetilde{\tau}_{j+1} - \widetilde{\tau}_j) \right| - |r_{k_1}(\tau(\theta) - \widetilde{\tau}_{k_1})| \right]
$$

$$
\geq E \left[|r_0|(\widetilde{\tau}_1 - \tau(\widetilde{\tau})) - \widetilde{r}_{\max}(\widetilde{\tau}_{k_1} - \widetilde{\tau}_1) - \widetilde{r}_{\max}(\tau(\theta) - \widetilde{\tau}_{k_1}) \right]
$$

$$
\geq E \left[|r_0|(\widetilde{\tau}_1 - \tau(\widetilde{\tau})) - \widetilde{r}_{\max}(\tau(\theta) - \widetilde{\tau}_1) \right],
$$

where $\widetilde{r}_{\max} = \max_j |r_j|$ (Recall that the values of r_j belong to a finite set since $\alpha(t)$ is a finite-state Markov chain). Then using similar estimates as in the case having one jump between $\tau(\widetilde{\theta})$ and $\tau(\theta)$, we obtain $E[\tau(\theta) - \tau(\widetilde{\theta})] \to 0$, which is again a contradiction to (28). $\qquad\square$

Proof of Lemma 3.7. In view of the interpolation and the choice of Δ_j^n, to derive the desired result, it suffices to consider the term $\widetilde{\zeta}_j^n$. In fact, we have

$$
\widetilde{\zeta}_j^n = \frac{1}{\beta_j} \sum_{i=n+q_j^n}^{n+q_{j+1}^n-1} 1_{\alpha(\tau(\theta_{n+q_j^n}))} D\widehat{\varphi}_i + \frac{1}{\beta_j} \sum_{i=n+q_j^n}^{n+q_{j+1}^n-1} [1_{\alpha(\tau(\theta_i))} - 1_{\alpha(\tau(\theta_{n+q_j^n}))}] D\widehat{\varphi}_i.
$$

Examining the last term above, using the projection and the boundedness of $E^{1/2}|D\widehat{\varphi}_i|^2$ (owing to the projection and the continuity of $D\widehat{\varphi}_i$ with respect to θ), by virtue of the Cauchy-Shwartz inequality, we have

$$
E \left| \frac{1}{\beta_j} \sum_{i=n+q_j^n}^{n+q_{j+1}^n-1} [1_{\alpha(\tau(\theta_i))} - 1_{\alpha(\tau(\theta_{n+q_j^n}))}] D\widehat{\varphi}_i \right|
$$

$$
\leq \frac{1}{\beta_j} \sum_{i=n+q_j^n}^{n+q_{j+1}^n-1} E|1_{\alpha(\tau(\theta_i))} - 1_{\alpha(\tau(\theta_{n+q_j^n}))}| |D\widehat{\varphi}_i|
$$

$$
\leq \frac{K}{\beta_j} \sum_{i=n+q_j^n}^{n+q_{j+1}^n-1} E^{1/2}|1_{\alpha(\tau(\theta_i))} - 1_{\alpha(\tau(\theta_{n+q_j^n}))}|^2
$$

$$
\leq \frac{K}{\beta_j} \sum_{i=n+q_j^n}^{n+q_{j+1}^n-1} \max_{1 \leq \ell \leq m} E^{1/2}|I_{\{\alpha(\tau(\theta_i))=\ell\}} - I_{\{\alpha(\tau(\theta_{n+q_j^n}))=\ell\}}|^2.
$$

Thus, it suffices to examine

$$E[I_{\{\alpha(\tau(\theta_i))=\ell\}} - I_{\{\alpha(\tau(\theta_{n+q_j^n}))=\ell\}}]^2, \quad \text{for } \ell = 1, \ldots, m.$$

It is readily seen

$$E[I_{\{\alpha(\tau(\theta_i))=\ell\}} - I_{\{\alpha(\tau(\theta_{n+q_j^n}))=\ell\}}]^2$$
$$= P(\alpha(\tau(\theta_i)) = \ell) - 2P(\alpha(\tau(\theta_i)) = \ell, \alpha(\tau(\theta_{n+q_j^n})) = \ell)$$
$$+ P(\alpha(\tau(\theta_{n+q_j^n})) = \ell)$$
$$= [P(\alpha(\tau(\theta_i)) = \ell) - P(\alpha(\tau(\theta_{n+q_j^n})) = \ell)$$
$$\times P(\alpha(\tau(\theta_i)) = \ell | \alpha(\tau(\theta_{n+q_j^n})) = \ell)]$$
$$+ P(\alpha(\tau(\theta_{n+q_j^n})) = \ell)[1 - P(\alpha(\tau(\theta_i)) = \ell | \alpha(\tau(\theta_{n+q_j^n})) = \ell)].$$

Using the defining relation (13) and the choice of the sequence $\{q_j^n\}$, $\theta_i - \theta_{n+q_j^n} \to 0$ as $n \to \infty$, for all i satisfying $n + q_j^n \leq i < n + q_{j+1}^n$, and moreover,

$$P(\alpha(\tau(\theta_i)) = \ell) - P(\alpha(\tau(\theta_{n+q_j^n})) = \ell) \to 0,$$
$$P(\alpha(\tau(\theta_i)) = \ell | \alpha(\tau(\theta_{n+q_j^n})) = \ell) \to 1.$$

Thus $E[I_{\{\alpha(\tau(\theta_i))=\ell\}} - I_{\{\alpha(\tau(\theta_{n+q_j^n}))=\ell\}}]^2 \to 0$, and we conclude that

$$E\left| \frac{1}{\beta_j} \sum_{i=n+q_j^n}^{n+q_{j+1}^n-1} [1_{\alpha(\tau(\theta_i))} - 1_{\alpha(\tau(\theta_{n+q_j^n}))}] D\widehat{\varphi}_i \right| \to 0.$$

The lemma is proved. $\square$

Completion of Proof of Theorem 3.2. To derive the limit dynamic system for $\theta(\cdot)$, we start with the interpolation process indexed by n. In view of (25), Lemma 3.7 implies that

$$\theta^n(t+s) - \theta^n(t) = \sum_{j:t_j^n \in [t,t+s)} \Delta_j^n \frac{1}{\beta_j} 1_{\{\alpha(\tau(\theta_{n+q_j^n}))\}} \sum_{i=n+q_j^n}^{n+q_{j+1}^n-1} D\widehat{\varphi}_i$$
$$+ \sum_{j:t_j^n \in [t,t+s)} \Delta_j^n \frac{1}{\beta_j} \sum_{i=n+q_j^n}^{n+q_{j+1}^n-1} z_i + o(1),$$

where $o(1) \to 0$ in probability uniformly in t.

Note that using the expansive form (see (17) and (19)), $\theta^n(t)$ can also be written as

$$(34) \qquad \theta^n(t) = \theta^n(0) + g^n(t) + \widetilde{b}^n(t) + \widetilde{\psi}^n(t) + Z^n(t),$$

where

$$g^n(t) = \sum_{j:t_j^n < t} \varepsilon_j \varphi_\theta(\theta_j),$$

$$\widetilde{b}^n(t) = \sum_{j:t_j^n < t} \varepsilon_j b_j,$$

$$\widetilde{\psi}^n(t) = \sum_{j:t_j^n < t} \varepsilon_j 1_{\alpha(\tau(\theta_j))} \frac{\psi(\theta_j, \xi_j)}{2\delta_j}.$$

Define
$$(35) \qquad M^n(t) = \theta^n(t) - \theta^n(0) - g^n(t) - Z^n(t).$$

It is readily seen that we also have

$$(36) \qquad M^n(t) = \widetilde{b}^n(t) + \widetilde{\psi}^n(t).$$

Using (36), for any positive integer κ_0, any $0 < t_\ell \le t \le t + s$, and any bounded and continuous functions $\rho_\ell(\cdot)$, with $\ell \le \kappa_0$,

$$E \prod_{\ell=1}^{\kappa_0} \rho_\ell(\theta^n(t_\ell), Z^n(t_\ell))[M^n(t+s) - M^n(t)]$$

$$- E \prod_{\ell=1}^{\kappa_0} \rho_\ell(\theta^n(t_\ell), Z^n(t_\ell))[\widetilde{b}^n(t+s) - \widetilde{b}^n(t)]$$

$$- E \prod_{\ell=1}^{\kappa_0} \rho_\ell(\theta^n(t_\ell), Z^n(t_\ell))[\widetilde{\psi}^n(t+s) - \widetilde{\psi}^n(t)] = 0.$$

We proceed to figure out the limit of each of the terms above. As $j \to \infty$, $\beta_j \to \infty$, and $\Delta_j^n \to 0$. In view of the definitions of $g^n(\cdot)$ and $\widetilde{b}^n(\cdot)$, we have that

$$\widetilde{b}^n(t+s) - \widetilde{b}^n(t) = \sum_{j:t_j^n \in [t,t+s)} \; \sum_{i=n+q_j^n}^{n+q_{j+1}^n - 1} \varepsilon_i b_i$$

$$= \sum_{j:t_j^n \in [t,t+s)} \Delta_j^n \frac{1}{\beta_j} \sum_{i=n+q_j^n}^{n+q_{j+1}^n - 1} b_i + o(1),$$

where $o(1) \to 0$ in probability uniformly in t. The smoothness of $\varphi(\theta)$ then yields that

$$(37) \qquad \prod_{\ell=1}^{\kappa_0} \rho_\ell(\theta^n(t_\ell), Z^n(t_\ell))[\widetilde{b}^n(t+s) - \widetilde{b}^n(t)] \to 0 \quad \text{as } n, j \to \infty.$$

Similarly, we obtain that

$$\widetilde{\psi}^n(t+s) - \widetilde{\psi}^n(t)$$

$$= \sum_{j:t_j^n \in [t,t+s]} \Delta_j^n \frac{1}{\beta_j} 1_{\{\alpha(\tau(\theta_{n+q_j^n}))\}} \sum_{i=n+q_j^n}^{n+q_{j+1}^n - 1} \frac{\psi(\theta_{n+q_j^n}, \xi_i)}{2\delta_i} + o(1).$$

Owing to the interpolation $\theta^n(t_j^n) = \theta_{n+q_j^n}$. For arbitrarily small $\eta > 0$, let $\{B_l^\eta : l \le m_0\}$ be a finite partition of the set $[\theta_l^1, \theta_u^1] \times [\theta_l^2, \theta_u^2] \times \cdots \times [\theta_l^m, \theta_u^m]$ such that each B_l^η satisfies $\operatorname{diam}(B_l^\eta) < \eta$. Let $\widetilde{\theta}_l^\eta$ be an arbitrary point in B_l^η. Then

$$\frac{1}{\beta_j} \sum_{i=n+q_j^n}^{n+q_{j+1}^n - 1} \frac{\psi(\theta_{n+q_j^n}, \xi_i)}{2\delta_i} = \frac{1}{\beta_j} \sum_{i=n+q_j^n}^{n+q_{j+1}^n - 1} \frac{\psi(\theta^n(t_j^n), \xi_i)}{2\delta_i}$$

$$= \frac{1}{\beta_j} \sum_{l=1}^{m_0} \sum_{i=n+q_j^n}^{n+q_{j+1}^n - 1} \frac{\psi(\theta^n(t_j^n), \xi_i)}{2\delta_i} I_{\{\theta^n(t_j^n) \in B_l^\eta\}}$$

$$= \frac{1}{\beta_j} \sum_{l=1}^{m_0} \sum_{i=n+q_j^n}^{n+q_{j+1}^n - 1} \frac{\psi(\widetilde{\theta}_l^\eta, \xi_i)}{2\delta_i} I_{\{\theta^n(t_j^n) \in B_l^\eta\}} + o(1).$$

For each of the $\widetilde{\theta}_l^\eta$ chosen above,

$$\frac{1}{\beta_j} \sum_{i=n+q_j^n}^{n+q_{j+1}^n - 1} E_{n+q_j^n} \frac{\psi(\widetilde{\theta}_l^\eta, \xi_i)}{2\delta_i} \to 0 \text{ in probability } \text{ as } n, j \to \infty.$$

Thus, we obtain

(38) $$E \prod_{\ell=1}^{\kappa_0} \rho_\ell(\theta^n(t_\ell), Z^n(t_\ell))[\widetilde{\psi}^n(t+s) - \widetilde{\psi}^n(t)] = 0.$$

Then the definition of $M^n(\cdot)$ together with (37) and (38) leads to

(39) $$E \prod_{\ell=1}^{\kappa_0} \rho_\ell(\theta^n(t_\ell), Z^n(t_\ell))[M^n(t+s) - M^n(t)] \to 0.$$

On the other hand, using (35), the weak convergence of $(\theta^n(\cdot), Z^n(\cdot))$ to $(\theta(\cdot), Z(\cdot))$ and the definition of $M^n(\cdot)$ imply that $M^n(\cdot)$ converges weakly to a process $M(\cdot)$ defined by

(40) $$M(t) = \theta(t) - \theta(0) - \int_0^t \varphi_\theta(\theta(u)) du - Z(t),$$

where $Z(t) = \int_0^t z(u)du$. Using (39) and (40), Theorem 7.4.1 in Kushner and Yin (2003) yields that $M(\cdot)$ is a martingale. Since $\theta(\cdot)$ and $Z(\cdot)$ have Lipschitz continuous sample paths w.p.1, Theorem 4.1.1 in Kushner and Yin (2003) implies that $M(t) \equiv$ constant w.p.1. Since $M(0) = 0$, $M(t) \equiv 0$ w.p.1. That is, $\theta(\cdot)$ is the solution of equation (22). Thus Theorem 3.2 is proved. $\square$

Proof of Corollary 3.3. For any $\widehat{T} > 0$, using exactly the same argument, we can establish the weak convergence of the sequence $\{\theta^n(\widehat{T}_n + \cdot), \theta^n(\widehat{T}_n - \widehat{T} + \cdot)\}$. Select a convergent subsequence with the limit denoted by $(\theta(\cdot), \theta_{\widehat{T}}(\cdot))$. Then $\theta(0) = \theta_{\widehat{T}}(\widehat{T})$. The set of all possible $\{\theta_{\widehat{T}}(0)\}$ for all $\widehat{T}$ and all convergent subsequence belongs to a tight set. The form of the function $\varphi(\cdot)$ implies that the maximizer θ_* is unique. The stability of the ODE then implies that for any $\eta > 0$ there is a $0 < \widehat{T}_\eta < \infty$ such that for all $\widehat{T} > \widehat{T}_\eta$, $P(\theta_{\widehat{T}}(\widehat{T}) \in N_\eta(\theta_*)) \geq 1 - \eta$, where $N_\eta(\theta_*)$ denotes an η-neighborhood of θ_* (i.e., $N_\eta(\theta_*) = \{\theta : |\theta - \theta_*| \leq \eta\}$). This yields the desired result. $\square$

Proof of Theorem 3.9. For simplicity, we work out the details for $\gamma_1 = 1$ and $\gamma_2 = 1/6$ (so $\gamma_3 = 1/3$). The proofs for the other cases are essentially the same. Without loss of generality, assume $\theta_* = 0$ and $\delta = 1$ henceforth in the proof. Using (19), straightforward calculation yields

$$E_n V(\theta_{n+1}) - V(\theta_n)$$
$$\leq -\lambda\varepsilon_n V(\theta_n) + \varepsilon_n V_\theta'(\theta_n)\left[b_n + 1_{\alpha(\tau(\theta_n))}\frac{\psi(\theta_n, \xi_n)}{2\delta_n} + z_n\right] + O(\varepsilon_n^2).$$

(41)

Based on the idea of perturbed Liapunov function (see Kushner and Yin (2003)), to facilitate the cancellation of the unwanted terms, we introduce several perturbations as follows:

$$V_1(\theta, n) = \sum_{j=n}^{\infty} \varepsilon_j a_{j|n} V_\theta'(\theta) b_j,$$

$$V_2(\theta, n) = \sum_{j=n}^{\infty} \varepsilon_j E_n V_\theta'(\theta) 1_{\alpha(\tau(\theta_j))}\frac{\psi(\theta, \xi_j)}{2\delta_j},$$

$$V_3(\theta, n) = \sum_{j=n}^{\infty} \varepsilon_j c_{j|n} V_\theta'(\theta) z_j,$$

326

where

$$a_{j|n} = \begin{cases} \prod_{k=n}^{j}(1 - 1/(k+1)^{1/3}), & j > n, \\ 1, & \text{otherwise,} \end{cases}$$

and

$$c_{j|n} = \begin{cases} \prod_{k=n}^{j}(1 - 1/(k+1)^{1/6}), & j > n, \\ 1, & \text{otherwise.} \end{cases}$$

By (A4), the smoothness of $\varphi(\cdot)$ together with the boundedness of $\{\theta_n\}$ (owing to the projection) implies $b_n = O(\delta_n) = O(n^{-1/6})$ w.p.1. Using

$$\sum_{j=n}^{\infty} \frac{1}{(j+1)^{1/3}} a_{j|n} \leq K \sum_{j=n}^{\infty} \frac{1}{(j+2)^{1/3}} (a_{j|n} - a_{j|n+1}) = O(1),$$

$$|V_1(\theta, n)| \leq \sum_{j=n}^{\infty} \varepsilon_j a_{j|n} |V_\theta'(\theta)| |b_j|,$$

so we obtain

(42) $$|V_1(\theta, n)| \leq K \sum_{j=n}^{\infty} j^{-5/6}(j^{-1/3} a_{j|n}) \leq O((n+1)^{-5/6}).$$

Similarly, by (A4),

(43)
$$E|V_2(\theta, n)| \leq K \sum_{j=n}^{\infty} j^{-5/6} E|E_n \psi(\theta, \xi_j)|$$
$$\leq O((n+1)^{-5/6}),$$

and

(44)
$$|V_3(\theta, n)| \leq K \sum_{j=n}^{\infty} j^{-5/6}(j^{-1/6} c_{j|n}) |V_\theta'(\theta)| |z_j|$$
$$\leq O((n+1)^{-5/6}).$$

Next, detailed calculation leads to

$$E_n V_1(\theta_{n+1}, n+1) - V_1(\theta_n, n)$$
$$= [E_n V_1(\theta_n, n+1) - V_1(\theta_n, n)] + E_n[V_1(\theta_{n+1}, n+1) - V_1(\theta_n, n+1)]$$
$$= -\varepsilon_n V_\theta'(\theta_n) b_n + \sum_{j=n}^{\infty} \varepsilon_j a_{j|n} E_n V_{\theta\theta}(\breve{\theta}_n)(\theta_{n+1} - \theta_n) b_j$$
$$= -\varepsilon_n V_\theta'(\theta_n) b_n + O((n+1)^{-5/3}),$$

where $\breve{\theta}_n$ is a point on the line segment joining θ_n and θ_{n+1}. Moreover, we also have

$$
\begin{aligned}
&E_n V_2(\theta_{n+1}, n+1) - V_2(\theta_n, n) \\
&= [E_n V_2(\theta_n, n+1) - V_2(\theta_n, n)] + E_n[V_2(\theta_{n+1}, n+1) - V_2(\theta_n, n+1)] \\
&= -\varepsilon_n V'_\theta(\theta_n) 1_{\alpha(\tau(\theta_n))} \frac{\psi(\theta_n, \xi_n)}{2\delta_n} \\
&\quad + \sum_{j=n+1}^{\infty} \varepsilon_j E_n[V'_\theta(\theta_{n+1}) - V'_\theta(\theta_n)] 1_{\alpha(\tau(\theta_j))} \frac{\psi(\theta_n, \xi_j)}{2\delta_j} \\
&\quad + \sum_{j=n+1}^{\infty} \varepsilon_j E_n V'_\theta(\theta_{n+1}) 1_{\alpha(\tau(\theta_j))} \frac{1}{2\delta_j} \\
&\qquad\qquad \times \int_0^1 [\psi(\theta_n + s(\theta_{n+1} - \theta_n), \xi_j) - \psi(\theta_n, \xi_j)] ds(\theta_{n+1} - \theta_n) \\
&= -\varepsilon_n V'_\theta(\theta_n) 1_{\alpha(\tau(\theta_n))} \frac{\psi(\theta_n, \xi_n)}{2\delta_n} + O((n+1)^{-5/3}),
\end{aligned}
$$

and

$$
E_n V_3(\theta_{n+1}, n+1) - V_3(\theta_n, n) = -\varepsilon_n V'_\theta(\theta_n) z_n + O((n+1)^{-5/3}).
$$

Define

$$
\widetilde{V}(\theta, n) = V(\theta) + V_1(\theta, n) + V_2(\theta, n) + V_3(\theta, n).
$$

Then using (41), and the estimate on $E_n V_i(\theta_{n+1}, n+1) - V_i(\theta_n, n)$ for $i = 1, 2, 3$,

$$
\begin{aligned}
E[\widetilde{V}(\theta_{n+1}, n+1) - \widetilde{V}(\theta_n, n)] &\leq -\varepsilon_n \lambda E V(\theta_n) + O((n+1)^{-5/3}) \\
&\leq -\varepsilon_n \lambda E \widetilde{V}(\theta_n, n) + O((n+1)^{-5/3}).
\end{aligned}
$$

The last inequality above follows from (42), (43), and (44). Iterating on $E[\widetilde{V}(\theta_{n+1}, n+1) - \widetilde{V}(\theta_n, n)]$, we arrive at

$$
\begin{aligned}
E\widetilde{V}(\theta_{n+1}, n+1) &\leq \prod_{j=0}^{n}(1 - \lambda\varepsilon_j) E\widetilde{V}(\theta_0, 0) \\
&\quad + \sum_{j=0}^{n} \varepsilon_j(1 - \lambda\varepsilon_j) O((j+1)^{-2/3}) \\
&= O((n+1)^{-2/3}).
\end{aligned}
$$

By virtue of (42) and (43), we also have $EV(\theta_{n+1}) = O((n+1)^{-2/3})$. The desired order of estimation error then follows. $\square$

328

Proof of Corollary 3.10. By virtue of Theorem 3.9, (again assuming $\theta_* = 0$) the Markov inequality leads to

$$P(V(\theta_n) \geq K_1 n^{-2/3}) \leq \frac{EV(\theta_n)}{K_1 n^{-2/3}} \leq \frac{K_2}{K_1},$$

for some $K_2 > 0$. This implies that $\{n^{2/3}V(\theta_n)\}$ is tight. The local quadratic form of the Liapunov function $V(\cdot)$ further yields the tightness of $\{n^{1/3}(\theta_n - \theta_*)\}$. $\quad\square$

References

G. Barone-Adesi and R. Whaley, Efficient analytic approximation of American option values, *J. Finance*, **42** (1987), 301–320.

J. Buffington and R.J. Elliott, American options with regime switching, *Internat. J. Theoretical Appl. Finance*, **5** (2002), 497–514.

P. Billingsley, *Convergence of Probability Measures*, J. Wiley, New York, NY, 1968.

G.B. Di Masi, Y.M. Kabanov and W.J. Runggaldier, Mean variance hedging of options on stocks with Markov volatility, *Theory Probab. Appl.*, **39** (1994), 173–181.

J.P. Fouque, G. Papanicolaou, and K.R. Sircar, *Derivatives in Financial Markets with Stochastic Volatility*, Cambridge University Press, 2000.

M.C. Fu, S.B. Laprise, D.B. Madan, Y. Su, and R. Wu, Pricing American options: A comparison of Monte Carlo simulation approaches, *J. Comput. Finance*, Vol. 4 (2001), 39–88.

M.C. Fu, R. Wu, G. Gürkan, and A. Y. Demir, A note on perturbation analysis estimators for American-style options, *Probab. Eng. Informational Sci.*, **14** (2000), 385–392.

P. Glasserman, *Monte Carlo Methods in Financial Engineering*, Springer-Verlag, New York, 2003.

X. Guo and Q. Zhang, Closed-form solutions for perpetual American put options with regime switching, *SIAM J. Appl. Math.*, **64** (2004), 2034-2049.

Y.C. Ho and X.R. Cao, *Perturbation Analysis of Discrete Event Dynamic Systems*, Kluwer, Boston, MA, 1991.

J.D. Hamilton, A new approach to the economic analysis of nonstationary time series, *Econometrica*, **57** (1989), 357–384.

J.C. Hull and A. White, The pricing of options on assets with stochastic volatilities, *J. Finance*, **42** (1987), 281–300.

H. J. Kushner, *Approximation and Weak Convergence Methods for Random Processes, with applications to Stochastic Systems Theory*, MIT Press, Cambridge, MA, 1984.

H.J. Kushner and G. Yin, *Stochastic Approximation and Recursive Algorithms and Applications*, 2nd Ed., Springer-Verlag, New York, 2003.

F. A. Longstaff and E. S. Schwartz, Valuing American Options by Simulation: A simple least-squares approach, *Rev. Financial Studies*, **14** (2001), 113–147.

R.C. Merton, Lifetime portfolio selection under uncertainty: The continuous time case, *Rev. Economics Statist.*, **51** (1969), 247–257.

D.D. Yao, Q. Zhang, and X. Zhou, A regime-switching model for European option pricing, in *this volume* (2006).

G. Yin, R.H. Liu, and Q. Zhang, Recursive algorithms for stock Liquidation: A stochastic optimization approach, *SIAM J. Optim.*, **13** (2002), 240–263.

G. Yin, J.W. Wang, Q. Zhang, and Y.J. Liu, Stochastic optimization algorithms for pricing American put options under regime-switching models, to appear in *J. Optim. Theory Appl.* (2006).

Q. Zhang, Stock trading: An optimal selling rule, *SIAM J. Control Optim.*, **40**, (2001), 64–87.

Q. Zhang and G. Yin, Nearly optimal asset allocation in hybrid stock-investment models, *J. Optim. Theory Appl.*, **121** (2004), 197–222.

X.Y. Zhou and G. Yin, Markowitz mean-variance portfolio selection with regime switching: A continuous-time model, *SIAM J. Control Optim.*, **42** (2003), 1466–1482.

Chapter 16

OPTIMAL PORTFOLIO APPLICATION WITH DOUBLE-UNIFORM JUMP MODEL

Zongwu Zhu

Department of Mathematics, Statistics, and Computer Science
University of Illinois at Chicago

zzhu@math.uic.edu

Floyd B. Hanson

Department of Mathematics, Statistics, and Computer Science
University of Illinois at Chicago

hanson@math.uic.edu

Dedicated to Suresh P. Sethi on his 60th birthday for his fundamental contributions to optimal portfolio theory.

Abstract   This paper treats jump-diffusion processes in continuous time, with emphasis on the jump-amplitude distributions, developing more appropriate models using parameter estimation for the market in one phase and then applying the resulting model to a stochastic optimal portfolio application in a second phase. The new developments are the use of double-uniform jump-amplitude distributions and time-varying market parameters, introducing more realism into the application model – a log-normal diffusion, log-double-uniform jump-amplitude model. Although unlimited borrowing and short-selling play an important role in pure diffusion models, it is shown that borrowing and shorting is limited for jump-diffusions, but finite jump-amplitude models can allow very large limits in contrast to infinite range models which severely restrict the instant stock fraction to [0,1]. Among all the time-dependent parameters modeled, it appears that the interest and discount rate have the strongest effects.

Keywords: Optimal portfolio with consumption, portfolio policy, jump-diffusion, double-uniform jump-amplitude

1. Introduction

The empirical distribution of daily log-returns for actual financial instruments differs in many ways from the ideal pure diffusion process with its log-normal distribution as assumed in the Black-Scholes-Merton option pricing model [4, 27]. The log-returns are the log-differences between two successive trading days, representing the logarithm of the relative size. The most significant difference is that actual log-returns exhibit occasional large jumps in value, whereas the diffusion process in Black-Scholes [4] is continuous. Statistical evidence of jumps in various financial markets is given by Ball and Torous [3], Jarrow and Rosenfeld [18] and Jorion [19]. Hence, some jump-diffusion models were proposed including Merton's pioneering log-normal [28] (also [29, Chap. 9]), Kou and Wang's log-double-exponential [21, 22] and Hanson and Westman's log-uniform [13, 15] jump-diffusion models.

Another difference is that the empirical log-returns are usually negatively skewed, since the negative jumps or crashes are likely to be larger or more numerous than the positive jumps for many instruments, whereas the normal distribution associated with the diffusion process is symmetric. Thus, the coefficient of skew [5] is negative,

$$\eta_3 \equiv M_3/(M_2)^{1.5} < 0, \qquad (16.1)$$

where M_2 and M_3 are the 2nd and 3rd central moments of the log-return distribution here. A third difference is that the empirical distribution is usually leptokurtic since the coefficient of kurtosis [5] satisfies

$$\eta_4 \equiv M_4/(M_2)^2 > 3, \qquad (16.2)$$

where the value 3 is the normal distribution kurtosis value and M_4 is the fourth central moment. Qualitatively, this means that the tails are fatter than a normal with the same mean and standard deviation, compensated by a distribution that is also more slender about the mode (local maximum). A fourth difference is that the market exhibits time-dependence in the distributions of log-returns, so that the associated parameters are time-dependent.

For option pricing with jump-diffusions, in 1976 Merton [28] (see also [29, Chap. 8]) introduced Poisson jumps with independent identically distributed random jump-amplitudes with fixed mean and variances into the Black-Scholes model, but the ability to hedge the volatilities as with the Black-Sholes options model was not possible. Also for option pricing, Kou [21, 22] used a jump-diffusion model with a double exponential (Laplace) jump-amplitude distribution, having leptokurtic and negative skewness properties. However, it is difficult to see the empirical justification for this or any other jump-amplitude distribution due to the problem of separating the outlying jumps from the diffusion (see Aït-Sahalia [1]), although separating out the diffusion is a reasonable task.

For optimal portfolio with consumption theory Merton in another pioneering paper, prior to the Black-Scholes model, [25, 26] (see also [29, Chapters 4-6]) analyzed the optimal consumption and investment portfolio with geometric Brownian motion and examined an example of hyperbolic absolute risk-aversion (HARA) utility having explicit solutions. Generalizations to jump-diffusions consisting of Brownian motion and compound Poisson processes with general random finite amplitudes are briefly discussed. Earlier in [24] ([29, Chapter 4]), Merton also examined constant relative risk-aversion problems.

In the 1971 Merton paper [25, 26] there are a number of errors, in particular in boundary conditions for bankruptcy (non-positive wealth) and vanishing consumption. Some of these problems are directly due to using a general form of the HARA utility model. These errors are very thoroughly discussed in a seminal collection assembled by Suresh P. Sethi [32] from his papers and those of his coauthors. Sethi in his introduction [32, Chapter 1]) thoroughly summarizes these errors and subsequent generalizations. In particular, basic papers of concern here are the *KLSS* paper with Karatzas, Lehoczhy, Shreve [20] (reprint [32, Chapter 2]) for exact solutions in the infinite horizon case and with Taksar [33] (reprint [32, Chapter 2]) pinpointing the errors in Merton's [25, 26] work.

Hanson and Westman [10, 16] reformulated an important external events model of Rishel [31] solely in terms of stochastic differential equations and applied it to the computation of the optimal portfolio and consumption policies problem for a portfolio of stocks and a bond. The stock prices depend on both scheduled and unscheduled jump external events. The complex computations were illustrated with a simple log-bi-discrete jump-amplitude model, either negative or positive jumps, such that both stochastic and quasi-deterministic jump magnitudes were estimated. In [11], they constructed a jump-diffusion model with marked Poisson jumps that had a log-normally distributed jump-amplitude and rigorously derived the density function for the diffusion and log-normal-jump stock price log-return model. In [12], this financial model is applied to the optimal portfolio and consumption problem for a portfolio of stocks and bonds governed by a jump-diffusion process with log-normal jump amplitudes and emphasizing computational results. In two companion papers, Hanson and Westman [13, 14] introduce the log-uniform jump-amplitude jump-diffusion model, estimate the parameter of the jump-diffusion density with weighted least squares using the S&P500 data and apply it to portfolio and consumption optimization. In [15], they study the time-dependence of the jump-diffusion parameter on the portfolio optimization problem for the log-uniform jump-model. The appeal of the log-uniform jump model is that it is consistent with the stock exchange introduction of *circuit breakers* [2] in 1988 to limit extreme changes, such as in the crash of 1987, in stages. On the contrary, the normal and double-exponential jump models have an infinite domain, which is not a problem for the diffusion part of the jump-diffusion distribution since the contribution in the dynamic programming formulation is local appearing only in partial derivatives. However, the influence of the jump part in dynamic programming is global through integrals with integrands that have shifted arguments. This has important consequences for the choice of jump distribution since the portfolio wealth restrictions will depend on the range of support of the jump density.

In this paper, the log-double-uniform jump-amplitude, jump-diffusion asset model is applied to the portfolio and consumption optimizaition problem. In Section 2, the jump-diffusion density is rigorously derived using a modification of the prior theorem [11]. In Section 3, the time dependent parameters for this log-return process are estimated using this theoretical density and the S&P500 Index daily closing data for 16 years. In Section 4, the optimal portfolio and consumption policy application is presented and then solved computationally. Also, in this section, the big difference in borrowing and short-selling limits is formulated in a lemma. Concluding remarks are given in Section 5.

2. Log-Double-Uniform Amplitude Jump-Diffusion Density for Log-Return

Let $S(t)$ be the price of a single financial asset, such as a stock or mutual fund, governed by a Markov, geometric jump-diffusion stochastic differential equation (SDE) with time-dependent coefficients,

$$dS(t) = S(t)\left(\mu_d(t)dt + \sigma_d(t)dG(t) + \sum_{k=1}^{dP(t)} J(T_k^-, Q_k)\right), \quad (16.3)$$

with $S(0) = S_0$, $S(t) > 0$, where $\mu_d(t)$ is the mean appreciation return rate at time t, $\sigma_d(t)$ is the diffusive volatility, $dG(t)$ is a continuous Gaussian process with zero mean and dt variance, $dP(t)$ is a discontinuous, standard Poisson process with jump rate $\lambda(t)$, with common mean-variance of $\lambda(t)dt$, and associated jump-amplitude $J(t, Q)$ with log-return mark Q mean $\mu_j(t)$ and variance $\sigma_j^2(t)$. The stochastic processes $G(t)$ and $P(t)$ are assumed to be Markov and pairwise independent. The jump-amplitude $J(t, Q)$, given that a Poisson jump in time occurs, is also independently distributed, at pre-jump time T_k^- and mark Q_k. The stock price SDE (16.3) is similar in prior work [11, 12], except that time-dependent coefficients introduce more realism here. The Q_k are IID random variables with Poisson amplitude mark density, $\phi_Q(q; t)$, on the mark-space $\mathcal{Q}$.

The infinitesimal moments of the jump process are

$$\mathrm{E}[J(t, Q)dP(t)] = \lambda(t)dt \int_{\mathcal{Q}} J(t, q)\phi_Q(q; t)dq$$

and

$$\mathrm{Var}[J(t, Q)dP(t)] = \lambda(t)dt \int_{\mathcal{Q}} J^2(t, q)\phi_Q(q; t)dq.$$

The differential Poisson process is a counting process with the probability of the jump count given by the usual Poisson distribution,

$$p_k(\lambda(t)dt) = \exp(-\lambda(t)dt)(\lambda(t)dt)^k/k!, \tag{16.4}$$

$k = 0, 1, 2, \ldots$, with parameter $\lambda(t)dt > 0$.

Since the stock price process is geometric, the common multiplicative factor of $S(t)$ can be transformed away yielding the SDE of the stock price log-return using the stochastic chain rule for Markov processes in continuous time,

$$d[\ln(S(t))] = \mu_{ld}(t)dt + \sigma_d(t)dG(t) + \sum_{k=1}^{dP(t)} \ln(1 + J(T_k^-, Q_k)), \tag{16.5}$$

where $\mu_{ld}(t) \equiv \mu_d(t) - \sigma_d^2(t)/2$ is the log-diffusion drift and $\ln(1 + J(t, q))$ is the stock log-return jump-amplitude or the logarithm of the relative post-jump-amplitude. This log-return SDE (16.5) is the model that will be used for comparison to the S&P500 log-returns. Since jump-amplitude coefficient $J(t, q) > -1$, it is convenient to select the mark process to be the log-jump-amplitude random variable,

$$Q = \ln(1 + J(t, Q)), \tag{16.6}$$

on the mark space $Q = (-\infty, +\infty)$, so $J(t, Q) = e^Q - 1$ in general. Although this is a convenient mark selection, it implies the independence of the jump-amplitude in time, but not of the jump-amplitude distribution.

Since market jumps are rare and limited, while the tails are relatively fat, a reasonable approximation is the log-double-uniform (duq) jump-amplitude distribution with density ϕ_Q on the finite, time-dependent mark interval $[a(t), b(t)]$ as in [15]. However, since the optimistic strategies that play a role in rallies should be different from the pessimistic strategies used for crashes, it would be better to decouple the positive from the negative jumps giving rise to the log-double-uniform jump-amplitude model. The double-uniform density is the juxtaposition of two uniform densities, $\phi_1(q; t) = I_{\{a(t) \leq q \leq 0\}}/|a|(t)$ on $[a(t), 0]$ and $\phi_2(q; t) = I_{\{0 \leq q \leq b(t)\}}/b(t)$ on $[0, b(t)]$, such that $a(t) < 0 < b(t)$ and I_S is the indicator function for set S. The double-uniform density can be written,

$$\phi_Q(q; t) \equiv \begin{cases} 0, & -\infty < q < a(t) \\ p_1(t)/|a|(t), & a(t) \leq q < 0 \\ p_2(t)/b(t), & 0 \leq q \leq b(t) \\ 0, & b(t) < q < +\infty \end{cases} , \quad (16.7)$$

essentially undefined or doubly defined at $q = 0$, except $p_1(t)$ is the probability of a negative jump and $p_2(t)$ is the probability of a non-negative jump, conserving probability by assigning the null jump to the uniform sub-distribution with the positive jumps. Otherwise, $\phi_Q(q; t)$ is undefined as the derivative of the double-uniform distribution for the point of jump discontinuity at 0, but the distribution

$$\Phi_Q(q; t) = p_1(t) \frac{q - a(t)}{|a|(t)} I_{\{a(t) \leq q < 0\}} + \left(p_1(t) + p_2(t) \frac{q}{b(t)} \right) I_{\{0 \leq q < b(t)\}}$$
$$+ I_{\{b \leq q < \infty\}}$$

is well-defined and continuous since points of zero measure do not contribute. The assumption that $a(t) < 0 < b(t)$ is to make sure that both negative jumps (including crashes) and positive jumps (including rallies) are represented. The form of this double-uniform model was motivated by Kou's [21] double-exponential model.

The density $\phi_Q(q; t)$ yields the mean

$$E_Q[Q] = \mu_j(t) = (p_1(t)a(t) + p_2(t)b(t))/2$$

and variance

$$\text{Var}_Q[Q] = \sigma_j^2(t) = (p_1(t)a^2(t) + p_2(t)b^2(t))/3 - \mu_j^2(t)$$

which define the basic log-return jump-amplitude moment parameters. The third and fourth central moments are, respectively,

$$M_3^{(\text{duq})}(t) \equiv E_Q[(Q - \mu_j(t))^3]$$
$$= (p_1(t)a^3(t) + p_2(t)b^3(t))/4 - \mu_j(t)(3\sigma_j^2(t) + \mu_j^2(t))$$

and

$$M_4^{(\text{duq})}(t) \equiv E_Q[(Q - \mu_j(t))^4]$$
$$= (p_1(t)a^4(t) + p_2(t)b^4(t))/5 - 4\mu_j(t)M_3^{(\text{duq})}(t) - 6\mu_j^2(t)\sigma_j^2(t) - \mu_j^4(t).$$

The log-double-uniform distribution is treated as time-dependent in this paper, so $a(t)$, $b(t)$, $\mu_j(t)$ and $\sigma_j^2(t)$ all depend on t.

The difficulty in separating out the small jumps about the mode or maximum of real market distributions is explained by the fact that a diffusion approximation for small marks can be used for the jump process that will be indistinguishable from the continuous Gaussian process anyway.

The first four moments of the difference form stock log-return,

$$\Delta \ln(S(t)) \equiv \ln(S(t + \Delta t)) - \ln(S(t)),$$

assuming that a sufficiently close approximation of the double-uniform jump-diffusion (dujd) by (16.5), i.e.,

$$\Delta \ln(S(t)) \simeq \mu_{ld}(t)\Delta t + \sigma_d(t)\Delta G(t) + \sum_{k=1}^{\Delta P(t)} Q_k$$
$$= (\mu_{ld}(t) + \lambda(t)\mu_j(t))\Delta t + \sigma_d(t)\Delta G(t) \qquad (16.8)$$
$$+ \mu_j(t)(\Delta P(t) - \lambda(t)\Delta t) + \sum_{k=1}^{\Delta P(t)}(Q_k - \mu_j(t)),$$

the latter in a more convenient zero-mean and independent terms form, are

$$M_1^{(\text{dujd})} \equiv E[\Delta \ln(S(t))] = (\mu_{ld}(t) + \lambda(t)\mu_j(t))\Delta t, \qquad (16.9)$$

$$M_2^{(\text{dujd})} \equiv \text{Var}[\Delta \ln(S(t))] = \left(\sigma_d^2(t) + \lambda(t)\left(\mu_j^2(t) + \sigma_j^2(t)\right)\right)\Delta t, (16.10)$$

$$M_3^{(\text{dujd})}(t) \equiv E\left[\left(\Delta[\ln(S(t))] - M_1^{(\text{dujd})}(t)\right)^3\right]$$
$$= (p_1(t)a^3(t) + p_2(t)b^3(t))\lambda(t)\Delta t/4, \qquad (16.11)$$

$$M_4^{(\text{dujd})}(t) \equiv E\left[\left(\Delta[\ln(S(t))] - M_1^{(\text{dujd})}(t)\right)^4\right]$$
$$= (p_1(t)a^4(t) + p_2(t)b^4(t))\lambda(t)\Delta t/5 \qquad (16.12)$$
$$+ 3(\sigma_d^2(t) + \lambda(t)(\mu_j^2(t) + \sigma_j^2(t)))^2(\Delta t)^2.$$

The $M_4^{(\mathrm{dujd})}(t)$ moment calculation, in particular, needs a lemma from [9, Chapter 5] for the fourth power of partial sums of zero-mean IID random variables X_i, i.e.,

$$\mathrm{E}\left[\left(\sum_{i=1}^{n}X_i\right)^4\right] = n\mathrm{E}\left[X_i^4\right] + 3n(n-1)\left(\mathrm{E}\left[X_i^2\right]\right)^2.$$

The log-double-uniform jump-diffusion density can be found by basic probabilistic methods following a slight modification to time-dependent coefficients from the constant coefficients assumption used in the theorem of Zhu [36],

Theorem 1 (Probability Density). *The log-double-uniform jump-amplitude jump-diffusion log-return difference, written as*

$$\Delta\ln(S(t)) = \mathcal{G}(t) + \sum_{k=1}^{\Delta P(t)} Q_k$$

specified in the SDE (16.8) with non-standard Gaussian $\mathcal{G}(t) = \mu_{ld}\Delta t + \sigma_d \Delta G(t)$*, has a probability density given by*

$$\phi_{\Delta\ln(S(t))}^{(\mathrm{dujd})}(x) \simeq \sum_{k=0}^{\infty} p_k(\lambda(t)\Delta t)\phi_{\mathcal{G}(t)+\sum_{i=1}^{k}Q_i}^{(\mathrm{dujd})}(x)$$

$$\equiv \sum_{k=0}^{\infty} p_k(\lambda(t)\Delta t)\phi_k^{(\mathrm{dujd})}(x), \tag{16.13}$$

for sufficiently small Δt *and* $-\infty < x < +\infty$*, where* $p_k(\lambda(t)\Delta t)$ *is the Poisson distribution (16.4) with parameter* $\lambda(t)\Delta t$ *with multiple-convolution, Poisson coefficients*

$$\phi_k^{(\mathrm{dujd})}(x) = \left(\phi_{\mathcal{G}(t)}\prod_{i=1}^{k}(*\phi_{Q_i})\right)(x). \tag{16.14}$$

In the case of the corresponding normalized second order approximation,

$$\phi_{\Delta\ln(S(t))}^{(\mathrm{dujd},2)}(x) = \sum_{k=0}^{2} p_k(\lambda(t)\Delta t)\phi_k^{(\mathrm{dujd})}(x) / \sum_{k=0}^{2} p_k(\lambda(t)\Delta t), \tag{16.15}$$

where the density coefficients are given by

$$\phi_0^{(\mathrm{dujd})}(x) = \phi^{(n)}\left(x; \mu, \sigma^2\right), \tag{16.16}$$

for $k = 0$*, where* $\phi^{(n)}(x; \mu, \sigma^2)$ *is the normal distribution with mean* μ *and variance* σ^2*, while here* $(\mu, \sigma^2) = (\mu_{ld}, \sigma_d^2)\Delta t$*, for* $k = 1$*,*

$$\phi_1^{(\mathrm{dujd})}(x) = +\frac{p_1(t)}{|a|(t)}\Phi^{(n)}(a(t), 0; x - \mu, \sigma^2)$$

$$+\frac{p_2(t)}{b(t)}\Phi^{(n)}(0, b(t); x - \mu, \sigma^2), \tag{16.17}$$

where $\Phi^{(n)}(a, b; \mu, \sigma^2)$ is the normal distribution on (a, b) with density $\phi^{(n)}(x; \mu, \sigma^2)$, and for $k = 2$,

$$
\begin{aligned}
\phi_2^{(\text{dujd})}(x) = \sigma^2 \Big(& (p_1(t)/a(t) + p_2(t)/b(t))^2 \phi^{(n)}(0, *) \\
& + (p_1(t)/a(t))^2 \phi^{(n)}(2a(t), *) + (p_2(t)/b(t))^2 \phi^{(n)}(2b(t), *) \\
& - 2\left((p_1(t)/a(t))^2 + p_1(t)p_2(t)/(a(t)b(t))\right) \phi^{(n)}(a(t), *) \\
& - 2\left((p_2(t)/b(t))^2 + p_1(t)p_2(t)/(a(t)b(t))\right) \phi^{(n)}(b(t), *) \\
& + 2p_1(t)p_2(t)/(a(t)b(t)) \phi^{(n)}(a(t) + b(t), *) \Big) \\
& + (p_1(t)/a(t))^2 \Big((x - 2a(t) - \mu)\Phi^{(n)}(2a(t), a(t), *) \\
& - (x - \mu)\Phi^{(n)}(a(t), 0, *)\Big) \\
& + (2p_1(t)p_2(t)/(a(t)b(t))) \Big((x - \mu)\Phi^{(n)}(0, b(t), *) \\
& - (x - a(t) - \mu)\Phi^{(n)}(a(t), a(t) + b(t), *)\Big) \\
& + (p_2(t)/b(t))^2 \Big((x - \mu)\Phi^{(n)}(0, b(t), *) \\
& - (x - 2b(t) - \mu)\Phi^{(n)}(b(t), 2b(t), *)\Big) \\
& - 2(p_1(t)p_2(t)/a(t))\Phi^{(n)}(a(t) + b(t), b(t), *)
\end{aligned}
\tag{16.18}
$$

where the symbol $*$ means that the common parameter argument $x - \mu, \sigma^2$ has been suppressed.

Proof. The sum in (16.13) is merely an expression of the law of total probability [9, Chapters 0 and 5] and the multiple or nested form (16.14) follows from a convolution theorem [9]. When $k = 0$ there are no jumps and $\Delta \ln(S(t)) = \mathcal{G}(t)$, the purely Gaussian term, so the distribution is normal and is given in (16.16). Note in this case $\sum_{i=1}^{0} Q_i \equiv 0$ by convention.

When $k = 1$ jump, consider the double sum of IID random variables $\Delta \ln(S(t)) = \mathcal{G}(t) + Q_1$ near the jump for sufficiently small Δt and letting $(\mu, \sigma^2) = (\mu_{ld}, \sigma_d^2)\Delta t$,

$$
\begin{aligned}
\phi_1^{(\text{dujd})}(x) = \left(\phi_{\mathcal{G}(t)} * \phi_{Q_1}\right)(x) &= \int_{-\infty}^{+\infty} \phi^{(n)}(x - q; \mu, \sigma^2)\phi_{Q_1}(q; t)dq \\
&= \left(\frac{p_1(t)}{|a|(t)} \int_{a(t)}^{0} + \frac{p_2(t)}{b(t)} \int_{0}^{b(t)}\right) \phi^{(n)}(x - q; \mu, \sigma^2)dq \\
&= \frac{p_1(t)}{|a|(t)} \Phi^{(n)}(a(t), 0; x - \mu, \sigma^2) + \frac{p_2(t)}{b(t)} \Phi^{(n)}(0, b(t); x - \mu, \sigma^2),
\end{aligned}
$$

verifying (16.17) by the normal argument-mean shift identity [9, Chapter 0], $\phi^{(n)}(x - q; \mu, \sigma^2) = \phi^{(n)}(q; x - \mu, \sigma^2)$. The density $\Phi^{(n)}(\xi, \eta; \mu, \sigma^2)/(\eta -$

340

ξ) is a **secant-normal density** as the secant approximation to the derivative to the normal distribution [9, Chapter 5].

For $k = 2$ jumps, we consider the triple IID random variables $\mathcal{G}(t) + Q_1 + Q_2$, first treating the sum of the two double-uniform IID RVs,

$$
\begin{aligned}
(\phi_{Q_1} * \phi_{Q_2})(x) &= \int_{-\infty}^{+\infty} \phi_{Q_2}(x-q;t)\phi_{Q_1}(q;t)dq \\
&= \tfrac{p_1^2(t)}{a^2(t)} \int_{a(t)}^{0} I_{\{a(t)\leq x-q<0\}}dq + \tfrac{p_2^2(t)}{b^2(t)} \int_{0}^{b(t)} I_{\{0\leq x-q\leq b(t)\}}dq \\
&\quad + \tfrac{2p_1(t)p_2(t)}{b(t)|a|(t)} \int_{a(t)}^{0} I_{\{0\leq x-q\leq b(t)\}}dq \\
&= \tfrac{p_1^2(t)}{a^2(t)} \min(x-2*a(t),-x)I_{\{a(t)\leq x<0\}} \\
&\quad + \tfrac{p_2^2(t)}{b^2(t)} \min(x,2*b(t)-x)I_{\{0\leq x\leq b(t)\}} \\
&\quad + \tfrac{2p_1(t)p_2(t)}{b(t)|a|(t)} \min(x-a(t),\min(|a|(t),b(t)),b(t)-x)I_{\{a\leq x\leq b(t)\}},
\end{aligned}
$$

comprising two triangular densities [9, Chapter 5] plus one trapezoidal density. On substituting this density composite and again using the argument-mean normal shift identity again into the double convolution leads to

$$
\begin{aligned}
\phi_2^{(\mathrm{dujd})}(x) &= \left(\phi_{\mathcal{G}(t)} * (\phi_{Q_1} * \phi_{Q_2})\right)(x) \\
&= \tfrac{p_1^2(t)}{a^2(t)} \left(\int_{2a(t)}^{a(t)} (q-2a(t))\phi^{(n)}(q;*)dq + \int_{a(t)}^{0}(-q)\phi^{(n)}(q;*)dq \right) \\
&\quad + \tfrac{p_2^2(t)}{b^2(t)} \left(\int_{0}^{b(t)} q\phi^{(n)}(q;*)dq + \int_{b(t)}^{2b(t)}(2a(t)-q)\phi^{(n)}(q;*)dq \right) \\
&\quad + \tfrac{2p_1(t)p_2(t)}{b(t)|a|(t)} \left(\int_{a(t)}^{\min(a(t)+b(t),0)} (q-a(t))\phi^{(n)}(q;*)dq \right. \\
&\quad + \min(|a|(t),b(t)) \int_{\min(a(t)+b(t),0)}^{\max(a(t)+b(t),0)} \phi^{(n)}(q;*)dq \\
&\quad \left. + \int_{\max(a(t)+b(t),0)}^{b(t)} (b(t)-q)\phi^{(n)}(q;*)dq \right),
\end{aligned}
\qquad (16.19)
$$

where again the symbol $*$ denotes $x - \mu, \sigma^2$. The last equation follows from using the following normal integral identity,

$$
\pm\int_{\alpha}^{\beta}(q-\gamma)\phi^{(n)}(q;*)dq = \pm(x-\mu-\gamma)\Phi^{(n)}(\alpha,\beta;*) \mp \sigma^2\left(\phi^{(n)}(\beta;*)-\phi^{(n)}(\alpha;*)\right).
$$

Finally after some analysis for two cases: $a(t)+b(t) < 0$ or $a(t)+b(t) >= 0$, the equation (16.19) for $\phi_2^{(\mathrm{dujd})}(x)$ can be recollected and simplified as the form in (16.18). However, there are also practical computational considerations since some naive collections of terms lead to *exponential catastrophic cancellation problems* which are detected by checking a form for $\phi_2^{(\mathrm{dujd})}(x)$ for conservation of probability since $\phi_2^{(\mathrm{dujd})}(x)$ must be a proper density. The problem arises for the double-uniform jump-amplitude model coupled with difficulty of computing normal distributions with very small variances and combining similar exponential as well as distribution terms. Corrections to this problem require a very robust normal density integrator like the MATLAB™ [23] basic *erfc* complementary error function and a proper collection of terms. Note that the two forms (16.18) and (16.19) of $\phi_2^{(dujd)}(x)$ are analytically equivalent in infinite precision, but not computationally in finite precision.   □

Using the log-normal jump-diffusion log-return density in (16.13), the third and fourth central moment formulas (16.11,16.12) can be confirmed [36].

3.   Jump-Diffusion Parameter Estimation

Given the log-normal-diffusion, log-double-uniform jump density (16.13), it is necessary to fit this theoretical model to realistic empirical data to estimate the parameters of the log-return model (16.5) for $d[\ln(S(t))]$. For realistic empirical data, the daily closings of the S&P500 Index during the years from 1988 to 2003 are used from data available on-line [35]. The data consists of $n^{(\text{sp})} = 4036$ daily closings. The S&P500 (sp) data can be viewed as an example of one large mutual fund rather than a single stock. The data has been transformed into the discrete analog of the continuous log-return, i.e., into changes in the natural logarithm of the index closings, $\Delta[\ln(SP_i)] \equiv \ln(SP_{i+1}) - \ln(SP_i)$ for $i = 1, \ldots, n^{(\text{sp})} - 1$ daily closing pairs. For the period, the mean is $M_1^{(\text{sp})} \simeq 3.640 \times 10^{-4}$ and the variance is $M_2^{(\text{sp})} \simeq 1.075 \times 10^{-4}$, the coefficient of skewness is

$$\eta_3^{(\text{sp})} \equiv M_3^{(\text{sp})}/(M_2^{(\text{sp})})^{1.5} \simeq -0.1952 < 0,$$

demonstrating the typical negative skewness property, and the coefficient of kurtosis is

$$\eta_4^{(\text{sp})} \equiv M_4^{(\text{sp})}/(M_2^{(\text{sp})})^2 \simeq 6.974 > 3,$$

demonstrating the typical leptokurtic behavior of many real markets.

The S&P500 log-returns, $\Delta[\ln(SP_i)]$ for $i = 1 : n^{(\text{sp})}$ data points, are partitioned into 16 yearly (spy) data sets, $\Delta[\ln(SP_{j_y,k}^{(\text{spy})})]$ for $k = 1 : n_{y,j_y}^{(\text{sp})}$ yearly data points for $j_y = 1 : 16$ years, where $\sum_{j_y=1}^{16} n_{y,j_y}^{(\text{sp})} = n^{(\text{sp})}$. For each of these yearly sets, the six parameters

$$\mathbf{y}_{j_y} = \left(\mu_{ld,j_y}, \sigma_{d,j_y}^2, \mu_{j,j_y}, \sigma_{j,j_y}^2, p_{1,j_y}, \lambda_{j_y} \right),$$

are estimated for each year j_y to specify the jump-diffusion log-return distribution by *maximum likelihood estimation objective*,

$$f(\mathbf{y}_{j_y}) = - \sum_{k=1}^{n_{y,j_y}^{(\mathrm{sp})}} \log\left(\phi_{\Delta\ln(S(t))}^{(\mathrm{dujd},2)}(x_k; \mathbf{y}_{j_y}) \right). \qquad (16.20)$$

The time step $\Delta t = \Delta T_{j_y}$ is the reciprocal of the number of trading days per year, close to 252 days, but varies a little for $j_y = 1 : 16$ years used here for parameter estimation. The maximum likelihood estimation is performed for convenience directly on the set

$$\widehat{\mathbf{y}}_{j_y} = \left(\mu_{ld,j_y}, \sigma_{d,j_y}^2, a_{j_y}, b_{j_y}, p_{1,j_y}, \lambda_{j_y} \right),$$

since it is easier to get the pair $\{\mu_{j,j_y}, \sigma_{j,j_y}^2\}$ from $\{a_{j_y}, b_{j_y}\}$, rather then the other way around which would require a quadratic inversion.

Thus, we have a six dimensional global minimization problem for a highly complex discretized jump-diffusion density function (16.5). Due to the high level of flexibility with six free parameters, *barrier techniques* using large values in excluded regions are adopted to avoid negative variances σ_{d,j_y}^2, non-positive a_{j_y}, negative b_{j_y}, $p_{1,j_y} \notin [0,1)$ and negative λ_{j_y}. The analytical complexity indicates that a general global optimization method that does not require derivatives would be useful. For this purpose, such a method, the *Nelder-Mead downhill simplex* method [30], implemented in MATLAB™ [23] as the function fminsearch is used, since simple techniques are desirable in financial engineering. The method is quite efficient since it requires only one new function evaluation for each successive step to test for the best new search direction from the old simplex.

The jump-diffusion estimated yearly parameter results in the present log-double-uniform-jump amplitude case are summarized in the Figures 16.1 and 16.2. The graphs are piecewise linear interpolations of the yearly averages when the averages are assigned to the mid-year. Perhaps cubic splines or moving averages would portray the parameters better, but since a total of eight or more time dependent parameters are needed for the optimal portfolio and consumption application that follows, the piecewise linear interpolation is more convenient due to time constraints. Figure 16.1 displays the time-variation from 1988 to mid 2004 for the diffusion mean $\mu_d(t)$ and variance $\sigma_d(t)$ in Subfigure 16.1(a) and the jump mean $\mu_j(t)$ and variance $\sigma_j(t)$ in Subfigure 16.1(b). The average values of $(\mu_d, \sigma_d, \mu_j, \sigma_j)$ are (0.1654, 0.1043, 3.110e-4, 8.645e-3), respectively. In Figure 16.2, more jump parameters are displayed, but the three double-uniform distribution parameters $p_1(t)$, $a(t)$ and $b(t)$ determine the jump mean $\mu_j(t)$ and variance $\sigma_j(t)$. The biggest interest here is that the jump rate $\lambda(t)$, scaled by 500 to keep on the same graph, with both $\lambda(t)$ and $p_1(t)$ similarly variable in Subfigure 16.2(a). The double-uniform bounds, $a(t)$ and $b(t)$, vary quite a bit as would be expected from the variability of $\mu_j(t)$ and $\sigma_j(t)$.

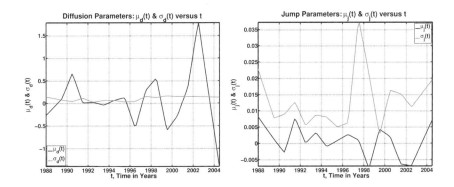

(a) Diffusion parameters: $\mu_d(t)$ and $\sigma_d(t)$.   (b) Jump parameters: $\mu_j(t)$ and $\sigma_j(t)$.

Figure 16.1. Jump-diffusion mean and variance parameters, $(\mu_d(t),\ \sigma_d(t)\ )$ and $(\mu_j(t),\ \sigma_j(t))$ on $t \in [1988, 2004.5]$, represented as piecewise linear interpolation of yearly averages assigned to the mid-year.

The *fminsearch* tolerances are $tolx = 0.5\text{e-}6$ and $toly = 0.5\text{e-}6$. All yearly iterations are converged in a range from 399 to 750 steps each. The time needed for the yearly estimations is in a range from 2 to 5 seconds using a Dual 2GHz PowerPC G5 computer processor.

In Figure 16.3 a sample comparison is made for the empirical S&P500 histogram on the left for the year of 2000 with the corresponding theoretical jump-diffusion histogram on the right using the fitted, optimized parameters and the same number of centered bins on the domain. The jump-diffusion histogram is a very idealized version of the empirical distribution, with the asymmetry of the tails clearly illustrated.

For reference, the summaries of the coefficients of skewness and kurtosis are given in Figure 16.4 for both the estimated theoretical jump-diffusion model and the empirical S&P500 data to facilitate comparison. The jump-diffusion skewness values $\eta_3^{(\text{dujd})}$ in Subfigure 16.4(a) are in the range of -308% to +204% of the empirical S&P500 values, except in the case of the year 2000 when the empirical value is near zero and the relative error is ill-defined. Note that contrary to the legendary long-

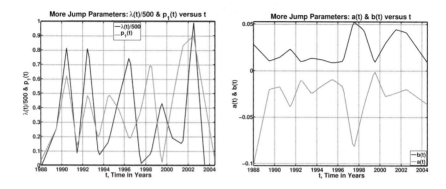

(a) Jump parameters: $\lambda(t)/500$ and $p_1(t)$.   (b) Jump parameters: $a(t)$ and $b(t)$.

Figure 16.2. More jump parameters, $(\lambda(t)/500,\ p_1(t)\ )$ and $(a(t),\ b(t))$ on $t \in [1988, 2004.5]$, represented as piecewise linear interpolation of yearly averages assigned to the mid-year.

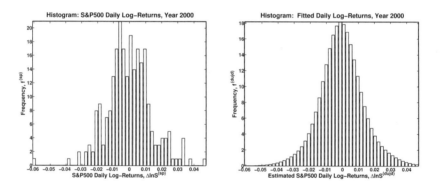

(a) Histogram sample of S&P500 log returns for year 2000.

(b) Histogram sample of S&P500 log returns for fitted jump-diffusion in the same year.

Figure 16.3. Sample comparison for year 2000 of the empirical S&P500 histogram on the left with the corresponding fitted theoretical log-double-uniform jump-diffusion histogram on the right, using 50 bins.

term negative market skewness, the skewness for some years is positive and the change in sign of the skewness is reflected in the large differences from the empirical results. The jump-diffusion kurtosis values $\eta_4^{(dujd)}$ in

Subfigure 16.4(b) are in the range of -19% to +24% with a mean of 3.2% of the empirical values, which is very good considering the difficulty of accurately estimated fourth moments and the results are very much better than that the skewness results. Any discrepancy between the estimated theoretical and observed data for kurtosis is likely due to the relative smallness of the yearly sample as well as the bin size and the fixed yearly double-uniform domain. The concept that the market data is usually leptokurtic refers to long term data and not to shorter term data.

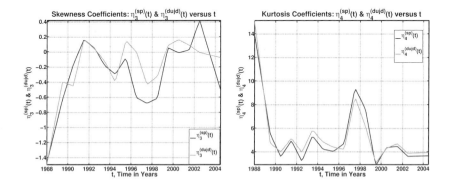

(a) Skewness coefficients: $\eta_3^{(\mathrm{sp})}$ and $\eta_3^{(\mathrm{dujd})}$.

(b) Kurtosis coefficients : $\eta_4^{(\mathrm{sp})}$ and $\eta_4^{(\mathrm{dujd})}$.

Figure 16.4. Comparison of skewness and kurtosis coefficients for both the S&P500 data and the estimated double-uniform jump diffusion values on $t \in [1988, 2004.5]$, represented as piecewise linear interpolation of yearly averages assigned to the mid-year.

The main purpose of this parameter estimation has been to have an estimate of the many time-dependence parameters . Hence, we use the simple piecewise linear interpolation to fit the jump-diffusion parameters in time assigning the estimate yearly averages to the mid-year as interpolation points.

4. Application to Optimal Portfolio and Consumption Policies

Consider a portfolio consisting of a riskless asset, called a bond, with price $B(t)$ dollars at time t years, and a risky asset, called a stock, with price $S(t)$ at time t. Let the instantaneous portfolio change fractions be $U_0(t)$ for the bond and $U_1(t)$ for the stock, so that the total satisfies $U_0(t) + U_1(t) = 1$. This does not necessarily imply bounds for $U_0(t)$ and $U_1(t)$, as will be seen later that their bounds depend on the jump-amplitude distribution in the presence of a non-negative of wealth (no bankruptcy) condition.

The bond price process is deterministic exponential,

$$dB(t) = r(t)B(t)dt , \quad B(0) = B_0 . \tag{16.21}$$

where $r(t)$ is the bond rate of interest at time t. The stock price $S(t)$ has been given in (16.3). The portfolio wealth process changes due to changes in the portfolio fractions less the instantaneous consumption of wealth $C(t)dt$,

$$\begin{aligned} dW(t) = {} & W(t)\left(r(t)dt + U_1(t)\left((\mu_d(t) - r(t))dt\right.\right. \\ & \left.\left. + \sigma_d(t)dG(t) + \sum_{k=1}^{dP(t)}\left(e^{Q_k} - 1\right)\right)\right) - C(t)dt , \end{aligned} \tag{16.22}$$

such that, consistent with non-negative constraints Sethi and Taksar [33] show are needed, $W(t) \geq 0$ and that the consumption rate is constrained relative to wealth $0 \leq C(t) \leq C_{\max}^{(0)}W(t)$. In addition, the stock fraction is bounded by fixed constants. $U_{\min}^{(0)} \leq U_1(t) \leq U_{\max}^{(0)}$, so borrowing and short-selling is permissible, and $U_0(t) = 1 - U_1(t)$ has been eliminated [12].

. The investor's portfolio objective is to maximize the conditional, expected current value of the discounted utility $\mathcal{U}_f(w)$ of terminal wealth at the end of the investment terminal time t_f and the discounted utility of instantaneous consumption $\mathcal{U}(c)$, i.e.,

$$\begin{aligned} v^*(t, w) = {} & \max_{\{u,c\}}\left[\mathrm{E}\left[e^{-\beta(t,t_f)}\mathcal{U}_f(W(t_f))\right.\right. \\ & \left.\left. + \int_t^{t_f} e^{-\beta(t,s)}\mathcal{U}(C(s))\, ds\,\middle|\, \mathcal{C}\right]\right] , \end{aligned} \tag{16.23}$$

conditioned on the state-control set $\mathcal{C} = \{W(t) = w, U_1(t) = u, C(t) = c\}$, where the time horizon is assumed to be finite, $0 \leq t < t_f$, and $\beta(t, s)$ is the cumulative time discount over time in (t, s) with $\beta(t, t) = 0$ and discount rate $\widehat{\beta}(t) = \partial\beta/\partial s(t, t)$ at time t. In order to avoid Merton's [25] problems with utility functions, $\mathcal{U}'(C) \to +\infty$ as $C \to 0^+$ will be assumed for the utility of consumption, while a similar form will be used for the final bequest $\mathcal{U}_f(W)$. Thus, the instantaneous consumption $c = C(t)$ and stock portfolio fraction $u = U_1(t)$ serve as control variables, while the wealth $w = W(t)$ is the single state variable.

Absorbing Boundary Condition at Zero Wealth: Eq. (16.23) is subject to zero wealth absorbing natural boundary condition (avoids arbitrage as pointed out by Karatzas, Lehoczky, Sethi, Shreve and Taksar ([20] or [32, Chapter 2] and [33] or [32, Chapter 3]) that it is necessary to enforce non-negativity feasibility conditions on both wealth and consumption. They formally derived explicit solutions for consumption-investment dynamic programming models with a time-to-bankruptcy horizon that qualitatively corrects the results of Merton [25, 26] ([29, Chapter 6]). See also Sethi and Taksar [33] and much more in the *Sethi volume* [32], which includes Sethi's very broad and excellent summary [32, Chapter 1].

Here the Merton correction [29, Chap. 6]) is used,

$$v^*(t, 0^+) = \mathcal{U}_f(0)e^{-\beta(t, t_f)} + \mathcal{U}(0) \int_t^{t_f} e^{-\beta(t, s)} ds, \qquad (16.24)$$

where the terminal wealth condition, $v^*(t_f, w) = \mathcal{U}_f(w)$, has been applied, following from the fact that the consumption must be zero when the wealth is zero.

Portfolio Stochastic Dynamic Programming: Assuming the optimal value $v^*(t, w)$ is continuously differentiable in t and twice continuously differentiable in w, then the stochastic dynamic programming equation (see [12]) follows from an application of the (Itô) stochastic chain rule to the principle of optimality,

$$
\begin{aligned}
0 = {}& v_t^*(t, w) - \widehat{\beta}(t)v^*(t, w) + \mathcal{U}(c^*(t, w)) \\
&+ [(r(t) + (\mu_d(t) - r(t))u^*(t, w))w - c^*(t, w)] v_w^*(t, w) \\
&+ \tfrac{1}{2}\sigma_d^2(t)(u^*)^2(t, w)w^2 v_{ww}^*(t, w) + \lambda(t)\left(\tfrac{p_1(t)}{|a|(t)} \int_{a(t)}^0 + \tfrac{p_2(t)}{b(t)} \int_0^{b(t)} \right) \\
&\cdot \left(v^*(t, (1 + (e^q - 1)u^*(t, w))w) - v^*(t, w) \right) dq,
\end{aligned}
\qquad (16.25)
$$

where $u^* = u^*(t, w) \in [U_{\min}^{(0)}, U_{\max}^{(0)}]$ and $c^* = c^*(t, w) \in [0, C_{\max}^{(0)}w]$ are the optimal controls if they exist, while $v_w^*(t, w)$ and $v_{ww}^*(t, w)$ are the partial derivatives with respect to wealth w when $0 \le t < t_f$.

Non-Negativity of Wealth and Jump Distribution: The non-negativity of wealth implies an additional consistency condition for the control since the jump in wealth argument $(1 + (e^q - 1)u^*)w$ in the stochastic dynamic programming equation (16.25) requires $\kappa(u, q) \equiv 1 + (e^q - 1)u \ge 0$ on the support interval of the jump-amplitude mark density $\phi_Q(q; t)$. Hence, it will make a difference in the optimal portfolio stock fraction u^* bounds if the support interval $[a(t), b(t)]$ is finite or if the support interval is $(-\infty, +\infty)$, i.e., had infinite range. Our results will be restricted to the usual case when $a(t) < 0 < b(t)$, i.e., when both crashes and rallies are modeled.

Lemma 1. Bounds on Optimal Stock Fraction due to Non-Negativity of Wealth Jump Argument

If the support of $\phi_Q(q;t)$ is the finite interval $q \in [a(t), b(t)]$ with $a(t) < 0 < b(t)$, then $u^*(t, w)$ is restricted by (16.25) to

$$\frac{-1}{\left(e^{b(t)} - 1\right)} \leq u^*(t, w) \leq \frac{1}{\left(1 - e^{a(t)}\right)}, \qquad (16.26)$$

but if the support of $\phi_Q(q)$ is fully infinite, i.e., $(-\infty, +\infty)$, then $u^*(t, w)$ is restricted by (16.25) to

$$0 \leq u^*(t, w) \leq 1. \qquad (16.27)$$

Proof. Since $\kappa(u, q) = 1 + (e^q - 1)u$ and it is necessary that $\kappa(u, q) \geq 0$ so that $\kappa(u, q)w \geq 0$ when the wealth and its jump argument need to be non-negative. The most basic instantaneous stock fraction case is when $u = 0$, so $\kappa(0, q) = 1 > 0$.

First consider the case when the support is the finite $a(t) \leq q \leq b(t)$. When $u > 0$, then

$$0 \leq 1 - \left(1 - e^{a(t)}\right)u \leq \kappa(u, q) \leq 1 + \left(e^{b(t)} - 1\right)u.$$

Since $e^{a(t)} < 1 < e^{b(t)}$, the worse case for enforcing $\kappa(u, q) \geq 0$ is on the left, so

$$u \leq \frac{+1}{\left(1 - e^{a(t)}\right)}.$$

When $u < 0$, then

$$0 \leq 1 - \left(e^{b(t)} - 1\right)(-u) \leq \kappa(u, q) \leq 1 + \left(1 - e^{a(t)}\right)(-u).$$

The worse case for enforcing $\kappa(u, q) \geq 0$ is again on the left so upon reversing signs,

$$u \geq \frac{-1}{e^{b(t)} - 1},$$

completing both sides of the finite case (16.26).

In the infinite range jump model case when $-\infty < q < +\infty$, then $0 < e^q < \infty$. Thus, when $u > 0$,

$$0 \leq 1 - u < \kappa(u, q) < \infty,$$

so $u \leq 1$. However, when $u < 0$, then

$$-\infty < \kappa(u, q) < 1 - u,$$

so $u < 0$ leads to a contradiction since $\kappa(u, q)$ is unbounded on the left and $u \geq 0$, proving (16.27), which is just the limiting case of (16.26). $\square$

Remark 1. *This lemma gives the constraints on the instantaneous stock fraction $u^*(t, w)$ that limits the jumps to the jumps that at most just wipe out the investor's wealth. Unlike the case of pure diffusion where the functional terms have local dependence on the wealth mainly through partial derivatives, the case of jump-diffusion has global dependence through jump integrals over finite differences with jump modified wealth arguments, leading to additional constraints under non-negative wealth conditions that do not appear for pure diffusions. The additional constraint comes not from the current wealth or nearby wealth but from the new wealth created by a jump. The more severe restrictions on the optimal stock fraction in the non-finite support case for the jump-amplitude models compared to the compact support case such as the double uniform model gives further justification for the uniform type models.*

Note that the compact support bounds can be rewritten in terms of the original jump-amplitude coefficient

$$-1/J(t, b(t)) \leq u^*(t, w) \leq -1/J(t, a(t)).$$

In the case of the fitted log-double-uniform jump-amplitude model, the range of the jump-amplitude marks $[a(t), b(t)]$ is covered by the estimated interval

$$[a_{\min}, b_{\max}] = \left[\min_t(a(t)), \max_t(b(t))\right] \simeq [-8.470e\text{-}2, 5.320e\text{-}2]$$

over the whole period from 1988-2003. The corresponding overall estimated range of the optimal instantaneous stock fraction $u^*(t, w)$ is then

$$[u_{\min}, u_{\max}] = \left[\frac{-1}{(e^{b_{\max}} - 1)}, \frac{+1}{(1 - e^{a_{\min}})}\right] \simeq [-18.30, +12.31] \quad (16.28)$$

in large contrast to the highly restricted infinite range models where $[\min(u^*(t, w)), \max(u^*(t, w))] = [0, 1]$ is fixed for any t.

Regular Optimal Control Policies: In absence of control constraints, then the maximum controls are the regular optimal controls $u_{\text{reg}}(t, w)$ and $c_{\text{reg}}(t, w)$, which are given implicitly, provided they are attainable and there is sufficient differentiability in c and u, by the dual critical conditions,

$$\mathcal{U}'(c_{\text{reg}}(t,w)) = v_w^*(t,w) ,\tag{16.29}$$

$$\sigma_d^2(t)w^2 v_{ww}^*(t,w)u_{\text{reg}}(t,w) = -(\mu_d(t) - r(t))wv_w^*(t,w)$$
$$-\lambda(t)w\left(\frac{p_1(t)}{|a|(t)}\int_{a(t)}^0 + \frac{p_2(t)}{b(t)}\int_0^{b(t)}\right)(e^q - 1)v_w^*(t,\kappa(u_{\text{reg}}(t,w),q)w)\,dq ,\tag{16.30}$$

for the optimal consumption and portfolio policies with respect to the terminal wealth and instantaneous consumption utilities (16.23). Note that (16.29-16.30) define the set of regular controls implicitly.

CRRA Utility and Canonical Solution Reduction: Assuming the investor is risk adverse, the utilities will be the constant relative risk-aversion (CRRA) power utilities [29, 10], with the same power for both wealth and consumption,

$$\mathcal{U}(x) = \mathcal{U}_f(x) = x^\gamma/\gamma , \quad x \ge 0 , \quad 0 < \gamma < 1 .\tag{16.31}$$

The CRRA utility designation arises since the relative risk aversion is the negative of the local change in the marginal utility $(\mathcal{U}''(x))$ relative to the average change in the marginal utility $(\mathcal{U}'(x)/x)$, or here

$$R(x) \equiv -\mathcal{U}''(x)/(\mathcal{U}'(x)/x) = (1 - \gamma) > 0,$$

i.e., a constant, and is a special case of the more general HARA utilities.

The CRRA power utilities for the optimal consumption and portfolio problem lead to a *canonical reduction* of the stochastic dynamic programming PDE problem to a simpler ODE problem in time, by the separation of wealth and time dependence,

$$v^*(t,w) = \mathcal{U}(w)v_0(t),\tag{16.32}$$

where only the time function $v_0(t)$ is to be determined. The regular consumption control is a linear function of the wealth,

$$c_{\text{reg}}(t,w) \equiv w \cdot c_{\text{reg}}^{(0)}(t) = w/v_0^{1/(1-\gamma)}(t),\tag{16.33}$$

using (16.29) and $\mathcal{U}'(x) = x^{\gamma-1}$ in (16.31). The regular stock fraction u in (16.30) is a wealth independent control, but is given in implicit form:

$$u_{\text{reg}}(t,w) = u_{\text{reg}}^{(0)}(t)$$
$$= \frac{1}{(1-\gamma)\sigma_d^2(t)}\left[\mu_d(t) - r(t) + \lambda(t)I_1\left(u_{\text{reg}}^{(0)}(t)\right)\right],\tag{16.34}$$

$$I_1(u) \doteq \left(\tfrac{p_1(t)}{|a|(t)} \int_{a(t)}^0 + \tfrac{p_2(t)}{b(t)} \int_0^{b(t)} \right) (e^q - 1)\kappa^{\gamma-1}(u, q) dq, \qquad (16.35)$$

The wealth independent property of the regular stock fraction is essential for the separability of the optimal value function (16.32). Since (16.34) only defines $u_{\text{reg}}^{(0)}(t)$ implicitly in fixed point form, $u_{\text{reg}}^{(0)}(t)$ must be found by an iteration such as Newton's method, while the Gauss-Statistics quadrature [34] can be used for jump integrals (see [12]). The optimal controls, when there are constraints, are given in piecewise form as $c^*(t, w)/w = c_0^*(t) = \max[\min[c_{\text{reg}}^{(0)}(t), C_{\text{max}}^{(0)}], 0]$, provided $w > 0$, and $u^*(t, w) = u_0^*(t) = \max[\min[u_{\text{reg}}^{(0)}(t), U_{\text{max}}^{(0)}], U_{\text{min}}^{(0)}]$, is independent of w along with $u_{\text{reg}}^{(0)}(t)$. Substitution of the separable power solution (16.32) and the regular controls (16.33-16.34) into the stochastic dynamic programming equation (16.25), leads to an apparent Bernoulli type ODE,

$$0 = v_0'(t) + (1 - \gamma) \left(g_1(t, u_0^*(t)) v_0(t) + g_2(t) v_0^{\frac{\gamma}{\gamma-1}}(t) \right), \qquad (16.36)$$

$$g_1(t, u) \equiv \tfrac{1}{1-\gamma} \left[-\widehat{\beta}(t) + \gamma \left(r(t) + u(\mu_d(t) - r(t)) \right) \right. \\ \left. - \tfrac{\gamma(1-\gamma)}{2} \sigma_d^2(t) u^2 + \lambda(t)(I_2(t, u) - 1) \right], \qquad (16.37)$$

$$g_2(t) \equiv \frac{1}{1 - \gamma} \left[\left(\frac{c_0^*(t)}{c_{\text{reg}}^{(0)}(t)} \right)^\gamma - \gamma \left(\frac{c_0^*(t)}{c_{\text{reg}}^{(0)}(t)} \right) \right], \qquad (16.38)$$

$$I_2(t, u) \equiv \left(\frac{p_1(t)}{|a|(t)} \int_{a(t)}^0 + \frac{p_2(t)}{b(t)} \int_0^{b(t)} \right) \kappa^\gamma(u, q) \, dq, \qquad (16.39)$$

for $0 \le t < t_f$. The coupling of $v_0(t)$ to the time dependent part of the consumption term $c_{\text{reg}}^{(0)}(t)$ in $g_2(t)$ and the relationship of $c_{\text{reg}}^{(0)}(t)$ to $v_0(t)$ in (16.33) means that the differential equation (16.36) is implicitly highly nonlinear and thus (16.36) is only of Bernoulli type formally. The apparent Bernoulli equation (16.36) can be transformed to an apparent linear differential equation by using $\theta(t) = v_0^{1/(1-\gamma)}(t)$, to obtain,

$$0 = \theta'(t) + g_1(t, u_0^*)\theta(t) + g_2(t),$$

whose general solution can be inverse transformed to the general solution for the separated time function,

$$v_0(t) = \theta^{1-\gamma}(t)$$

$$= \left[e^{-g_1(t, u_0^*(t))(t_f - t)} \left(1 + \int_t^{t_f} g_2(\tau) e^{g_1(t, u_0^*(t))(t_f - \tau)} d\tau \right) \right]^{1-\gamma}, \qquad (16.40)$$

352

given implicitly.

In order to illustrate this stochastic application, a computational approximation of the solution is presented. The main computational changes from the procedure used in [12] are that the jump-amplitude distribution is now double-uniform and the portfolio parameters as well as the jump-amplitude distribution are time-dependent. Parameter time-dependence is approximated by piecewise linear interpolation over the years from 1988-2003. The terminal time is taken to be $t_f = 2004.5$, one half year beyond this range.

For this numerical study, the economic rates are taken to be federal funds historical rates [6] from the U.S. Federal Reserve Bank, because they are readily available. For feasibility of the computation, the daily rates, $r(t)$ for interest and $\widehat{\beta}(t)$ for discounting, are transformed into approximate piecewise linear interpolation representations of the yearly averages of daily rates over the period 1988-2003. As for other time-dependent parameters the yearly averages are assigned to the mid-years as interpolation points. The federal funds rates are shown in Figure 16.5. Note that the economic rates are much more variable that the stock mar-

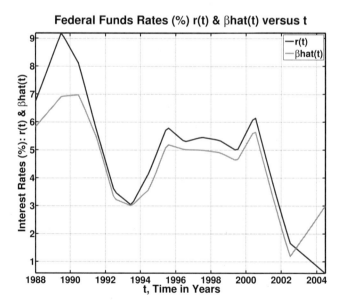

Figure 16.5. Federal funds rate [6] for interest $r(t)$ and discounting $\widehat{\beta}(t)$ on a daily bases, represented by piecewise linear interpolation with yearly averages assigned to the midpoint of each year for $t = 1988.5 : 2003.5$.

ket parameters displayed early. Also, the typical approximate ordering of interest and discount rates, $\widehat{\beta}(t) \leq r(t)$, is not valid in the recent anomalous low interest period, 2002-present.

The portfolio stock fraction constraints are chosen so that there is at least one active constraint within the time horizon,

$$[U_{\min}^{(0)}, U_{\max}^{(0)}] = [-18, +12],$$

since in a realistic trading environment there would be some bounds on the extremes of borrowing and short-selling, but not as severe as constraining the control to [0,1] as in (16.27). Also, the bound on consumption relative to wealth are assumed to be

$$C_{\max}^{(0)} = 0.75,$$

meaning that the investor cannot consume more that 75% of the wealth in the portfolio and $0 \leq c(t, w) \leq C_{\max}^{(0)} w$.

Subfigure 16.6(a) shows the regular or unconstrained optimal instantaneous portfolio stock fraction. Although the $u_{\mathrm{reg}}(t)$ results appear to be out of the conservative range of $[u_{\min}, u_{\max}]$ in (16.28) using $[a_{\min}, b_{max}]$, the results are consistent with the worst case scenario range

$$[\widetilde{u}_{\min}, \widetilde{u}_{\max}] \simeq [\text{-1.150e+02, 1.940e+12}]$$

using the tighter distribution range $[a_{\max}, b_{min}]$ of [-5.156e-13, 8.658e-03] in (16.26). In Subfigure 16.6(b), the optimal portfolio stock fraction $u^*(t)$ is displayed. The portfolio policy is not monotonic in time and the maximum control constraint at $U_{\max}^{(0)}$ is active during the interval just prior to the end of the time horizon $t \in [0, t_f]$, while the minimum constraint $U_{\min}^{(0)}$ remains unused since the stock fraction remains mostly in the borrowing range with the corresponding bond fraction negative, $1 - u^*(t) < 0$. The $u^*(t)$ non-monotonic behavior is very interesting compared to the constant behavior in the constant parameter model in [12] or Merton's [25] mainly pure diffusion results.

In Figure 16.7 on the left, the optimal, expected, discounted utility of terminal wealth and cumulative consumption, $v^*(t, w)$, is displayed in three dimensions. The behavior of $v^*(t, w)$ for fixed time t reflects the CRRA utility of function $\mathcal{U}(w)$ template of the separable canonical solution form in (16.32), while the decay in time toward the final time $t_f = 16.5$ and final value $v^*(t_f, w) = 0$ for fixed wealth w derives from the separable time function $v_0(t)$. The optimal value function $v^*(t, w)$ results, and the following optimal consumption policy $c^*(t, w)$ results in

354

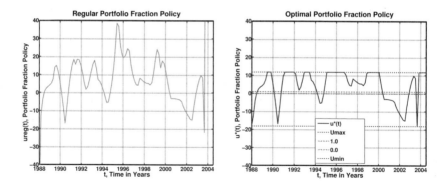

(a) Regular stock fraction policy $u_{\mathrm{reg}}(t)$. (b) Optimal stock fraction policy, $u^*(t)$.

Figure 16.6. Regular and optimal portfolio stock fraction policies, $u_{\mathrm{reg}}(t)$ and $u^*(t)$ on $t \in [1988, 2004.5]$, the latter subject to the control constraints set $[U_{\mathrm{min}}^{(0)}, U_{\mathrm{max}}^{(0)}] = [-18, 12]$.

Fig. 16.7 on the right, in this computational example, are qualitatively similar to that of the time-independent log-normal jump parameter case in [12] and the time-independent log-uniform jump parameter case in [15] computational results. Note that the wealth grid uses a specially constructed transformation tailored to the CRRA utility to capture the non-smooth behavior as $w \to 0^+$.

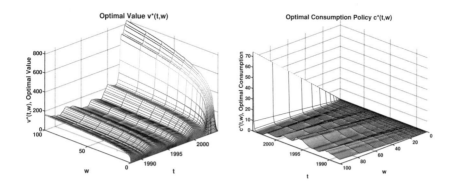

(a) Optimal portfolio value $v^*(t, w)$. (b) Optimal consumption policy $c^*(t, w)$.

Figure 16.7. Optimal portfolio value $v^*(t, w)$ and optimal consumption policy $c^*(t, w)$ for $(t, w) \in [1988, 2004.5] \times [0, 100]$.

5.   Conclusions

The main contributions of this work are the introduction of the log-double-uniformly distributed jump-amplitude into the jump-diffusion stock price model and the development of time-dependent jump-diffusion parameters. In particular, a significant effect on the variation of the instantaneous stock fraction policy is seen to be due to variations in the interest and discount rates. The double-uniformly distributed jump-amplitude feature of the model is a reasonable assumption for rare, large jumps, crashes or buying-frenzies, when there is only a sparse population of isolated jumps in the tails of the market distribution. Additional realism in the jump-diffusion model is given by the introduction of time dependence in the distribution and in the associated parameters. Finally, the large difference in the severity of the limits on borrowing and short-selling is made clear for the bounds on the instantaneous stock fraction with respect to compact support and non-compact support models of jump-amplitudes.

Further improvements, but with greater computational complexity, would be to estimate the double-uniform distribution limits $[a, b]$ by fitting the theoretical distribution to real market distributions, using longer and overlapping (moving-average) partitioning of the market data to reduce the effects of small sample sizes.

Acknowledgement

This work is supported in part by the National Science Foundation under Grant DMS-02-07081. Any conclusions or recommendations expressed in this material are those of the authors and do not necessarily reflect the views of the National Science Foundation.

References

[1] Aït-Sahalia, Y. (2004). "Disentangling Diffusion from Jumps," *J. Financial Economics*, vol. 74, pp. 487–528.

[2] Aourir, C. A. , D. Okuyama, C. Lott and C. Eglinton. (2002). *Exchanges - Circuit Breakers, Curbs, and Other Trading Restrictions*, http://invest-faq.com/articles/exch-circuit-brkr.html .

[3] Ball, C. A., and W. N. Torous. (1985). "On Jumps in Common Stock Prices and Their Impact on Call Option Prices," *J. Finance*, vol. 40 (1), pp. 155–173.

[4] Black, F., and M. Scholes. (1973). "The Pricing of Options and Corporate Liabilities," *J. Political Economy*, vol. 81, pp. 637–659.

[5] Evans, M., N. Hastings, and B. Peacock. (2000). *Statistical Distributions*, 3rd edn., New York: John Wiley.

[6] Federal Reserve Bank. (2005). "FRB: Federal Reserve Statistical Release H.15 - Historical Data", `http://www.federalreserve.gov/releases/h15/data.htm` .

[7] Feller, W. (1971). *An Introduction to Probability Theory and Its Application*, vol. 2, 2nd edn., New York: John Wiley.

[8] Forsythe, G. E., M. A. Malcolm and C. Moler. (1977). *Computer Methods for Mathematical Computations*, Englewood Cliffs, NJ: Prentice-Hall.

[9] Hanson, F. B. (2005). *Applied Stochastic Processes and Control for Jump-Diffusions: Modeling, Analysis and Computation*, SIAM Books, Philadelphia, PA, to appear 2006; Preprint: `http://www2.math.uic.edu/~hanson/math574/#Text` .

[10] Hanson, F. B., and J. J. Westman. (2001). "Optimal Consumption and Portfolio Policies for Important Jump Events: Modeling and Computational Considerations," *Proc. 2001 American Control Conference*, pp. 4556–4561.

[11] Hanson, F. B., and J. J. Westman. (2002). "Stochastic Analysis of Jump–Diffusions for Financial Log–Return Processes," *Proceedings of Stochastic Theory and Control Workshop, Lecture Notes in Control and Information Sciences*, vol. 280, B. Pasik-Duncan (Editor), New York: Springer–Verlag, pp. 169–184.

[12] Hanson, F. B., and J. J. Westman. (2002). "Optimal Consumption and Portfolio Control for Jump-Diffusion Stock Process with Log-Normal Jumps," *Proc. 2002 American Control Conference*, pp. 4256–4261; corrected paper: `ftp://ftp.math.uic.edu/pub/Hanson/ACC02/acc02webcor.pdf` .

[13] Hanson, F. B., and J. J. Westman. (2002). "Jump-Diffusion Stock Return Models in Finance: Stochastic Process Density with Uniform-Jump Amplitude," *Proc. 15th International Symposium on Mathematical Theory of Networks and Systems*, 7 CD pages.

[14] Hanson, F. B., and J. J. Westman. (2002). "Computational Methods for Portfolio and Consumption Optimization in Log-Normal Diffusion, Log-Uniform Jump Environments," *Proc. 15th International Symposium on Mathematical Theory of Networks and Systems*, 9 CD pages.

[15] Hanson, F. B., and J. J. Westman. (2002). "Portfolio Optimization with Jump–Diffusions: Estimation of Time-Dependent Parameters and Application," *Proc. 41st Conference on Decision and Control*, pp. 377–382; partially corrected paper: `ftp://ftp.math.uic.edu/pub/Hanson/CDC02/cdc02web.pdf` .

[16] Hanson, F. B., and J. J. Westman. (2004). "Optimal Portfolio and Consumption Policies Subject to Rishel's Important Jump Events Model: Computational Methods," *Trans. Automatic Control*, vol. 48 (3), Special Issue on Stochastic Control Methods in Financial Engineering, pp. 326–337.

[17] Hanson, F. B., J. J. Westman and Z. Zhu. (2004). "Multinomial Maximum Likelihood Estimation of Market Parameters for Stock Jump-Diffusion Models," *AMS Contemporary Mathematics*, vol. 351, pp. 155-169.

[18] Jarrow, R. A., and E. R. Rosenfeld. (1984). "Jump Risks and the Intertemporal Capital Asset Pricing Model," *J. Business*, vol. 57 (3), pp. 337–351.

[19] Jorion, P. (1989), "On Jump Processes in the Foreign Exchange and Stock Markets," *Rev. Fin. Studies*, vol. 88 (4), pp. 427–445.

[20] Karatzas, I., J. P. Lehoczky, S. P. Sethi and S. E. Shreve. (1986). "Explicit Solution of a General Consumption/Investment Problem," *Math. Oper. Res.*, vol. 11, pp. 261–294. (Reprinted in Sethi [32, Chapter 2].)

[21] Kou, S. G. (2002). "A Jump Diffusion Model for Option Pricing," *Management Science*, vol. 48, pp. 1086–1101.

[22] Kou, S. G. and H. Wang. (2004). "Option Pricing Under a Double Exponential Jump Diffusion Model," *Management Science*, vol. 50 (9), pp. 1178–1192.

[23] Moler, C., et al. (2000). *Using MATLAB*, vers 6., Natick, MA: Mathworks.

[24] Merton, R. C. (1969). *Lifetime Portfolio Selection Under Uncertainty: The Continuous-Time Case, Rev. Econ. and Stat.*, vol. 51, 1961, pp. 247-257. (Reprinted in Merton [29, Chapter 4].)

[25] Merton, R. C. (1971). "Optimal Consumption and Portfolio Rules in a Continuous-Time Model," *J. Econ. Theory*, vol. 4, pp. 141-183. (Reprinted in Merton [29, Chapter 5].)

[26] Merton, R. C. (1973). "Eratum," *J. Econ. Theory*, vol. 6 (2), pp. 213-214.

358

[27] Merton, R. C. (1973). "Theory of Rational Option Pricing," *Bell J. Econ. Mgmt. Sci.*, vol. 4, 1973, pp. 141-183. (Reprinted in Merton [29, Chapter 8].)

[28] Merton, R. C. (1976). "Option Pricing When Underlying Stock Returns are Discontinuous," *J. Financial Economics*, vol. 3, pp. 125–144. (Reprinted in Merton [29, Chapter 9].)

[29] Merton, R. C. (1990). *Continuous-Time Finance*, Cambridge, MA: Basil Blackwell.

[30] Nelder, J. A. and R. Mead. (1965). "A simplex method for function minimization," *Computer Journal*, vol. 7, pp. 308–313.

[31] Rishel, R. (1999). "Modeling and Portfolio Optimization for Stock Prices Dependent on External Events," *Proc. 38th IEEE Conference on Decision and Control*, pp. 2788–2793.

[32] Sethi, Suresh P. (1997). *Optimal Consumption and Investment with Bankruptcy*, Boston: Kluwer Academic Publishers.

[33] Sethi, Suresh P., and M. Taksar. (1988). "A Note on Merton's "Optimal Consumption and Portfolio Rules in a Continuous-Time Model"", *J. Econ. Theory*, vol. 46 (2), pp. 395–401. (Reprinted in Sethi [32, Chapter 3].)

[34] Westman, J. J., and F. B. Hanson. (2000). "Nonlinear State Dynamics: Computational Methods and Manufacturing Example," *International Journal of Control*, vol. 73, pp. 464–480.

[35] Yahoo! Finance. (2001). *Historical Quotes, S & P 500 Symbol ^SPC*, http://chart.yahoo.com/ .

[36] Zhu, Z. (2005). *Option Pricing and Jump-Diffusion Models*, Ph. D. Thesis in Mathematics, Dept. Math., Stat., and Comp. Sci., University of Illinois at Chicago, 17 October 2005.

Index

Early Titles in the
INTERNATIONAL SERIES IN
OPERATIONS RESEARCH & MANAGEMENT SCIENCE
Frederick S. Hillier, Series Editor, *Stanford University*

* A list of the more recent publications in the series is at the front of the book *